Food Security and Climate-Smart Food Systems

Mohamed Behnassi · Mirza Barjees Baig ·
Mohamed Taher Sraïri · Abdlmalek A. Alsheikh ·
Ali Wafa A. Abu Risheh
Editors

Food Security and Climate-Smart Food Systems

Building Resilience for the Global South

Editors
Mohamed Behnassi
Center for Environment, Human Security
and Governance (CERES)
Université Ibn Zohr
Agadir, Morocco

Mohamed Taher Sraïri
Department of Animal Production
and Biotechnology
Hassan II Agronomy and Veterinary
Medicine Institute (IAV)
Rabat, Morocco

Ali Wafa A. Abu Risheh
The Prince Sultan Institute for
Environmental, Water and Desert Research
King Saud University
Riyadh, Saudi Arabia

Mirza Barjees Baig
The Prince Sultan Institute for
Environmental, Water and Desert Research
King Saud University
Riyadh, Saudi Arabia

Abdlmalek A. Alsheikh
The Prince Sultan Institute for
Environmental, Water and Desert Research
King Saud University
Riyadh, Saudi Arabia

ISBN 978-3-030-92737-0   ISBN 978-3-030-92738-7   (eBook)
https://doi.org/10.1007/978-3-030-92738-7

© The Editor(s) (if applicable) and The Author(s), under exclusive license to Springer Nature Switzerland AG 2022
This work is subject to copyright. All rights are solely and exclusively licensed by the Publisher, whether the whole or part of the material is concerned, specifically the rights of translation, reprinting, reuse of illustrations, recitation, broadcasting, reproduction on microfilms or in any other physical way, and transmission or information storage and retrieval, electronic adaptation, computer software, or by similar or dissimilar methodology now known or hereafter developed.
The use of general descriptive names, registered names, trademarks, service marks, etc. in this publication does not imply, even in the absence of a specific statement, that such names are exempt from the relevant protective laws and regulations and therefore free for general use.
The publisher, the authors and the editors are safe to assume that the advice and information in this book are believed to be true and accurate at the date of publication. Neither the publisher nor the authors or the editors give a warranty, expressed or implied, with respect to the material contained herein or for any errors or omissions that may have been made. The publisher remains neutral with regard to jurisdictional claims in published maps and institutional affiliations.

This Springer imprint is published by the registered company Springer Nature Switzerland AG
The registered company address is: Gewerbestrasse 11, 6330 Cham, Switzerland

Foreword: Reversing Climate Change and Restoring Ecosystems as Pre-requisites for a Food-Secure and Resilient Future

Climate-induced impacts and disasters such as drought, heat stress, flooding, and crop pests have led to a decline in crop yields and livestock productivity in many regions, while severely affecting the livelihood and security of vulnerable communities, including small-holder farmers. These impacts and disasters are also critically undermining water security through changes in rainfall patterns, drying-up of rivers, and receding bodies of water. Many global assessments predict that conflicts over natural resources, especially water, may be a growing cause of violent tensions in the future, leading among other implications to mass displacements within and between countries. In this broad context, environmental degradation, climate change, biodiversity loss, and resource scarcity are increasingly impacting food systems and security, thus adding urgency to our fight against poverty and disease.

Yet science and technology now exist to reverse such trends and ensure sustainability and resilience against risks. Some evolving applications have the potential to make crops more viral- and drought-resistant in a changing climate. The Clustered Regularly Interspaced Short Palindromic Repeats (CRISPR) technology, for instance, provides an alternative approach for improving the genetic traits of plants that is easier and generally cheaper than traditional breeding techniques. Many scientific advances have produced new genes that can fortify crops to withstand natural calamities such as pests and diseases and provide higher nutritional value. Innovations such as smart agriculture, environmental conservation, healthcare and disease prevention, big data-driven bioinformatics, and industrial biotechnology have the potential to help maintain productivity while improving human wellbeing and resilience to shocks. All these pathways can sustainably help meet the food needs of a growing global population. In the same vein, the broad elements of diverse diets, seed and crop diversity, improvements in seed and crop delivery and cultivation, and the maintenance of our agrobiodiversity must be optimized to ensure a healthy and prosperous future for the humanity. Evidence of the wisdom of such investments is strong and growing; therefore, we must continue to invest in in order to enhance the sustainability of food systems, thus meeting a myriad of Sustainable Development Goals.

At the same time as we modernize, we must support local and traditional knowledge systems, such as those related to sustainable farming. Among the features of this preservation are enlightened agriculture and trade policies, intellectual property rights on the conservation and sustainable use of biological resources, the empowerment of women as guardians of these systems, and the equitable sharing of benefits across sectors, genders, and communities. We have as well a special obligation to build a rhetorical bridge between our environment and scientific research. The vital threats of environmental degradation, climate change, and biodiversity loss are themselves nested in the drive to advance human health and wellbeing. And the effectiveness of driving human health and wellbeing is correspondingly leveraged by a continued investment in research and innovation. Our very existence sits at the nexus of not just the interdependent, transdisciplinary nature of scientific research, but in a broader context, at the intersection of nutrition, health, agriculture, economy, environment, and governance.

These investments require, however, sustained operational funding and capital support and the capacity to engage successfully with funders, governments, policymakers, and communities. Significant investment in environmental preservation and climate resilience for sustainable farming systems and food and health security should be simultaneously combined with public-private investment in basic and applied research, building access to sustainable resources, creating appropriate and inclusive legal and policy frameworks, and enhancing innovative capacity-building.

It is true that the world is continuously trying to tackle the challenges of decelerating climate change and restoring ecosystems, but the trade-offs that they require inevitably pit one country's interests against another's. We must thus make use of all possible capacities to jointly fight climate and environmental change, thus creating a liveable and secure planet for all. This challenge is not easy: reaching it depends on the contributions of all, especially highly skilled individuals from everywhere in the world, political courage and foresight, and creative, ambitious innovation.

But it *can* be done, because it *must* be done. Our ability to create a sustained future for ourselves isn't optional: it's existential.

This timely contributed volume, *Food Security and Climate-Smart Food Systems—Building Resilience for the Global South*, falls under this prospect. It aims to provide guiding inputs to the different issues highlighted above. Under the editorship of specialized and insightful scholars, the volume advocates for a rethinking of agricultural development models through, among others, the enhancement of climate-smart practices, extension systems' innovation, and digitalization. This will help accelerate the transformational change of current food systems towards sustainability, adaptation to, and mitigation of climate change, therefore enhancing the resilience of a food-secure future. By doing this, this volume contributes to the advancement of science, knowledge, and governance settings and inspires change processes.

Dr. Ameenah Gurib-Fakim
Distinguished Professor in Women Leadership
John Wesley School of Leadership
Carolina University
North Carolina
USA & Former President of the Republic of Mauritius

Acknowledgements

This contributed volume, as a part of a series of CERES publications, is the outcome of an international cooperation between 50 authors—scientists, junior researchers, experts, and practitioners—from many countries, disciplines, and professional areas. As mentioned above, the core idea of the volume emerged upon completion of a related volume, which had as objective the identification and analysis of the main emerging challenges to food production and security in Asia, Middle-East, and Africa, especially climate risks and resource scarcity. Building on the findings and conclusions, the current volume aims at identifying and exploring the best ways to overcome such challenges and to build resilience in almost the same regions of the Global South.

I have been honored to share the editorship of this volume with my colleagues: Dr. Mirza Barjees Baig, Prince Sultan Institute for Environmental, Water, and Desert Research, King Saud University, Saudi Arabia; Dr. Mohamed Taher Sraïri, Department of Animal Production and Biotechnology and Head of the School of Agricultural Sciences, Hassan II Agronomy and Veterinary Medicine Institute, Rabat, Morocco; and Dr. Abdlmalek A. Alsheikh and Dr. Ali Wafa A. Abu Risheh, Prince Sultan Institute for Environmental, Water, and Desert Research, King Saud University (KSU), Riyadh, Saudi Arabia. I seize this opportunity to warmly thank all of them for their collaboration and support during the publishing process. Their professionalism, expertise, and intellectual capacity made the editing process an exciting and instructive experience and definitely contributed to the quality of this publication.

I would also like to seize this opportunity as well to pay tribute to all chapters' authors without whom this valuable and original publication could not have been produced. Their collaboration, reactivity, and engagement during the process were very remarkable and impressive.

The chapters published in this volume are also the result of the invaluable contribution made by peer-reviewers, who generously gave their time and energy to provide insight and expertise regarding the volume's content. On behalf of my co-editors, I would specifically like to acknowledge, with sincere and deepest thanks, the following peer-reviewers:

- Dr. Abdelaaziz Aitali, Senior Economist, Policy Center for the New South, Morocco
- Dr. Abdelmajid Saidi, Department of Economics, University Moulay Ismail, Morocco
- Dr. Aftab Alam, AliGarh Muslim University, India
- Dr. Ashrafuzzaman Chowdhury, Assistant Professor, Department of Economics, J.B. College, India
- Dr. Azra Musavi AliGarh Muslim University, India
- Dr. B. G. N. Sewwandi, University of Kelaniya, Sri Lanka
- Dr. Bandara Wanninayake, University of Kelaniya, Sri Lanka
- Dr. Boubaker Dhehibi, Senior Agricultural Resource Economist, Sustainable Intensification and Resilient Production Systems Program (SIRPS International), Center for Agricultural Research in the Dry Areas (ICARDA), Jordan
- Dr. Dastgir Alam, AliGarh Muslim University, India
- Dr. Duk-Byeong Park, Department of Regional Development, Kongju National University, South Korea
- Dr. Ekaterina Arabska, Vice-Dean, University of Agribusiness and Rural Development, Bulgaria
- Dr. Fatima Ezzahra Mengoub, Economist, Policy Center for the New South, Morocco
- Dr. Foued Amari, Senior Research Associate, Ohio State University, United States
- Dr. Guillaume Baggio, Research Associate, UNU-INWEH, Hamilton, Canada
- Dr. H. S. R. Rosairo, Sabaragamuwa University of Sri Lanka, Sri Lanka
- Dr. Jan W Hopmans, Distinguished Professor Emeritus, Soil Science and Irrigation Water Management, University of California Davis, USA
- Dr. Khaled Abbas, Research Director, National Institute of Agronomic Research (INRA), Algeria
- Dr. Larbi Toumi, Chief Engineer/Agro Socio Economist, General Secretariat, Ministry of Agriculture, Fisheries, Rural Development, Water, and Forests, Morocco
- Dr. M. Kamal Sheikh, Pakistan Agriculture Research Council (PARC), Pakistan
- Dr. M. M. M. Najim, Vice-Chancellor, South Eastern University of Sri Lanka, Professor in Environmental Conservation and Management, University of Kelaniya, Sri Lanka
- Dr. Manzoor Ahmad Malik, Director (R), Pakistan Council of Research in Water Resources (PARC), Pakistan
- Dr. Maqsood Anwar, Pakistan Pir Mehr Ali Shah Arid University
- Dr. Michael R. Reed, Emeritus Professor, Agricultural Economics, University of Kentucky, United States
- Dr. Mohammad Aftab, Department of Economic Development, Jobs, Transport and Resources, Government of Victoria, Australia
- Dr. Mohammed Bashir Umar, Federal University, Gashua, Yobe State, Nigeria
- Dr. Muhammad Ashfaq, Institute of Agricultural and Resource Economics, Faculty of Social Sciences, University of Agriculture, Pakistan
- Dr. Muhammad Munir, Pakistan Agriculture Research Council (PARC), Pakistan

- Dr. Muhammad Qaiser Alam, Associate Professor, Department of Economics, D.S (Post Graduate) College, Aligarh (Uttar Pradesh), India
- Dr. Peyman Falsafi, Assistant Professor, Agricultural Extension Education, Head of Halal Food Commission, Ministry of Jihad-e-Agriculture, Islamic Republic of Iran
- Dr. R. Kirby Barrick, Emeritus Professor and former Dean, Agricultural Education, University of Florida, United States
- Dr. Shaikh Shamim Hasan, Head, Department of Agricultural Extension and Rural Development, Bangabandhu Sheikh Mujibur Rahman Agricultural University (BSMRAU), Bangladesh
- Dr. Syed Ghazanfar Abbas, National Consultant (Innovation, E-Agriculture & E-Learning), Food and Agriculture Organization of the United Nations (FAO), Pakistan
- Dr. Usman Haruna, Federal University Dutse, Nigeria
- Dr. Usman Mustafa, Pakhtunkhwa Economic Policy Research Institute (PEPRI), AWK University, Pakistan
- Dr. Zakir Hussain, Ex-Vice-Chancellor/Adjunct Professor, Metropolitan College, New York Wilmington, United States.

Agadir, Morocco					Mohamed Behnassi

About the Publishing Institutions

The Center for Environment, Human Security and Governance (CERES), Morocco

CERES, previously the North-South Center for Social Sciences (NRCS), 2008–2015, is an independent and not-for-profit research institute founded by a group of Moroccan researchers and experts in 2015 and joined by many partners worldwide. It aspires to play the role of a leading think tank in the Global South and to serve as a reference point for relevant change processes. Since its creation, CERES managed to build a robust network involving various stakeholders such as researchers, experts, Ph.D. Students, decision-makers, practitioners, and journalists from different spheres and scientific areas. These achievements are being rewarded by the invitation of CERES members to contribute to global and regional assessments and studies (especially Ipbes, Medecc, EuroMeSco, etc.) and the invitation of the Center to become a member of the MedThink 5 + 5, which aims at shaping relevant research and decision agendas in the Mediterranean Basin. The Center has organized so far five international conferences and several training/building capacity workshops, provided expertise for many institutions, and published numerous books, scientific papers, and studies which are globally distributed and recognized. These events and publications cover many emerging research areas mainly related to the human-environment nexus from a multidimensional, multiscale, interdisciplinary, and policy-making perspectives. Through its initiatives, the CERES attempts to provide expertise, to advance science and its applications, and to contribute to effective science and policy interaction.

The Prince Sultan Institute for Environmental, Water and Desert Research (PSIEWDR), King Saud University, Riyadh, Kingdom of Saudi Arabia

PSIEWDR was established in 1986 to conduct scientific research related to environmental issues and water resources. It also engages with vital issues related to the problem of aridity and the desert environment. It conducts development initiatives for the country's desert areas, particularly programs for combating desertification in the Arabian Peninsula. PSIEWDR designed and carried out two major water harvesting and storage programs, including the construction of purpose-built infrastructure, throughout the Kingdom of Saudi Arabia using novel techniques and equipment. The Institute actively applies remote sensing technologies using advanced satellite image processing systems and GIS to study the country's environment and natural resources. In 2007, the Institute published *The Space Image Atlas of the Kingdom of Saudi Arabia*, and it is currently developing *The Environmental Atlas of the Kingdom of Saudi Arabia*. The Institute has been the primary sponsor of the biennial International Conference on Water Resources and Arid Environments (ICWRAE) held in Riyadh, Saudi Arabia since 2004. The institute hosts the General Secretariat of the Prince Sultan Bin Abdulaziz International Prize for Water (PSIPW) which honors scientists all over the world for their innovative water-related research. PSIPW, in turn, has many agreements with various international water associations as well as a close partnership with the United Nations. PSIPW and the United Nations Office of Outer Space Affairs (UNOOSA) jointly produce and maintain the International Space4Water Portal, an online hub for all stakeholders involved in utilizing space technologies for water resources applications.

Contents

1 **Food Security and Climate-Climate Smart Food Systems—An Introduction** 1
Mohamed Behnassi, Mirza Barjees Baig, Mohamed Taher Sraïri, Abdlmalek A. Alsheikh, and Ali Wafa A. Abu Risheh

2 **A Co-Evolving Governance Perspective of Climate-Smart Agriculture and Some Important Yet Unaddressed Elements for Integrated Gains** 15
Mohamed Behnassi, Gopichandran Ramachandran, and Gireesh Chandra Tripathi

3 **Rethinking the Agricultural Development Model in Post-COVID-19 Era Based on Scientific Knowledge: The Moroccan Case** 33
Mohamed Taher Sraïri

4 **Impacts of Climate Change on Agricultural Labour Force and Food Security in Pakistan: The Importance of Climate-Smart Agriculture** 51
Syed Ghazanfar Abbas, Mirza Barjees Baig, and Gary S. Straquadine

5 **Impacts of Climate Change on Horticultural Crop Production in Sri Lanka and the Potential of Climate-Smart Agriculture in Enhancing Food Security and Resilience** 67
W. M. Wishwajith W. Kandegama, Rathnayake Mudiyanselage Praba Jenin Rathnayake, Mirza Barjees Baig, and Mohamed Behnassi

6	**Managing Climate Change Impacts on Food Production and Security in Sri Lanka: The Importance of Climate-Smart Agriculture**	99
	Mohamed Mujithaba Mohamed Najim, V. Sujirtha, Muneeb M. Musthafa, Mirza Barjees Baig, and Gary S. Straquadine	
7	**Impacts of Covid-19 Pandemic on Agriculture Industry in Sri Lanka: Overcoming Challenges and Grasping Opportunities Through the Application of Smart Technologies**	117
	Prasanna Ariyarathne S. M. W., W. M. Wishwajith W. Kandegama, and Mohamed Behnassi	
8	**Integrated Transition Toward Sustainability: The Case of the Water-Energy-Food Nexus in Morocco**	141
	Afaf Zarkik and Ahmed Ouhnini	
9	**Food Security in the Kingdom of Saudi Arabia Face to Emerging Dynamics: The Need to Rethink Extension Service**	157
	Mirza Barjees Baig, Khodran H. AlZahrani, Abdulmalek A. Al-shaikh, Ali Wafa A. Abu Risheh, Gary S. Straquadine, and Ajmal M. Qureshi	
10	**Role of Agriculture Extension in Ensuring Food Security in the Context of Climate Change: State of the Art and Prospects for Reforms in Pakistan**	189
	Abdullah Bin Kamal, Muhammad Kamal Sheikh, Bismah Azhar, Muhammad Munir, Mirza Barjees Baig, and Michael R. Reed	
11	**Role of Agricultural Extension in Building Climate Resilience of Food Security in Ethiopia**	219
	Burhan Ozkan, Ahmed Kasim Dube, and Michael R. Reed	
12	**The Nexus of Climate Change, Food Security, and Agricultural Extension in Islamic Republic of Iran**	241
	Peyman Falsafi, Mirza Barjees Baig, Michael R. Reed, and Mohamed Behnassi	
13	**Better Crop-Livestock Integration for Enhanced Agricultural System Resilience and Food Security in the Changing Climate: Case Study from Low-Rainfall Areas of North Africa**	263
	Mina Devkota, Aymen Frija, Boubaker Dhehibi, Udo Rudiger, Veronique Alary, Hatem Cheikh M'hamed, Nasreddine Louahdi, Zied Idoudi, and Mourad Rekik	
14	**Realizing Food Security Through Agricultural Development in Sudan**	289
	Sharafeldin B. Alaagib, Imad Eldin A. Yousif, Khaled N. Alrwis, Mirza Barjees Baig, and Michael R. Reed	

15	**Agriculture in Fragile Moroccan Ecosystems in the Context of Climate Change: Vulnerability and Adaptation** .... Larbi Aziz	303
16	**Digitalization and Agricultural Development: Evidence from Morocco** .... Hayat Lionboui, Abdelghani Boudhar, Youssef Lebrini, Abdelaziz Htitiou, Fouad Elame, Rachid Hadria, and Tarik Benabdelouahab	321
17	**Natural and Regulatory Underlying Factors of Food Dependency in Algeria** .... Amel Bouzid, Messaoud Lazereg, Slimane Bedrani, Mohamed Behnassi, and Mirza Barjees Baig	339
18	**Boosting Youth Participation in Farming Activities to Enhance Food Self-Sufficiency: A Case Study from Nigeria** .... Omowumi A. Olowa and Olatomide W. Olowa	363

Postface 379

Biographical Notes of Authors 383

About the Editors

Dr. Mohamed Behnassi is a Full Professor and Head of Public Law in French Department at the College of Law, Economics, and Social Sciences of Agadir, Ibn Zohr University, Morocco. He is a Senior Researcher of international law and politics of environment and human security. He holds a Ph.D. in International Environmental Law and Governance (2003), a MSc. in Political Sciences (1997), and a B.A. of Administration (1995) from Hassan II University of Casablanca. He obtained a Diploma in International Environmental Law and Diplomacy from the University of Eastern Finland and UNEP, 2015. He is also an Alumnus of the International Visitors Leadership Program of the Department of State, United States of America. Dr. Behnassi is currently the Founding Director of the Center for Environment, Human Security and Governance (CERES). From 2015 to 2018, he was the Director of the Research Laboratory for Territorial Governance, Human Security and Sustainability (LAGOS). Recently, he was appointed as Expert Evaluator for the National Center for Scientific and Technical Research (CNRST/Morocco), and selected twice (2019–2024) as an Assessment Scoping Expert and a Review Editor by the Intergovernmental Science-Policy Platform on Biodiversity and Ecosystem Services (IPBES) and Member of the Mediterranean Experts on Climate and Environmental Change (MEDECC). He was also selected by The Intergovernmental Panel on Climate Change (IPCC) as Expert Reviewer of the 1st Order Draft of the Synthesis Report (SYR) of the IPC—VI Assessment Report (AR6). Accordingly, he was among the Lead Authors of the 1st Assessment

Report (MAR1): Climate and Environmental Change in the Mediterranean Basin - Current Situation and Risks for the Future (MEDECC, 2021). Dr. Behnassi has published considerable number of scientific papers and book chapters in addition to 20 books, including recent ones on: Social-Ecological Systems in the Era of Risks and Insecurity—Pathways to Viability and Resilience (Springer, 2021); Building Resilience for Food and Water Security Face to Climate Change and Biodiversity Decline - Perspectives from Asia, Middle-East and Africa (Springer, 2021); and Human and Environmental Security in the Era of Global Risks (Springer International Publishing, 2019). Dr. Behnassi serves as a reviewer for many global publishers (such as Routledge and Springer) and scientific journals with high impact factor. He has organized many international conferences covering the above research areas, managed many research and expertise projects, and is regularly requested to provide scientific expertise nationally and internationally. Other professional activities include social compliance auditing and consultancy by monitoring human rights at work and the sustainability of the global supply chain.

Dr. Mirza Barjees Baig is a Professor at the Prince Sultan Institute for Environmental, Water, and Desert Research, King Saud University, Saudi Arabia. He earned his MS degree in International Agricultural Extension in 1992 from the Utah State University, Logan, Utah, USA, and was placed on the "Roll of Honor". He completed his Ph.D. in Extension for Natural Resource Management from the University of Idaho, USA, and was honored with the "1995 outstanding graduate student award". Dr. Baig has published extensively on the issues associated with natural resources in national and international journals. He has also presented issues in agriculture and natural resources and the role of extension education at various international conferences. Food waste, water management, degradation of environment and natural resources, and their relationship with society/community are his areas of interest. He has attempted to develop strategies to conserve natural resources, promote environment, and develop sustainable communities. Dr. Baig started his scientific career in 1983 as a researcher at the Pakistan

Agricultural Research Council, Islamabad, Pakistan. He served at the University of Guelph, Ontario, Canada, as the Special Graduate Faculty from 2000 to 2005. He served as a Foreign Professor at the Allama Iqbal Open University (AIOU), Pakistan, through Higher Education Commission from 2005 to 2009. He also served as a Professor of Agricultural Extension and Rural Society at the King Saud University, Saudi Arabia, from 2009 to 2020. He is a member of the editorial boards of many international journals and professional organizations.

Dr. Mohamed Taher Sraïri is a Senior Lecturer in Animal Science at the Hassan II Agronomy and Veterinary Medicine Institute (IAV), Rabat, Morocco. He is currently the Head of the School of Agricultural Sciences at IAV. Dr. Sraïri completed his Ph.D. in 2004 from the Gembloux Agro-Bio Tech University, Belgium. He has an experience of 26 years in research on topics related to crop-livestock systems in Morocco with a focus on dairy production. His research activities have allowed him to publish more than 60 peer-reviewed articles in international journals, as well as the participate in several conferences dedicated to livestock and agricultural sciences. He has also been associated with reviewing several manuscripts submitted to international scientific journals. He is mainly interested in crop-livestock integration, with an emphasis on resources use efficiency, particularly land, water, and work. As a consequence, he had achieved on-farm research in semi-arid to arid areas on water productivity in crop-livestock systems, in contexts where farms are in majority small-holder units, with a focus on the origin of water (whether rainfall or irrigation from surface or groundwater sources). This has necessitated an accurate characterization of the resources uses (volumes of water and hours of work) and the yields and profitability they allowed by crops and livestock products. He is currently involved in a global research network dealing with work uses in farming systems and their consequences on the attractiveness of agriculture, particularly for young generations. He is also involved in research activities with a transdisciplinary team in the oases areas, where global changes (climate, social, etc.) are severely impacting farming systems, putting at risk their resilience, in a context of severe resource scarcity.

Dr. Abdlmalek A. Alsheikh is the Director of the Prince Sultan Institute for Environmental, Water, and Desert Research at the King Saud University (KSU) in Riyadh, Saudi Arabia. He earned his Ph.D. in Arid Land Studies from the University of Arizona, USA, in 1983. He also serves as the General Secretary of the Prince Sultan bin Abdulaziz International Prize for Water (PSIPW). The Prize is awarded bi-annually at the United Nations Headquarters. Being the Director and Research Chair, he oversees the Research Programs at the Institute. He is the team leader of many research projects concerning the environment, desertification, climate change, and water harvesting in Saudi Arabia including the King Fahd Project for Water Harvesting and Storage in Saudi Arabia, Prince Sultan Project for Villages and Hamlets Rehabilitation in Saudi Arabia, and the Space Images Atlas of Saudi Arabia and the Environmental Atlas of Saudi Arabia. Being the Chairman of the international conferences on water resources and arid environments, he successfully organized them at the King Saud University. He has brought world fame water scientists to Saudi Arabia and published 9 proceedings of organized international conferences. He has published extensively in journals of international repute.

Eng. Ali Wafa A. Abu Risheh earned his M.Sc. Degree in Applied Remote Sensing from Cranfield University in the UK in 1987. He is the Head of the Department of Remote Sensing and GIS at the Prince Sultan Institute for Environmental, Water, and Desert Research at King Saud University in Riyadh. He is also the Executive Director of the Prince Sultan Bin Abdulaziz International Prize for Water and a Researcher for the Prize's dedicated research chair at KSU. He has overseen a number of projects involving remote sensing, including the Space Image Atlas of Saudi Arabia, the King Fahd Project for Water Harvesting and Storage in Saudi Arabia, and recently the Atlases of the Environmental Systems and the Sustainable Development of Saudi Arabia.

Abbreviations and Acronyms

AAV	Agricultural Added Value
ADCs	Agricultural Development Corporations
AES	Agricultural Extension Services
AESA	Agriculture Extension in South Asia
AGDP	Agricultural Gross Domestic Product
AI	Artificial Intelligence
AMEE	Moroccan Agency for Energy Efficiency
AOAD	Arab Organization for Agricultural Development
AREEO	Agricultural Research, Education and Extension Organization
ASCs	Agricultural Services Centers
ATVET	Agricultural Technical and Vocational Training
AVRDC	Asian Vegetable Research and Development Center
BDS	Basic Democracies System
CA	Conservation Agriculture
CABI	Centre for Agricultural Bioscience International
CD	Crop Diversification
CDIAC	Carbon Dioxide Information Analysis Center
CDM	Clean Development Mechanisms
CERES	Center for Environment, Human Security and Governance
CF	Collective Farms
CGIAR	Consultative Group for International Agricultural Research
CGMS	Crop Growth Monitoring System
CIAT	International Centre for Tropical Agriculture
CIMMYT	International Maize and Wheat Improvement Center
CLCA	Crop-Livestock under Conservation Agriculture
CLI	Crop-Livestock Integration
COTUGRAIN	Compagnie Grainière Tunisienne
CR	Climate Resilience
CRGE	Climate-Resilient Green Economy
CRP Livestock	CGIAR Research Program on Livestock
CSA	Climate-Smart Agriculture

CSP	Concentrating Solar Power
CV	Coefficient of Variance
CWR	Crop-Water Relationships
DEA	Data Envelopment Analysis
DOA	Department of Agriculture
DSR	Design Science Research
DZ	Dry Zone
EACCE	International Market Monitoring System
EIA	Environmental Impact Assessment
EPAS	Electronic Partner for Agro Services
FAO	Food and Agriculture Organization
FBOs	Farmer-based Organizations
FDA	Agricultural Development Fund
FFEM	French Facility for Global Environment
FFS	Farmer Field School
FGLS	Feasible Generalized Least Square
FSA	Food Security Assessment Survey
FTCs	Farmer Training Centers
GAM	General Census of Agriculture (Algeria)
GAP	Good Agriculture Practices
GCC	Golf Council Countries
GCF	Green Climate Fund
GCRI	Global Climate Risk Index
GDA	Groupements de Développement Agricole
GDP	Gross Domestic Product
GEF	Global Environment Facility
GEI	Group of Economic Interest Around Saffron
GHG	Greenhouse Gas
GMP	Green Morocco Plan
GoP	Government of Pakistan
GPS	Global Positioning System
GTP	Growth and Transformation Plan (Ethiopia)
ha	Hectare
ICARDA	International Center for Agricultural Research in the Dry Areas
ICARDA	International Centre for Agricultural Research in Dry Areas
ICIMOD	International Centre for Integrated Mountain Development
ICLS	Integrated Crop-Livestock Systems
ICRISAT	International Crops Research Institute for the Semi-Arid Tropics
ICT	Information and Communications Technology
ICT	Information Communication Technology
IEA	International Energy Agency
IEE	Initial Environmental Examination
IF	Individual Farms
IFAD	International Fund for Agricultural Development
IFPRI	International Food Policy Institute

ILRI	International Livestock Research Centre
INGC	Institut National des Grandes Cultures
INRA	National Institute of Agronomic Research (Morocco)
INRAT	Institut National de Recherche Agronomique de Tunisie
INRGREF	Institut National de la Recherche en Génie Rural, Eaux et Forêts
IoTs	Internet of Things
IPC	Integrated Food Security Phase Classification
IPCC	Intergovernmental Panel on Climate Change
IRDP	Integrated Rural Development Program
IRDP	Integrated Rural Development Program' (Pakistan)
IRESA	Institution de la Recherche et de l'Enseignement Supérieur Agricoles
IRRI	International Rice Research Institute
ITGC	Institut Technique des Grandes Cultures
ITU	International Telecommunication Union
IVRs	Interactive Voice Responses
IWMI	International Water Management Institute
IZ	Intermediate Zone
KPK	Khyber Pakhtoon Khwa
KSA	Kingdom of Saudi Arabia
LDL	Low-Density Lipoprotein
MASEN	Moroccan Agency for Sustainable Energy
MEL	Monitoring, Evaluation and Learning
MEWA	Ministry of Environment, Water and Agriculture (KSA)
ML	Machine Learning
MODIS	Moderate Resolution Imaging Spectroradiometer
MTH	Mega Trading Hub
NA	North Africa
NADP	National Agricultural Development Program (Algeria)
NAES	National Agricultural Extension Service
NARS	National agricultural research system
NCCP	First National Climate Change Policy (Pakistan)
NCDs	Non-Communicable Diseases
NCP	National Water Plan (Morocco)
NDA	National Designated Authority
NFPP	National Food Production Program
NGOs	Non-Governmental Organizations
NLP	Natural Language Processing
OEP	Office de l'Elevage et des Pâturages
ONEE	Office National de l'Electricité et de l'Eau Potable
ORMVAO	Ouarzazate Regional Agricultural Development Office
PAD&SC	Punjab Agricultural Development and Supply Corporation
PARC	Pakistan Agricultural Research Council
PASDEP	Plan for Accelerated and Sustainable Development to End Poverty (Ethiopia)

PICCPMV	Project for the Integration of Climate Change in the implementation of Morocco Green Plan
PITB	Punjab Information Technology Board
PMP	Positive Mathematical Planning
PSIEWDR	Prince Sultan Institute for Environmental, Water, and Desert Research
PSIPW	Prince Sultan bin Abdulaziz International Prize for Water
Q&A	Questions and Answers
R&D	Research and Development
R4D	Research for Development
SAAB	Saudi Arabian Agricultural Bank
SABA	System of Agricultural Aid and Subsidies
SAGO	Saudi Grains Organization
SD	Standard Deviation
SDGs	Sustainable Development Goals
SDPRP	Sustainable Development and Poverty Reduction Program (Ethiopia)
SMSA	Société Mutuelle de Services Agricoles
TND	Tunisien Dinar (1 TND=0.36 US$ - average Jan July 2021)
TRC	Telecommunication Regulatory Commission
UAA	Utilized Agricultural Area
UAL	Utilized Agricultural Land
UAVs	Unmanned Aerial Vehicles or Drones
UNDP	United Nations Development Program
UNFCCC	United Nations Framework Convention on Climate Change
UNFPA	United Nations Population Fund
UNOOSA	United Nations Office of Outer Space Affairs
UNPRI	United Nations Principles for Responsible Investing
US$	United States Dollar
USAID	United States Aid Agency
USDA	United States Department of Agriculture
USGS	United States Geological Survey
V-AID	Village Agricultural and Industrial Development Programme
VDSR	Very Deep Super-Resolution
WEF	Water-Food-Energy
WFP	World Food Program
WHO	World Health Organization
WWF	World Wildlife Fund
WZ	Wet Zone

List of Figures

Chapter 2

Fig. 1	Africa CSA vision 25 × 25—Africa's strategic approach for food security and nutrition in the face of climate change. *Source* NEPAD (2014: 9)	26

Chapter 3

Fig. 1	Regional average rainfall levels in Morocco	37
Fig. 2	Renewable water availability *per capita* per year in the Middle East and North Africa (MENA) countries. *Source* The World Bank (2017)	38
Fig. 3	Inter-annual variability of the output of cereal crops (1981–2019). *Source* Adapted from ONICL (2020)	40
Fig. 4	Evolution of the annual raw milk output in Morocco (1961–2019). *Source* FAOSTAT (2020)	43

Chapter 4

Fig. 1	Pakistan population trend (1950–2050). *Source* World Population Review—Live	55
Fig. 2	Area under wheat crop in Pakistan (2018–2019)	58
Fig. 3	Scenario of the world economy in 2050 with variables like GDP, population, capital, skilled and unskilled labour	59

Chapter 5

Fig. 1	Agro-ecological regions in Sri Lanka	70
Fig. 2	GDP of agriculture in Sri Lanka (2005–2019). *Source* https://data.worldbank.org/indicator	76
Fig. 3	Export performance of fruits and vegetables sector (2011–2021). *Source* Export Development Board (2021)	79

Fig. 4	Increasing demand for value-added production in Sri Lanka. *Source* World Bank (2021)	80
Fig. 5	Agricultural losses due to Natural Disasters from 1974–2008. *Source* UNDP, Sri Lanka	83
Fig. 6	Comparison of carbon sequestration potential of tea lands among different tea growing regions with different densities of shade trees relating to different levels of compliance with TRI recommendations. HS: High Shade; MS: Medium Shade; LC: Low Country; MC: Mid Country; UC: Up Country. *Source* Wijeratne et al. (2014)	92

Chapter 6

Fig. 1	Conceptual framework—Adoption of CSA practices among small-scale farmers	111

Chapter 7

Fig. 1	Three cycle view of design science research. *Source* Hevner (2007: 2)	119
Fig. 2	Cellular mobile telephone subscriptions (2000–2019). *Source* Sri Lanka telecommunication regulatory commission	124
Fig. 3	Conceptual Agriculture product management system. *Source* Ginige et al. (2016)	129
Fig. 4	Segments of Present agriculture trading network in Sri Lanka. *Source* Developed by the authors	130
Fig. 5	Sketch of the proposed EPAS technical architecture. *Source* Developed by the authors	131
Fig. 6	Proposed Modules of EPAS development. *Source* Developed by the authors	133

Chapter 8

Fig. 1	Percentage of yield reduction, according to scenarios A2 and B2, by 2100. *Source* Gommes et al. (2009)	145
Fig. 2	GDP growth rate and AAV (% GDP). *Source* World Bank Database	147
Fig. 3	GDP growth rate and household consumption. *Source* DEPF (2019)	148
Fig. 4	Changes in hydropower capacity factor in Morocco, 2020–2099, relative to the baseline 2010–2019. *Source* IEA (2020)	153

Chapter 9

Fig. 1	Saudi Arabia on the map. *Source* FAOSTAT (2021)	159
Fig. 2	Water usage by the agricultural, municipal and industrial sectors. *Source* Al-Subaiee (2018)	166

Fig. 3	Wheat imports and export. *Source* http://statistics.amis-out look.org/data/index.html#	172

Chapter 10

Fig. 1	Overview of Agriculture in Pakistan. *Source* Statistical Supplement of Pakistan Economic Survey 2019–2020 (GoP 2020)	192
Fig. 2	Current system of Agriculture Extension and Adaptive Research. *Source* Extension and Punjab (2020)	206
Fig. 3	The Organogram. *Source* Extension and Punjab (2020)	207
Fig. 4	Farmers' training session in progress. *Source* Extension and Punjab (2020)	209
Fig. 5	Three Plant Clinics in Progress. *Source* Extension and Punjab (2020)	210

Chapter 11

Fig. 1	Number of agricultural extension agents per 10,000 farmers in selected countries. *Source* Davis et al. (2010)	232

Chapter 12

Fig. 1	Map of Iran (left) and a picture on the climate diversity in Iran (right)	242
Fig. 2	Rural and urban population of Iran, 1990–2018. *Source* FAOSTAT (2019)	243
Fig. 3	Annual temperature of Iran, 1901–2016. *Source* FAOSTAT (2019)	243
Fig. 4	Iran's arable lands, under permanent crop and pastures lands, 1961–2015. *Source*: FAOSTAT (2019)	245
Fig. 5	Iran historical wheat production, 2001–2012. *Source*: FAOSTAT (2019)	248
Fig. 6	A picture of Khuzestan flood in southern Iran in 2019 (left) and Map of flood-prone areas of Iran (right)	251
Fig. 7	FAO map of the prevalence of desert migratory locusts in different countries, 2019. *Source*: Dadashi (2020)	252
Fig. 8	Organizational chart of the Agricultural Research, Education, and Extension Organization (AREEO)	254
Fig. 9	International cooperation of AREEO	254
Fig. 10	Applying ICT-based instruments in farmers' lands	257

Chapter 13

Fig. 1	Sheep grazing stubble in Fernana region—North West Tunisia. *Photo* Zied Idoudi (ICARDA)	269

Fig. 2	Relationship between biomass of residues (%) on soil surface and grazing duration (day). (Author's own work)	270
Fig. 3	Clustering Crop-Livestock Integration (CLI) options based on the scale of implementation and resource-orientations. Graphic Design: Zied Idoudi and Mourad Rekik (ICARDA)	272
Fig. 4	Traditional and manual seed cleaning by woman farmer. Photo: Zied Idoudi (ICARDA)	273
Fig. 5	Mobile seed cleaning and treatment unit. *Photo* Zied Idoudi (ICARDA)	274
Fig. 6	Training in use of mobile feed grinder. *Photo* Udo Rudiger (ICARDA)	275
Fig. 7	Partnership for upscaling crop-livestock integration. *Note* NARES = national agricultural research and extension systems. *Source* Frija and Idoudi (2020)	280

Chapter 14

Fig. 1	GDP growth rate during the period 2000–2017	290
Fig. 2	Forest Area in Sudan, 1990–2015	294
Fig. 3	Cereal production in Sudan (000 tons), 1990–2017	294
Fig. 4	Animal numbers during 1990–2018 (1000 heads)	295
Fig. 5	Sudan employment in agriculture, 1991–2017	298

Chapter 15

Fig. 1	Localisation of the two study sites	305

Chapter 16

Fig. 1	Socio-economic importance of agriculture in Africa. Average 1991–2020 calculated based on the World Bank data. *Source* World Bank (2021a)	323
Fig. 2	Agricultural land monitoring steps using remote sensing. *Source* Developed by the authors	325
Fig. 3	Schematic presentation of agricultural land classification steps	327
Fig. 4	The disaggregation levels of the proposed water management modeling framework	330
Fig. 5	Average estimated yield from 2001 to 2016. *Source* Developed by the authors	332
Fig. 6	Trend of wheat production in the main agricultural areas (2001 to 2016). *Source* Developed by the authors	333
Fig. 7	Risk of losses in wheat production value in agricultural areas	334

Chapter 17

Fig. 1	Trend for indicator values—Algeria. *Source* Global Hunger Index (2020)	342
Fig. 2	Evolution of the Gross Domestic product per capita. *Source* Elaborated based on FAO's data (2020)	342
Fig. 3	Prevalence of severe food insecurity. *Source* Elaborated based on FAO's data (2020)	343
Fig. 4	Volume growth rate of agricultural production (2000–2015). *Source* Office for National Statistics (2017)	350
Fig. 5	Food Import Dependency rate 1997–2017, percentage points. *Source* The Economist (May 9th, 2020)	351

Chapter 18

Fig. 1	Map of Ogun State ADP zones and blocks showing study location	366

List of Tables

Chapter 3

Table 1	Main imports of food commodities in Morocco, 2018 and 2019 ($\times 10^6$ US $)	41

Chapter 4

Table 1	Share of agriculture in total employment in Pakistan (%)	54
Table 2	Agriculture growth percentages (2011–2012 to 2017–2018)	57
Table 3	Wheat area by province, 2018–2019	58
Table 4	Rice growing zone of Pakistan	58
Table 5	Baseline development (percent change)	59
Table 6	Current wheat and rice production, consumptions and the future requirements (2030/2050)	60
Table 7	Projection on Pakistan's total, agricultural, and agricultural labor force (2030/2050)	61
Table 8	Energy, cost and time factors in minimum and conventional cultivation	63

Chapter 5

Table 1	Relief of vegetation recommended in agro-ecological regions for crop diversification with paddy lands	74

Chapter 7

Table 1	Agricultural export revenue in Sri Lanka, 2015–2019	123
Table 2	Distribution of cellular mobile subscriptions in provinces as of year 2019	124

Chapter 9

Table 1	Cereal production in Saudi Arabia	161
Table 2	Water Demand (million Cubic meters/year)	167

Chapter 10

Table 1	Agriculture Growth (Base = 2005–2006) (%)	193
Table 2	Food and Nutrition Insecurity Situation in Pakistan	195
Table 3	Budgetary allocations and expenditures incurred during last 12 years	211

Chapter 11

Table 1	Shares of crop, livestock, forestry, and fishery in AGDP (percent)	221
Table 2	Summary of some common CSA practices in Ethiopia	228

Chapter 13

Table 1	Summary of the impact of different agronomic innovations (technology from the Crop–Livestock under Conservation Agriculture [CLCA] initiative) on improving sustainability indicators in Algeria and Tunisia under rainfed conditions	277

Chapter 14

Table 1	Total and rural population in Sudan during 2014–2018 (in millions)	291
Table 2	Forest Area in Sudan, 1990–2015	293
Table 3	Animal products in Sudan, 1990–2019 (1,000 tons)	296
Table 4	Sudan employment in agriculture, 1991–2017	297
Table 5	Production and domestic consumption of important food commodities in Sudan, 1990–2016 (1,000 tons)	298
Table 6	Trend in production and consumption of important agricultural commodities in Sudan, 1990–2016	299

Chapter 15

Table 1	The surveyed population in the two sites	307

Chapter 17

Table 1	Quantitative evolution and composition of average food intake per capita (in Kcal) in selected North-African countries	347
Table 2	The evolution of the protein and fat composition of the diet in selected North-African countries (gr/capita/day)	348

Table 3	The sources of supply of the national market (average1990/2017)	351
Table 4	Rural areas indicators	353
Table 5	Algerian yields compared to proximate countries (Spanish yield Index = 100)	355
Table 6	Evolution of irrigated areas (in hectares)	356

Chapter 18

Table 1	Descriptive Statistics and socio-economic characteristics of youth in the Study area	369
Table 2	Youths' distribution according to rate of participation in farming activities	370
Table 3	Farming activities in which youths participated	371
Table 4	Estimates of Logit Regression on factors determining the youth's decision to participate in farming activities	372
Table 5	Poisson estimates on determinants of hours spent by youth in farming activities	374

List of Boxes

Chapter 4

Box 1   Estimated Agricultural Labour Force Efficiency   53

Chapter 8

Box 1   National Program for Irrigation-Water Saving and the Solar Pumping Project (an example of the WEF nexus)   150

Chapter 9

Box 1   Greenhouse Crop Production   170

Chapter 1
Food Security and Climate-Climate Smart Food Systems—An Introduction

Mohamed Behnassi, Mirza Barjees Baig, Mohamed Taher Sraïri, Abdlmalek A. Alsheikh, and Ali Wafa A. Abu Risheh

Abstract A healthy planet provides the foundation for the sustainability and resilience of social-ecological systems. However, the recent evolution of human society has impacted the vital balance of such systems in dangerous ways, with some irreversible trends. The conventional growth models are to be blamed for current environmental and climate disruptions, perceived as the dynamics behind the onset of the Anthropocene era. Coinciding with this, profound changes in the way food is grown, processed, distributed, consumed, and wasted have resulted in food systems which accelerate global warming, biodiversity loss, shifts in nutrient cycles, and land-use change beyond planetary boundaries. Such systems are, in turn, severely impacted by the consequences of the ongoing environmental and climate disruptions. To date, the prevailing paradigm is still focusing on the 'feed the world' or the 'productivist' narrative, which has resulted in unsustainable, vulnerable, and exclusive food systems. Against this background, this introductory chapter outlines the overall framework on which the volume is shaped while presenting the scope, objectives, and key findings of subsequent chapters. It highlights the opportunities to accelerate the transformation of current food systems towards sustainability

M. Behnassi (✉)
International Politics of Environment and Human Security, College of Law, Economics and Social Sciences, Ibn Zohr University of Agadir, Agadir, Morocco
e-mail: m.behnassi@uiz.ac.ma

Center for Environment, Human Security and Governance (CERES), Agadir, Morocco

M. B. Baig
Prince Sultan Institute for Environmental, Water and Desert Research, King Saud University, Riyadh, Saudi Arabia
e-mail: mbbaig@ksu.edu

M. T. Sraïri
Department of Animal Production and Biotechnology, Head of the School of Agricultural Sciences, Hassan II Agronomy and Veterinary Medicine Institute, Rabat, Morocco
e-mail: mt.srairi@iav.ac.ma

A. A. Alsheikh · A. W. A. A. Risheh
King Saud University, Environmental, Water and Desert Research, Riyadh, Kingdom of Saudi Arabia
e-mail: aasheikh@ksu.edu.sa

© The Author(s), under exclusive license to Springer Nature Switzerland AG 2022
M. Behnassi et al. (eds.), *Food Security and Climate-Smart Food Systems*,
https://doi.org/10.1007/978-3-030-92738-7_1

and climate adaptation and mitigation, therefore enhancing the resilience of food security.

Keywords Social-ecological systems · Food systems · Climate change · Resilience · Sustainability

1 The Volume's Background, Scope, and Objectives

It is commonly agreed in the literature that a healthy planet provides the foundation for the sustainability and resilience of social-ecological systems. However, over the last decades, the evolution of many societies around the world—both in the Global North and South—has impacted the vital balance of ecosystems in a dangerous way, with some irreversible trends. As a consequence, many anthropogenic dynamics are already approaching several Earth system thresholds. The conventional growth models based on which many societies are shaping their organization, production and consumption patterns, lifestyle, and well-being are to be blamed for the exhaustion and depletion of natural resources, and environmental and climate disruptions, which have prompted researchers to announce the onset of the Anthropocene era.

Currently, climate change is widely accepted as the single most pressing issue facing society. According to the recently published Sixth Assessment Report (AR6) of the Intergovernmental Panel on Climate Change (IPCC), each of the last four decades has been successively warmer than any decade that preceded it since 1850 and world is going to face more climatic crises. Global surface temperature in the first two decades of the twenty-first century (2001–2020) was 0.99 [0.84–1.10] °C higher than 1850–1900. Observed warming is driven by greenhouse gas (GHG) emissions from human activities. Since 2011, GHG concentrations have continued to increase in the atmosphere, reaching in 2019 annual averages of 410 ppm for carbon dioxide (CO_2), 1866 ppb for methane (CH_4), and 332 ppb for nitrous oxide (N_2O). In 2019, atmospheric CO_2 concentrations were higher than at any time in at least 2 million years (high confidence), and concentrations of CH_4 and N_2O were higher than at any time in at least 800,000 years (very high confidence). Continued emissions of GHGs will cause further warming and changes in the climate system. Based on the assessment of multiple lines of evidence, global warming of 2 °C, relative to 1850–1900, would be exceeded during the twenty-first century under the high and very high GHG emissions scenarios. The human-induced climate change is already affecting many weather and climate extremes in every region across the globe. Evidence of observed changes in extremes such as heatwaves, heavy precipitation, droughts, and tropical cyclones, and in particular, their attribution to human influence, has strengthened over some years.

Agriculture is one of the important economic sectors of the global society. However, climate change is having a profound influence on agroecosystems, posing serious threats to food security, human health, and protection of environment. Of the total annual crop losses in world agriculture, many are due to increasing frequency and magnitude of weather and climate extremes such as droughts, flash floods, untimely rains, frost, hail, and severe storms. Agricultural impacts from natural events and disasters most commonly include: alteration of ecosystems, contamination of water bodies, loss of harvest or livestock, increased susceptibility to disease, and destruction of irrigation systems and other agricultural infrastructure.

Coinciding with these dynamics, profound changes in the way food is grown, processed, distributed, consumed, and wasted have resulted in agriculture and food systems which accelerate global warming, biodiversity loss, shifts in nutrient cycles, and land-use change beyond planetary boundaries.

In addition, the ongoing Covid-19 pandemic has unprecedently exposed the vulnerability of such systems and their inherent inequality, injustice, and interlinked nature. Medium- and long- term impacts are expected to severely affect agriculture and food systems and the human security of vulnerable segments of populations. Even before the outbreak of Covid-19, the state of global food security and nutrition was already alarming, with an estimated average of 821 million people undernourished and poor nutrition causing stunting which is responsible for nearly 45 percent of the deaths of children under five. Moreover, there are more than 2 billion people in overweight or obese, adding pressure on health systems and questioning the efficiency of food education and consumption patterns.

To date, the prevailing paradigm is still focusing on the 'feed the world' or the 'productivist' narrative, thus giving more attention to the quantity of food and calories produced rather than characterizing the efficiency and equity of natural resource use. It is based on the assumption that doubling food production by 2050, maximizing yields, and basing food production on export-oriented models as the sole viable trajectory. This has resulted in unsustainable, vulnerable, and exclusive agriculture and food systems which support the industrialization of food production, the excessive reliance on foreign food markets, the focus on a limited number of globally traded crops, and the concentration of power and influence in the hands of a few, as well as the dissemination of a single food consumption model.

Within agriculture and food systems, five major trends have enabled widespread industrialization, including the discovery of inorganic chemical inputs, mechanization, specialization, consolidation, and market concentration. Industrialization in itself is not systematically inappropriate since it has lifted countless people out of poverty and enabled a wide range of quality-of-life improvements. However, such an option should not be solely oriented by profit and over-consumerism while sidelining other options which are more efficient and sustainable. Therefore, conceiving solutions to address the disadvantages of any advancement must be system-based and capable of offering viable solutions that maintain sustainability, resilience, and economic advantage.

Produced crops are often transformed into animal feed and ultra-processed food or used as inputs for bio-energy production. Many biophysical, environmental, and disease trends confirm that current agriculture and food systems are at crossroads, and this has the potential to jeopardize future food security. Indeed, these systems are excessively dependent on fossil fuels and high external inputs' uses (fertilizers, pesticides, seeds, etc.), resulting in environmental degradation, resource depletion, and greenhouse gas emissions. Also, agriculture and food systems are responsible for the accelerated pace of natural resource degradation and scarcity and climate change at the same time that they are affected by the induced impacts. Meanwhile, food safety risks, hazards, pests, and emerging diseases have wide-ranging impacts on food security. In many cases, such systems are at the root of eroding livelihoods and important social, cultural, and spiritual traditions. They also promote an economic system that results in liabilities due to hidden costs, global trade vulnerabilities, declining rural economies, and increased inequality.

Efforts to tackle the social-ecological costs and negative externalities of current evolutions of agriculture and food systems are generally considered in this prevailing narrative but often perceived as marginally important compared to the purpose of increasing food production to cater to the needs of the global population, meeting industry demands, and achieving high levels of profit. This situation is no longer viable given the growing biophysical, environmental, social, and ethical imperatives. Therefore, agriculture and food systems need to undergo transformative changes while countering the prevailing powerful narratives that often guide research, policy, practice, investment, and production and consumption. The alternative narratives are geared towards shaping agriculture and food systems that can appropriately nourish a growing global population while ensuring social-ecological sustainability and resilience. Such systems should be simultaneously decoupled from excessive resource use, environmental degradation, carbon emissions, and cultural and social disruptions. They should be resilient to shocks and able to constantly adapt to emerging changes and risks. The inclusive and redistribution potential of such systems is currently a growing social demand in many settings to ensure that no one is left behind and to avoid the continuous exclusive concentration of power and influence.

In such perspectives, the focus should be made on the quality of food produced so that it contributes toward healthy, equitable, renewable, resilient, just, inclusive, and culturally diverse agriculture and food systems. This approach facilitates the delivery of global commitments such as the Sustainable Development Goals (SDGs), the Paris Climate Agreement, the Convention on Biological Diversity, and the UN Decade for Action for Nutrition. Fully embracing the new narrative to prioritize the social-ecological sustainability and resilience will mean challenging prevailing mainstream narratives and mindsets, and transforming the physical, technical, socioeconomic, and political contexts that shape current agriculture and food systems.

On the ground, it should be noted that the scientific debate to shape future agriculture and food systems is already ongoing from many perspectives. Moreover, there have been important shifts in policy approaches to food security and nutrition that are informed by the evolving understandings of food security and agriculture and food systems thinking, technology innovation, and lessons learned from the practice or from the cumulative successes and failures. Policies that embrace these shifts often: support radical transformations of food systems; appreciate agriculture and food system complexity and interactions with other sectors and systems; focus on a broader understanding of hunger and malnutrition; and develop diverse policy solutions to address context-specific problems and needs. The sustainability and resilience of agriculture and food systems—which embody qualities that support the various dimensions of food security—are increasingly perceived as an imperative both globally and locally. Sustainable and resilient agriculture and food systems are: productive and prosperous (to ensure the availability of sufficient food); equitable and inclusive (to ensure access for all people to food and to livelihoods within that system); empowering and respectful (to ensure agency for all people and groups, including those who are most vulnerable and marginalized to make choices and exercise voice in shaping such systems); resilient (to ensure stability in the face of shocks and crises); regenerative (to ensure sustainability in all its dimensions); and healthy and nutritious (to ensure nutrient uptake and utilization).

In addition, recent years have seen the evolution of many technology, innovation, and infrastructure trends. There is indeed a growing support for innovation to ensure sustainable agricultural production methods—such as agroecology, sustainable intensification, innovative extension service, and climate-smart farming—although some controversies still persist over which of these approaches should be applied, how, and in which contexts. Also, many solutions are being developed and tested to overcome the ongoing weaknesses and postharvest handling and storage infrastructure, which still present serious challenges, including high levels of food loss and waste.

Moreover, the resilience of agriculture and food systems to shocks, especially environmental and climate risks, is gaining growing importance both in research, policy, and practice. 'Resilience' has been commonly defined from a social-ecological perspective as the capacity of socioeconomic systems to withstand shocks through absorption, adaptation, and transformation. It has been applied in various contexts to understand whether and how social-ecological systems could become more robust to shocks. Since agriculture and food systems are highly vulnerable to various risks and shocks, enhancing their resilience is becoming imperative. Moreover, addressing resilience from a food security perspective is important and can help focus research by building on existing best practices. In the literature, resilience is increasingly measured as an indicator of food security; higher resilience scores are assumed to be indicative of a better food security status.

Against this background, the timing of this contributed volume seems crucial. Based on the principle that transformational change will not occur without a shift of narrative and mindsets, the volume contributes to the scientific debate about the above-mentioned issues while highlighting the opportunities to accelerate the transformation of current agriculture and food systems towards sustainability, adaptation to and mitigation of climate change, therefore enhancing the resilience of food security. In addition to some conceptual research, most chapters are based on empirical investigations, particularly done in climate-vulnerable and resource-constrained countries from the Global South (Algeria, Ethiopia, Iran, Kingdom of Saudi Arabia, Morocco, Nigeria, Pakistan, Sri Lanka, Sudan, India, Thailand, and Tunisia), with challenging issues related to human development and security, and provide research and policy-oriented inputs and recommendations to guide change processes at multiple scales.

2 The Volume's Content

This volume is composed of 18 chapters, which can be presented as follows:

In Chap. 2, *A Co-Evolving Governance Perspective of Climate-Smart Agriculture and some Important yet Unaddressed Elements for Integrated Gains*, Behnassi, Gopichandran, and Tripathi claim that climate change presents multiple challenges, especially for agriculture; however, it cannot be addressed effectively in silos. Specific science, technology, finance, and market elements of agriculture, in addition to stakeholder-specific and cross-cutting approaches relevant to governance should be increasingly considered. In this perspective, the governance of climate-smart agriculture (CSA) has evolved in form and function over recent years with a growing diversity of thrusts, including for instance the water-energy-food-ecosystems nexus, the variety of stakeholders, and environmental, social, and governance (ESG) criteria. Simultaneously, deeper insights about policy cycles that relate to the continuum of incremental and drastic changes have emerged and set the context to infer areas that need greater attention. Accordingly, the authors attempt in their eclectic synthesis to highlight the emerging calls for accelerated CSA due to a deeper understanding of ecosystem science, social contexts, and synergies/trade-offs between productivity, mitigation, adaptation, and resultant resilience that should be sustained. While doing it, the authors also highlight the governance constraints and leads by reference to regional- and state- level frameworks and elaborate recommendations to guide future change processes in the area of CSA.

In Chap. 3, *Rethinking the Agricultural Development Model in Post-Covid-19 Era Based on Scientific Knowledge: The Moroccan Case*, Sraïri recalls the strategic socio-economic importance of the agricultural sector for Morocco while questioning its capacity to withstand shocks and risks, such as the effects of the current Covid-19 pandemic, which is associated with growing uncertainties about market opportunities and workers' availability, given the drop in consumers' incomes and the requested social distancing. This crisis, according to the author, was an opportunity for Moroccan citizens to realize the importance of agriculture, as it ensured a regular and affordable food supply during the lockdown. It was also an opportunity to revise the constraints facing this sector while being more sensitive to the food sovereignty imperative given the positive and negative implications of both food import and export, especially in terms of water security. Given such dynamics, the author believes that the post-Covid-19 agricultural sector's growth should be different from its conventional form and, to be so, there is a need for a paradigm shift where rain-fed agriculture has to be a political priority, with significant financial allocations. Moreover, such a path will have to be coupled with a change in farmers and consumers' awareness about the positive effects of short circuits, to decrease fossil energy inputs, as well as to promote a low external inputs' agriculture with an insight on the sustainability of diets. In such a perspective, the author analyses these issues while considering that it is not only food production which is at stake while addressing post-Covid-19 farming, but there are also other relevant issues such as environmental preservation, rural development, and sustainable food systems. This comprehensive vision should be based on scientific knowledge, to ensure social inclusiveness and sustainable use of scarce resources, to reduce rural exodus, and to guarantee the attractiveness of farming activities to youths.

In Chap. 4, *Impacts of Climate Change on Agricultural Labour Force and Food Security in Pakistan: The Importance of Climate-Smart Agriculture*, Abbas et al. focus on the case of Pakistan, given its high vulnerability to climate change and induced disasters, in addition to the different socio-economic implications of such a vulnerability coupled with an insufficient adaptive capacity. Building on the available climate research findings relevant to the country, the authors show that the impacts of climate change will significantly influence the productivity of water-dependent sectors—such as agriculture and energy—and the agricultural labour force since field workers' efficiency is likely to be affected due to higher temperatures and fatigue. Therefore, the impacts of climate change are forcing a change in habits and habitats. Against this background, the authors assess the effects of climate change on agricultural labour force, the related implications for food production and security, as well as the technological developments in these areas in the last few decades. The main objective of the authors is to inform relevant decision-making processes about current dynamics, with an emphasis on Pakistan. In other terms, the objective is to better assess, adapt, and bring issues before policy makers so that climate change associated risks are promptly addressed. Climate-Smart Agriculture (CSA) has been used by the authors as a framework of such an analysis.

In Chap. 5, *Impacts of Climate Change on Horticultural Crop Production in Sri Lanka and the Potential of Climate-Smart Agriculture in Enhancing Food Security and Resilience*, Kandegama et al. focus on the case of Sri Lanka where dramatic climatic changes are severely affecting crop production, food supply chain, and farmers' livelihood. This is happening in a context where the annual production hardly fulfils domestic food demand, a situation making food security an increased political priority. The authors emphasized on the various impacts climate change has on horticultural crops, which result in a reduction of yield and nutritional quality, while the perishable nature of harvest often worsens the loss level. To reverse such trends, the authors adopt the climate-smart agriculture (CSA) as a suitable approach to shift to a climate-based cropping system, thus allowing a mitigation of climate change impacts on horticultural crops while fostering food security. Through their analysis based on the assessment of the available literature on the country's current adaptation efforts, various issues were identified, mainly the continuous decline in livelihood of rural farmers and crop productivity, postharvest losses, destruction of the supply chain, and a high investment in food export. In light of the findings, the authors claim that horticultural crops could be used in CSA to address climate change and achieve food security in Sri Lanka due to an increase in productivity and resilience and a decrease of emissions. Also, CSA related practices help fill knowledge gaps and serve as a guide for future investments and development initiatives. The authors, however, recommend further research in these areas to determine the magnitude of climate change impacts on horticultural crop productivity and how such impacts vary depending on the agricultural geographical zone, variety, and customer demand.

By referring to the same geographical scope and using the same framework of analysis, Najim et al. in Chap. 6, *Managing Climate Change Impacts on Food Production and Security in Sri Lanka: The Importance of Climate-Smart Agriculture*, report the case of Sri Lanka, which is increasingly impacted by climate change, especially in the areas of food production, food security, and livelihoods. Instead of focussing only on crop productivity (food availability) as the main impacted area, the authors adopt a food system model to obtain a better perspective on food security issues in the country (especially food access and use). The diminishing agricultural productivity, food loss and wastage along supply chains, low rural poor subsistence resilience, and the prevalence of high under-nutrition and infant malnutrition are among the issues addressed by this model. In this research, the authors indicate that ensuring food security should be guided by an integrated approach that can promote the stability of the entire food system face to climate risks. Developing a climate-smart agricultural framework to tackle all aspects of food security is a recommended option in this perspective. Accordingly, the authors identify a knowledge gap in Sri Lanka regarding the impacts of climate change on food systems—especially the productivity of a diverse range of food crops, livestock, and fisheries—and recommend further research in this regard, including the avenues of an environmentally-induced nutritional insecurity.

In to the same geographical context, Ariyarathne et al. report in Chap. 7, *Impacts of Covid-19 Pandemic on Agriculture Industry in Sri Lanka: Overcoming Challenges and Grasping Opportunities Through the Application of Smart Technologies*, the case of the agriculture industry sector where stakeholders, including consumers, have been coming across several challenges such as food scarcity, dramatic price fluctuations of commodities, and difficulties in identifying markets for both product buying and selling. On the community level, challenges related to the timely purchase of planting materials and other agro-inputs, loss of income, the inadequacy of reliable advices and directions, and market uncertainties have generated many concerns with no way forward. Such an impasse created a sudden imbalance of the entire value chain of agriculture industry affecting almost all stakeholders in Sri Lanka. To overcome such challenges, the authors believe that the promotion of smart practices in the agriculture sector is a viable solution. From this perspective, they propose a smart-agriculture support system—developed according to the Design Science Research (DSR) Methodology—with the potential to virtually connect relevant stakeholders of the agriculture value chain. Such a system intends to regulate the price for goods and services while organizing a balanced supply and demand in a more informative and intelligent manner, thus provisioning electronic financial accounting facilities for subscribers. Timely dissemination of knowledge, advice, financial services, and linking agro-input suppliers are also embedded into the proposed model.

In Chap. 8, *Integrated Transition Toward Sustainability: The Case of the Water-Energy-Food Nexus in Morocco*, Zarkik et Ouhnini refer to the case of Morocco, which is at the forefront of a disaster due to its current condition as a climate change hotspot. Indeed, climate change-induced impacts, especially increasing temperatures coupled with decreasing precipitations, directly threaten the country's social-ecological systems. In such a context, the authors focused on the complex and interlinked food, energy, and water systems. With an increased water scarcity due to both over-use, degradation, and climate change, it has now become essential to manage such a resource carefully. Accordingly, the authors outline in their chapter the physical limitations of available resources in Morocco, and determine physical areas of intersection between the energy, water, and food sectors. They also explore the existing sectoral policies and adaptation strategies employed in the three sectors in order to evaluate them and assess whether they are evolving separately or co-evolving as part of a nexus system approach. Finally, the authors investigate the potential of seawater desalination as a key economic alternative that the country can develop to face climate change-induced water scarcity and its food security implications.

In Chap. 9, *Food Security in the Kingdom of Saudi Arabia Face to Emerging Dynamics: The Need to Rethink Extension Service*, Baig et al. analyse the food security in KSA in a context marked by climate change impacts and induced disasters (such as prolonged droughts and devastating floods), and the risks associated with an excessive reliance on food import given the various constraints facing domestic agricultural production, especially water scarcity. In such a context, the authors claim that the climate change-food security nexus has to be a top political priority and a growing concern for all stakeholders. In this change process, rethinking the roles of extension system through reform and the upgrade will be a promising initiative to ensure its potential in simultaneously addressing the impacts of climate change on food systems and responding to the growing needs of farmers. To support this option, the authors provide ample evidence that strengthening the agriculture extension system through sound scientific curriculum will help realize food security and address climate change. Workable strategies are outlined by the authors as well to assist decision makers in reforming the extension system with the objective to make it more productive and efficient in today's context.

In the same perspective, Bin Kamal et al. report in Chap. 10, *Role of Agriculture Extension in Ensuring Food Security in the Context of Climate Change: State of the Art and Prospects for Reforms in Pakistan*, the case of Pakistan where agriculture is a key socio-economic sector, which enables the country to make significant progress in food production over the last several decades. However, this sector is being faced with many challenges, resulting in a declining performance, including a slow rate of technological innovation and inadequate extension services and technology transfer. Due to this low performance, the country is shifting to food import to meet its domestic demands despite being one of the world's large growers of wheat, rice, cotton, fruits, and livestock. Consequently, food security has remained a key challenge due to the high population growth, rapid urbanization, low purchasing power, high price fluctuations, erratic food production, and inefficient food distribution systems. Furthermore, the agriculture sector is expected to be severely impacted by climate change-induced disasters (especially floods and droughts), thus threatening food security. To reverse such trends, the authors highlight the importance of coping strategies, including the renovation and empowerment of extension services—through agricultural knowledge and good practices—given their potential in ensuring food security, dealing with the emerging challenges in agriculture, and enhancing productivity and rural incomes.

In Chap. 11, *Role of Agricultural Extension in Building Climate Resilience of Food Security in Ethiopia*, Ozkan et al. start with the fact that climate change and related disasters are currently a leading cause of global food insecurity, affecting all its dimensions at unprecedented levels, thus undermining current efforts to eradicate hunger and undernutrition. On the African level, despite that agriculture is the backbone of many sub-Saharan countries' economies, food production has failed to keep up with fast-rising demands and this gap may become wider in the future due to climate change. In this context, Ethiopia has been identified by the authors as one of the most vulnerable countries since the variability of rainfall and increasing temperature cause frequent droughts and famines, with disastrous implications for peoples' livelihood. To counter such trends, the authors believe that building climate resilience, in a way that ensures sustainable food security decoupled from further depletion of natural resources, is an imperative option. To do so, governance systems should be improved to ensure their capacity in managing risks in the future and people should be empowered to cope with current changes through appropriate adaptive capacities. This also requires long-term strategies that build on agro-ecological knowledge to enable smallholder farmers to counter the impacts of environmental and climatic changes. More specifically, ensuring greater access of farmers to technologies, markets, information, and credit to adapt their production practices, is highly recommended. In such a process, agricultural extension services, according to the authors, may provide an opportunity for strengthening the resilience of rural and farming households, hence the focus of their research on the ways through which extension services may play a critical role in promoting and improving the resilience of agricultural and rural development in Ethiopia.

In Chap. 12, *The Nexus of Climate Change, Food Security, and Agricultural Extension in Islamic Republic of Iran*, Falsafi et al. focus on the case of Iran where agriculture is one of the most important sectors, meeting domestic food, nutrition, and employment needs in a context of water scarcity and high dependence on climatic and environmental conditions. The authors recall that climate change impacts on agriculture, and subsequently on the development of this sector in Iran, vary depending on spatial and temporal scales. More specifically, due to its special geographical location and topographic characteristics, the country has a different climate in each region and due to severe fluctuations in rainfall, droughts have had several detrimental effects on agriculture and the economy. To face these challenges, the authors claim that planning and mobilization of facilities and resources focusing on agricultural extension can help ensure food security through an enhanced supply of basic foodstuffs and commodities. Currently, the agricultural extension system is experiencing a radical change to render it more effective in making farmers productive, profitable, and climate resilient. In this research, the authors, after presenting the current changes in climate and examining the food security situation in Iran, outline workable and viable strategies by redefining the vibrant role that the agricultural extension system may play in addressing related challenges.

In Chap. 13, *Better Crop-Livestock Integration for Enhanced Agricultural System Resilience and Food Security in the Changing Climate*, Devkota et al. base their research on the growing evidence that frequent droughts, declining soil fertility, and poor plant-animal-atmosphere interactions are threatening the sustainability of integrated crop-livestock systems in the rainfed drylands of North Africa. The authors found highly relevant for rainfed drylands the integration of crop and livestock activities within agricultural production systems, which is now widely approved by research given its potential to boost food productivity, soil health, and overall farm profitability. In the study area, especially Algeria and Tunisia, this practice did exist, but it is decreasing with negative consequences in terms of soil fertility depletion and an overall decrease of relative farm incomes. Based on this, the authors attempt to highlight alternative options for a better integration of the crop-livestock system into the region's long-existing cereal-based livestock farming system, in order to help boost food and nutrition security, farmers' income, and soil health. For a wider adoption of such options by smallholder farmers, the authors stress that it is important to consider different approaches such as: participatory evaluation, field visits, farmers field schools and the use of information and communications technology, along with improving farmers' capacity to access and use these tools.

In Chap. 14, *Realizing Food Security through Agricultural Development in Sudan*, Alaagib et al. recall the socio-economic importance of agriculture and livestock for this African country, which suffers from severe political instability during the last years. In addition, the economic crisis, compounded by seasonal hardship and heightened conflict, has led to a deterioration in food security and nutrition in addition to a high incidence of poverty in the country. Currently, the agriculture sector is facing many challenges such as the lack of finance, low productivity, little quality control, poor extension services, marketing problem, and heavy taxes. This situation is incompatible with the fact that Sudan has vast and diverse agricultural resources that may qualify agriculture to be the leading sector in the overall development of the country. Furthermore, the volatility and rise in food prices on global markets can be used to increase agricultural investment in Sudan and help the country benefit from modern technology. In this context, the authors investigate the potential of agricultural development in achieving domestic food security and reducing poverty using descriptive and analytical statistics. The analysis concludes by developing some public policy-oriented recommendations.

In Chap. 15, *Agriculture in Fragile Moroccan Ecosystems in the Context of Climate Change: Vulnerability and Adaptation*, Larbi considers climate disruption as one of the mains factors degrading agro-systems in Morocco, and the related effects are often felt by smallholder farmers, especially in oasis and mountain vulnerable ecosystems. For the author, farmers often deploy their reactive capacities to respond to such emerging effects as an imperative while maintaining the agricultural production for subsistence and income stabilization. To capture these dynamics, the author undertakes an empirical investigation in the Anti-Atlas Mountain and the oases of Ouarzazate, Morocco. Through his research, the vulnerability of populations and production systems were analyzed and the adaptive practices deployed by farmers to reduce the sensitivity of local systems to natural risks and increase their climate resilience were observed. The findings show that farmers have detected variations in the local climate with negative implications for their production systems. To cope with these changes, they have adopted different strategies according to their socioeconomic contexts, soil, and climatic conditions. Based on his investigation to understand the resilience mechanisms of local populations geared towards ensuring food security, the author concludes that building new knowledge often results from the interweaving of imported technical and traditional knowledges. Moreover, the findings show that adaptation strategies of local populations also translate their responses, in a more global way, to the development challenges of their territories.

Focusing on the same context, Lionboui et al. in Chap. 16, *Digitalization and Agricultural Development: Evidence from Morocco*, claim that advances induced by digital innovation have positively impacted the agriculture sector in Morocco and could have a considerable impact by reducing the risk of food insecurity. To support such a claim and highlight some of the opportunities that digital transformation can offer in agriculture, the authors report applied research experiences conducted in Morocco. Their approach consists of: first, approaching agricultural land management through remote monitoring of agricultural land and mapping of cropland using satellite imageries and machine learning; second, addressing the question of what the digital can offer for the management of agricultural water resources; and third, discussing the experiences on the importance of digital transformation in risk management with the objective to analyze the risks in a more informed way and to relate aspects that could not be connected before in the past, in an efficient and relevant manner. According to the authors, digital innovation allows precise monitoring of agricultural land and offers socio-economic conditions that are more advantageous for farmers. Overall, all of the experiences cited on the use of digital innovation in the agriculture sector can be extended to other contexts, particularly in Africa where sustainable agricultural development remains the ultimate goal.

In another North-African country, Bouzid et al. analyse in Chap. 17, *Natural and Regulatory Underlying Factors of Food Dependency in Algeria*, the situation of Algeria in terms of food policy and its potential in ensuring food security based on domestic production and supply. The authors observed that previous food policies implemented in the country since its independence in 1962 either focus on enhancing local food production or rather relying on food imports. Currently, the country combines these two options by supporting local producers while seeking new foreign suppliers. For the authors, the main shared objective of such policies was to meet a growing food demand generated by a rising population, improved incomes, and changing food consumption patterns. Through this research, the authors show that Algeria's food consumption has increased significantly in the last half-century both quantitatively and qualitatively, resulting in an increased dependency on foreign suppliers. This dependency can be explained, in addition to population growth and rising households' incomes, by the agricultural policies in place and inefficient economic governance. The authors find that yields are low compared to what could technically be achieved because they are constrained by insufficient use of productive inputs, lack of (or outdated) equipment, irrational use of irrigation water, poor access to loans, land fragmentation, and the decrease in rainfall.

In Chap. 18, *Boosting Youth Participation in Farming Activities to Enhance Food Self-Sufficiency: A Case Study from Nigeria*, Olowa et Olowa start by the fact that the food system in Nigeria is characterized by a demand–supply gap or a distorted balance between market and society, thus culminating in high food prices and endangering the much-touted food self-sufficiency. To unravel the possible underlying causes, the authors prefer assessing this situation through the lens of youth involvement in farming production, which has been given little attention in the literature. More specifically, the authors focus on factors determining youth decision to participate in farming in Nigeria using Ogun state as a case study. To do so, the authors used purposive and random sampling techniques to obtain data from 300 youths spread across the four agricultural zones of the state and collated data were analysed using descriptive, Logit, and Poisson regression. Based on the findings, and in order to motivate youth to intensify their participation in farming activities in the study area, the authors recommend, among others, the adoption of public measures to facilitate the access of youths to soft loans, tractor hiring services, and land acquisition, considered as positive determinants of participation.

This volume has been designed in complimentarity with another volume being published by Springer International Publishing under the title, *Emerging Challenges to Food Production and Security in Asia, Middle East, and Africa—Climate Risks and Resource Scarcity*, co-edited by Mohamed Behnassi, Mirza Barjees Baig, Mahjoub El Haiba, and Michael R. Reed. This volume provides a multi-regional and cross-sectoral analysis of food and water security, especially in the era of climate risks, biodiversity loss, pressure on scarce resources, increasing global population, and changing dietary preferences. It includes both conceptual and empirically-based research, which provide context-specific analyses and recommendations based on a variety of case studies from Africa, Middle East, and Asia. Building on this publication, and after identifying the various emerging challenges to food production and security, the current volume focuses on emerging practices to manage such challenges within a resilience perspective. More specifically, rethinking agricultural development models, enhancing climate-smart practices, extension systems' innovation, and digitalization have been identified by different chapters, among others, as key pathways given their potential in empowering food systems and enabling transformational changes in developing countries, especially in these times of crisis. We do hope that these two volumes contribute to the advancement of science and knowledge and impact change processes on the ground.

Chapter 2
A Co-Evolving Governance Perspective of Climate-Smart Agriculture and Some Important Yet Unaddressed Elements for Integrated Gains

Mohamed Behnassi, Gopichandran Ramachandran, and Gireesh Chandra Tripathi

Abstract Climate change presents multiple challenges and it cannot be addressed effectively in silos. Attention must be paid not only to specific science, technology, finances and market elements of agriculture, but also to governance through stakeholder-specific and cross-cutting approaches. In this perspective, the governance of climate-smart agriculture (CSA) and related farming systems has evolved in form and function over recent years. This is reflected in the growing diversity of thrusts, including for instance the water-energy-food-ecosystems nexus, the variety of stakeholders, and environmental, social, and governance (ESG) criteria. Deeper insights about policy cycles that relate to the continuum of incremental and drastic changes have emerged and set the context to infer areas that need greater attention. The present eclectic synthesis accordingly highlights the call for accelerated CSA. This relates to a deeper understanding of ecosystem science, social contexts and synergies/trade-offs between productivity, mitigation, adaptation, and resultant resilience that should be sustained. Governance constraints and leads have been highlighted by reference to regional- and state- level frameworks and recommendations made to guide future change processes in the area of CSA.

Keywords Food production and security · Climate-smart agriculture · Adaptation and mitigation · Alternative energy systems · Governance

---

M. Behnassi (✉)
International Politics of Environment and Human Security, College of Law, Economics and Social Sciences, Ibn Zohr University of Agadir, Agadir, Morocco
e-mail: m.behnassi@uiz.ac.ma

Center for Environment, Human Security and Governance (CERES), Agadir, Morocco

G. Ramachandran · G. C. Tripathi
NTPC School of Business, Noida, UP, India
e-mail: gopichandran@nsb.ac.in

G. C. Tripathi
e-mail: gireesh.tripathi@nsb.ac.in

© The Author(s), under exclusive license to Springer Nature Switzerland AG 2022
M. Behnassi et al. (eds.), *Food Security and Climate-Smart Food Systems*,
https://doi.org/10.1007/978-3-030-92738-7_2

1 Introduction

It is widely recognized that the current rate of environmental and climatic changes is accelerating. This is reportedly exacerbated by past and present risk-enhancing unsustainable growth models. Despite a growing body of empirical evidences and a deeper understanding of social and ecological problems caused by such pathways, these models predominantly continue to presume that a society can only develop by intensifying and expanding its use of resources, altering ecosystems, and increasing per capita consumption patterns. Such worldviews prevail, notwithstanding implicit long-term environmental disruptions, resource scarcity, biodiversity loss, and climate change. Agriculture is a typical case in point. Multiple risks affect agriculture and livestock-centered farming systems that are part of key socio-economic strategies for many developing countries.

It is important to reverse prevailing trends of growing impacts and vulnerability through significant paradigm and policy shifts. The framework of climate-smart agriculture (CSA) has emerged through such a perspective as an option to tackle the stated and related challenges. Several mutually reinforcing elements of such a framework are currently tested and prescribed by the scientific community, development agencies, and relevant multilateral negotiations to enhance sustainability and resilience. Some of the most important elements of such robust governance are co-evolving inclusiveness, consensus, safeguards to prevent backsliding through mutually reinforcing market-financial and institutional mechanisms across production and consumption, the precautionary principle and preventive management, empowerment and science-based decisions, that recognize the pervasiveness of impacts of climate change.

The present chapter accordingly highlights the call for accelerated CSA. This relates to a deeper understanding of ecosystem science, social contexts and synergies/trade-offs between productivity, mitigation, adaptation, and resultant resilience that should be sustained. Governance constraints and leads have been highlighted by reference to regional- and state- level frameworks and recommendations made to guide future change processes in the area of CSA.

2 The Growing Climate Vulnerability of Farming Systems and The Need for Concerted Mitigation and Adaption Strategies

Farming systems around the world are reportedly increasingly vulnerable to climate change. Some comprehensive insights about forms, scales, intensity and periodicities of such challenges in the interface of agriculture and related adaptation have been presented by a myriad of institutions. These include initiatives by the Adaptation

Fund,[1] the International Fund for Agricultural Development (IFAD),[2] the World Business Council for Sustainable Development (WBCSD),[3] the World Bank,[4] the FAO with special reference to CSA,[5] CARE International and three other partners' research project (Sustainable, Productive, Profitable, Equitable and Resilient (CSA-SuPER)),[6] the Consortium of International Agricultural Research Centres (CGIAR) Research Program on Climate Change, Agriculture and Food Security (CCASF),[7] the Development Alternatives, Inc. (DAI),[8] the European Union,[9] the United States Department of Agriculture (USDA),[10] the Indian Council for Agricultural Research (ICAR), India,[11] and the World Agro-Forestry[12] Initiative, Africa.

Recently Stephanie Jaquet[13] referred to climate risk profiling, the influence of investment plans across the globe and thrust areas for mitigation and adaptation across crops and livestock management. The World Bank[14] and the Climate Finance Lab[15] too focus on resilience rating to demonstrate that systems of governance to reduce vulnerabilities are evolving over time. These are substantiated by Crouch et al. (2017) through an economic modeling of agricultural productivity and resilience as a function of mitigation and other sustainability correlates. GGGI (2021) presents a *Compendium of practices in climate-smart agriculture and solar irrigation* considering more than a dozen modulators of outcomes. The impact of solar-powered tools has also been analyzed.

Climate change induced ecological, social, and income externalities are compounded further through locally incompatible chemical-intensive, trade considerations, and, often, mono-crop systems. Genetically modified crops are yet to be

---

[1] https://www.adaptation-fund.org/about/.

[2] https://www.ifad.org/en/.

[3] https://www.wbcsd.org/Programs/Food-and-Nature/Food-Land-Use/Scaling-Positive-Agriculture.

[4] https://www.worldbank.org/en/topic/climate-smart-agriculture.

[5] http://www.fao.org/3/i3325e/i3325e.pdf.

[6] https://careclimatechange.org/info-note-climate-smart-agriculture-and-super-approach/

[7] https://ccafs.cgiar.org/research/climate-smart-technologies-and-practices/publications.

[8] https://www.dai.com/uploads/climate-flyer-2021.pdf.

[9] https://ec.europa.eu/eip/agriculture/sites/default/files/eip_agri_brochure_climate-smart_agriculture_2021_en_web_final.pdf.

[10] https://www.usda.gov/media/press-releases/2021/05/20/usda-releases-90-day-progress-report-climate-smart-agriculture-and.

[11] http://www.nicra-icar.in/nicrarevised/.

[12] https://www.worldagroforestry.org/publication/what-evidence-base-climate-smart-agriculture-east-and-southern-africa-systematic-map.

[13] https://cgspace.cgiar.org/bitstream/handle/10568/113957/3%20_%20CCAFS%20situation%20analysis.pdf?sequence=3&isAllowed=y.

[14] https://openknowledge.worldbank.org/bitstream/handle/10986/35039/Resilience-Rating-System-A-Methodology-for-Building-and-Tracking-Resilience-to-Climate-Change.pdf?sequence=7&isAllowed=y.

[15] https://www.climatefinancelab.org/news/global-innovation-lab-for-climate-finance-launches-new-program-to-increase-impact-through-replication/.

comprehensively accepted by communities, partly due to ethical issues and incomplete understanding of sustainability related challenges. Agro-ecological approaches and the scope to recycle nutrients and energy from biomass through augments offer hope but have to be significantly up-scaled, duly recognizing political economy considerations. Other important governance outcomes are yet to be comprehensively secured in many parts of the world, including: changes in cultivation practices that can be sustained across medium- and long-term periods; sustained incomes with implications to fulfill related sustainable development goals; and safeguards to prevent backsliding.

Such questions as the outcome of the complex interplay of physical, chemical, and biological attributes of the productivity of both social and ecological systems, in addition to the concurrent uncertainties imposed by economic and health challenges, continue to perplex decision makers. The 5th Assessment Report of the IPCC (Denton et al. 2014) reinforces the considerations stated. It redefines goals for climate-resilient pathways; wherein adaptation as a strategy implies the ability to anticipate and cope with unavoidable impacts, especially evidenced in agriculture. This will be through appropriate system-specific risk management interventions that have to be expeditiously delivered, because with passing time, the spread of options could reduce significantly. This calls for a deeper understanding of approaches and consequences pertaining to incremental and transformational adaptation.

The governance framework of CCAFS links food security policies with poverty, gender equity, and landscape performance outcomes. Recently, it highlighted the role of two major drivers. They are growing consumption patterns over the next three decades and the safety-net imperative to simultaneously tackle productivity decline of ecosystems. The latter could enhance inequalities and, therefore, further reduce preparedness of communities to tackle existing challenges. Interestingly, the program proposes an appropriate mix of agro-ecological and landscape approaches and market governance to target consumption patterns.[16]

One of the most recent calls for integrated action to enhance CSA was from FAO (2021a), which aptly summarizes the predicament we continue to face. According to the report, "with only a decade left to achieve the Sendai Framework for Disaster Risk Reduction 2015–2030 (SFDRR) and Sustainable Development Goals (SDG), urgent efforts are necessary to build disaster-, disease-, and climate-resilient agricultural systems that will be capable of improving the nutrition and food security of present and future generations, even in the face of mounting threats" (FAO 2021a: 1). Therefore, it is not only about tackling prevailing inadequacies, it is also about the need to address increasing and abrupt threats communities are not familiar with. This raises such important public policy questions pertaining to the appropriateness, timeliness, and spread and depth of remedial and preventive measures and, most importantly, the scope for mid-course corrections.

---

[16] https://www.cgiar.org/research/program-platform/climate-change-agriculture-and-food-security/.

3 Co-Evolution of Adaptation and Resilience Needs Correlated with Mutually Reinforcing Challenges and the Governance Implications

The boundaries between conventions and protocols dealing with the three pillars of sustainability are increasingly dissolving and posing newer challenges of 'governance of evidence' (IFRC 2020; FAO 2021a) to justify action beyond political economy considerations. This is especially so of such mutually reinforcing domains like disaster assessments as the basis for disaster risk reduction and the implications for climate change adaptation. The latter highlights the link between actual loss in money terms and drop in potential impacted by disasters. Nutrient loss and related dynamics are at the centre of such spirals that equally impact ecosystems, crops, livestock, marine food stocks, and humans.

Interestingly, as part of technology-enabled and enhanced governance, the FAO report (2021a: 148) refers to the Modelling System for Agricultural Impacts of Climate Change (MOSAICC),[17] impact attribution approaches, and the ability to differentiate delayed and rapid induced responses in biotic and abiotic interactions. These were amplified earlier by Schmidhube and Tubiello (2007); establishing the persistence of challenges and opportunities for need-based approaches. They defined the predicament of landless and wage-dependent farmers, losing access to food, and the need for a robust insurance and entitlements centered on traditional rights. A comprehensive review by Vermeulen et al. (2012) reinforced the widespread, spatially and temporally complex, and variable nature of food security as influenced by socioeconomic conditions. They present a pre-production, production, and post-production continuum for policy interventions and their impacts on direct and indirect emissions. A unique policy thrust is about potential conflicts between adaptation and food security across long and short terms, respectively. They cite the cases of immediate increase in yields that may not be sustained and the implications of high-capital technologies that could influence farmers' budgets. These could be useful to assess the tradeoffs based on variability and uncertainty, with implications for technical, economic, and institutional aspects of mitigation and adaptation options.

In a recent FAO Study (2020), it was observed that many national legal frameworks still do not include laws and related governance measures specifically to holistically tackle impacts of climate change that affect crops, livestock, forestry, fisheries, and other bioresources. The referred study presented a comprehensive overview of the legal and institutional issues with implications for SDGs and other commitments; that should therefore be addressed on priority. Barasa et al. (2021) corroborated these with

---

[17] FAO's MOSAICC helps countries project potential crop loss. Historical yield time series adjusted for non-climatic variations (e.g. changes in production systems) can be analysed for correlations with climatic variables (average, minimum and maximum temperatures and precipitation) as well as soil water-related variables (evapotranspiration, soil water balance, etc.) derived from those same climatic variables. Based on these correlations, a performance function quantifies relative contributions of selected climatic and water variables to the yields, location by location, crop by crop (FAO 2021a: 148).

special reference to the nascent stage of CSA preparedness in Africa. Most countries in the continent do not have national CSA-investment plans and, therefore, miss the opportunity to infuse governance elements at all stages.

4 Addressing Crosscutting Elements of Production and Consumption Systems to Secure Comprehensive Resilience Gains

Porter et al. (2014) highlighted the preponderance of artificially induced negative impacts on food production and security systems on the basis of a drivers/response framework that differentiates impacts, vulnerabilities, and risks. This is true for natural processes and, at least, four types of stakeholders across the food access continuum. Such critical science-led policy calls were about the sensitivity of crops to daytime temperatures and tropospheric ozone in particular, inviting attention to the occurrence and distribution of pollutants of immediate relevance at the local level and consequences on feral hosts. Most importantly, the authors established the paucity of information about related food processing, packaging, transport, and storage, with implications for smart management of produce; essential for holistic resilience.

Five years later, Mbow et al. (2019) elaborated on the vulnerability of pastoral, fruit, and vegetable systems with implications for trade-related product quality and supply chains, income security, nutrition, equity, and gender facets. Pests, diseases, and pollinators featured prominently in this analysis, alongside infrastructure, markets, and sovereignty. The most important policy perspective was about incremental and transformational adaptation and related enabling circumstances, the typology of 'food system response options', and the links with SDGs. It is only logical that they will soon revisit the payments for ecosystems framework to reinforce community action and conservation.

The UNFCCC Secretariat reinforced the need to address post-harvest processing and transport as part of food value chains and the lack of synergies with policies and governance of consumer preferences. (UNFCCC Secretariat 2021). Importantly, the need for deeper insights about precision farming, locally adapted weather forecasting and climate modeling, integration of mitigation and adaptation outcomes, thrusts of the Adaptation Fund, the GEF, and the three pathways of the GCF about climate proofing production systems with alternative energy supply and energy efficiency improvements, was emphasized. Gupta et al. (2021) reinforce this perspective with special reference to India. They indicate that a significant focus is on green revolution-led farming, including organic farming; while detailed analyses of the scope and long-term consequences of available alternatives are missing. This extends to such related areas as ecosystem services and gender parity, with incentives for conservation, rather than input-based subsidies.

5 Selected Governance Mechanisms with the Potential to Enhance the Development and Implementation of CSA Practices

Interestingly, Nelson et al. (2010), a decade ago, presented a comprehensive modeling-based overview of emerging impacts through a 2050-horizon and the need for a holistic policy perspective to tackle challenges. They indicated that price-related trade safeguards and targeted resilience-enhancing investments could offset systemic impacts of climate change; that could otherwise neutralize economic security along with food security and human well-being. Recently, Kerr et al. (2021) raised interesting questions about the benefits of the agro-ecological framework through integrated and mixed crop and livestock management. This is an interesting case of uncertainties in policy framing. They refer to "food system governance based on human and social values" that addresses crop, food security, and nutrition across individual, family, and community scales of organization. Notably, such socio-economic aspects of equity and justice, derived through direct marketing and land and natural resource governance to sustain benefits, are rarely reported.

The OECD Global Forum on Agriculture (2021) has just acknowledged that the triggers of risks and vulnerabilities are ubiquitous on livelihoods and human security. A special emphasis was on shocks affecting domestic and international markets. Recommendations related to entrepreneurship skills, support for on-farm adaptation, and vulnerability reduction assessments improved access to information systems and incentives to transform risk management operations. Interestingly these were highlighted more than a decade earlier by Schmidhube and Tubiello (2009). The German Environment Agency caps these calls, as stated by Wunder et al. (2021), to reduce food losses/waste through food policies and accreditation programmes across the globe as part of the value chain.

The Food Farming and Countryside Commission (FFCC-UK) (2021) has presented as well a roadmap for agroecology-centered farming with certified practices. It raised important questions about the scale of output that can be sustained with implications for farming enterprises, qualitative and quantitative profiles of soils, and related ecosystems that can change through bio-mediations and related biodiversity. These are also seen within the larger framework of net-zero commitments. The ICRIER-OECD paper (2018) illustrated the success story of India's agriculture, which has been also interpreted also by Rosman and Singh (2021) based on policy reforms. Importantly, such reforms were designed to enable environmentally efficient infrastructure, sustainable use of fertilizers, extraction and use of groundwater, and watershed management. The latter also aims to correct perverse incentives that allow resource over-use.

Equally important is the focus on innovations to secure mitigation and adaptation gains. Such governance measures, as information and capacity building to use appropriate tools and techniques through agro-advisories, are embedded in this approach. Tolerant breeds of livestock and poultry with better fertility and mothering instincts, thermo-regulation, ability to walk long distances, ingest and digest low-quality feed, and capacity to resist diseases are chosen as part of this strategy. The larger context of the challenges and approaches across the global North and South has been highlighted by Miller et al. (2013). They asked for detailed holistic assessments of cultural, political, and crosscutting ecosystem-level changes to reduce food security limitations.

A typical challenge that evades comprehensive solutions pertains to insect plant interactions. The FAO (2021b) presented a comprehensive review of these aspects with special reference to pathogens, alternative hosts interactions with implications for agricultural landscape management. While bio-ecological and chemical ecological measures have to be system-specific, they have to be scaled up with policy measures that relate pest risk analyses to locally observed micro-climate changes.

Alvar-Beltran et al. (2021) present a framework of climate-risk screening for 'transformational development' that guides focused investments to tackle historically evidenced risks. Importantly, they take note of the individual and synergistic impacts of threats to define system-specific adaptive capacities as part of governance strategies. It is probably the most comprehensive framework that relates ecosystem functions across multiple trophic levels and nutrient flows with implications for rewards to communities in terms of amelioration outcomes. Six elements of governance enrich the framework with scope for inclusive decision making by a wide variety of stakeholders. This includes the role of market instruments, regulations, and institutional mechanisms to sustain impacts.

Willingham (2021) rightly highlights challenges faced in North America. Significant damage to farmlands compound challenges posed by changing pest and disease dynamics. The author calls for safeguards in lending to tackle climate-induced challenges with regulatory implications for disclosure of risks, public stake, and appropriate credit planning. Such elements of preventive governance, as climate resilience analysis to determine capital requirements, are also included, with additional implications for commodity prices as a function of changing crop patterns. Acevedo et al. (2020) discussed the Preferred Reporting Items for Systematic review and Meta-Analysis Protocols (PRISMA-P) approach to determine choice of crops. They relate several abiotic stresses mediated through exposure and sensitivity of crops; information about which is embedded in focused outreach activities that promote preventive options.

A detailed analysis of practices in South Asia, including the vulnerability framework and nine related parameters, highlights the reverse-flow impacts of growing populations, malnutrition and prevailing poverty addressed by the Action on Climate Today (ACT) initiative of the UK Department for International Development (Pound et al. 2018).

Local-level initiatives on risk management in Australia have well defined roles and responsibilities for stakeholders (Department of Agriculture, Water, and the Environment undated). The most important element of governance pertains to the role of government in managing risks to the natural environment and assets only to promote private adaptation. This, in turn, should be determined by stakeholder specific exposure to risk. Reviews and social inclusion are at the centre of this governance approach. This is also part of the transformative adaptation framework elaborated by Carter et al. (2018). Arguably, this helps avoid maladaptation, with well-designed financial support across all time horizons of cropping systems, and goes beyond incremental adaptation. This also calls for timely and targeted decision-support systems aligned with stakeholders' preparedness to implement them for collective good. The authors present as well an overview of adaptation funds with the scope to integrate transformative adaptation with national commitments.

UNESCO (2021) recently elaborated on a five-pronged adaptation action-oriented skill development approach with the involvement of private sector. aligned with nationally determined contributions (NDCs) by countries. This was in response to a felt need to expedite adaptation action at the local level. The further builds on the interface of governance with behaviour change and benefit sharing through conservation and enhancement of ecosystem services (IPE Global undated) and energy-centered coping strategies.

Agriculture features prominently in India's third biennial update report to the UNFCCC (MoEFCC 2021) as the sector that lends itself most to adaptation. It elaborates on the social and economic vulnerability of the sector and corresponding policy and plan measures taken by the government to concomitantly reduce burdens on people and ecosystems implemented through local-level institutions. Drought and heat tolerant/resistant varieties and other vulnerability reduction measures through a four-thrust strategy are unique to these interventions. These are rationalized through a deep understanding of emission profiles of agriculture, the sink function, and demand-side management. At the state level, several programmes in India for instance (Beaton et al. 2019) address the water-energy-food nexus on the basis of nation-wide policies. Solar pumps reportedly helped save money and secured higher yields compared to conventional practices. It is, however, equally important to infuse safeguards against un-intended overexploitation in the longer run.

In this context, it is useful to refer to the process to prioritize stakeholder engagement as presented by Khatri-Chhetria (2019). These include local-level government departments, outreach strategies, research and financial institutions, civil society organizations and farmers in particular. The objective is to enable strategic decisions to improve adaptability, increase farm productivity and income, and in particular reduce emissions. It is equally important the focus on the technical feasibility of options and remove barriers to the adoption of CSA practices. The most recent 'toolbox with participatory foresight methods, relevant for climate and food governance has been presented by van den Ende et al. (2021). It justifies the need to define processes and outcomes of citizen engagement/collaboration with policy-makers and other actors. They build on insights from many international programs on citizen engagement across three thrust areas, five approaches, degrees of participation, and the scope to deal with uncertainties, distributed across preparatory and delivery phases of action. These are essential elements of governance much needed to rise up to the unfinished agenda of CSA (Govt. of Madhya Pradesh 2017; Green Climate Fund 2020; ICRISAT 2016; UNDP 2020; UNFCCC 2016; Abhilash et al. 2021; USDA 2021).

6 The Emerging Funding Focus

The US Farmers and Ranchers in Action platform (2021) has recently deliberated on agro-ecosystems services and the means to secure them through innovative finance and technology interventions. This relates to the Climate-Smart Soil Technology Landscape and analysis to substantiate the case for climate-resilient management of nutrients, manure, cultivation, and grazing. The impact of these on carbon stocks was also established. The scope to reach out to investors to support climate-efficient transitions and set such mechanisms as green bonds was established, aligned with the growing importance of environmental, social, and governance (ESG) ratings. The initiative called for improved farm-management systems and logical frameworks that demonstrate risk/reward links of returns on investments. The above stated were corroborated by Mikolajczyk et al. (2021) in the context of Southeast Asia that has a high proportion of small-farm holders who should be assisted in adopting sustainable farming practices given their importance in decoupling production and environmental degradation.

This calls for business models uniquely adapted to scales and adaptive abilities of systems. Such correlates, like costs and benefits of sustainable land-use options as indictors of credit worthiness of farmers, are important to devise system-specific interventions. Technical assistance to farmers to comprehend and interpret outcomes periodically will be at the core of such initiatives. Interestingly, the most recent call for a sustainable finance initiative in Brazil (CCAFS, Biodiversity International and CIAT 2021) integrates these elements through seven action points. Earlier, CICERO and TERI (2019) presented a 10-country case analysis to integrate climate change considerations in agricultural finance. They established the need to enhance income-generating potential of farmers to, in turn, raise their credit profile and assist mitigation of production risk. The Climate Finance Leadership Initiative (CFLI), along with European Development Finance institutions (EDFI) and Global Infrastructure Facility (GIF) (2021) deliberate on the importance of climate-smart water and land use and the need to guide investments for sustainability.

An equally important strand of CSA governance is the status of institutional mechanisms that complement finances. One of the earliest arguments on these aspects was presented by Scherr et al.(2012) with special reference to co-evolving farming practices, resilience demands, and consequences on social, economic, and ecological correlates. These gravitate towards much needed multi-stakeholder planning essential for governance, with implications of resource tenure and investments to meet goals of resilience. Stephanie et al. (2021) present the case of Ghana's development plans to mainstream CSA. Significant challenges prevail despite a general awareness of climate change impacts on agriculture. This is attributed to data gaps about local-level change trends and impacts and issues of ownership that determine sustained action. This is true also of financing for mitigation and adaptation and, therefore, the need to revise guidelines to develop and implement local development plans. These, in turn, impact socio-ecological transformations.

Earlier in 2016, Zougmoré et al. argued for a greater spread and depth of system-specific information on impacts even of West Africa to guide the development of mitigation and adaptation plans. One of the most recent calls for a deeper understanding of regulatory insights that enable mitigation and adaptation was by Vidya (2021). The ADRC (2021) highlighted the need for deeper climate-informed planning to guide budgets and investments with a special focus on risk assessments. This was derived through its CARE South Asia project that also focused on infrastructure and safety considerations. An equally important needs—assessment was reported by Anschell and Salamanca (2020) at the ASEAN level through a multilateral framework. This was with special reference to climate-resilient land use and that despite the wide use of guidelines, studies on the lifespan of thematic guidelines are uncommon; duly recognizing the importance of crop insurance, related investments, and gender dynamics.

7 Some Specific Leads About Governance Challenges to CSA from National and Regional Levels

Mizik (2021) defines the characteristics of adopting CSA and draws attention to the niche occupied by small-farm systems that should target short-term benefits, with payments for ecosystem services and sustainable intensification as augments. Muthee et al. (2021) too emphasize the ecosystem-based adaptation practices on the basis of a sixteen -parameters matrix and pathways that substantiate outcomes. Golam and Nilhari (2021) superimpose three criteria and a four-step policy process dynamic to rationalize decisions and improve policy coherence. These relate well with the framework already elaborated by FAO (2014).

The International Finance Corporation (IFC) (2017) presented an excellent business perspective to CSA with examples from South Asia through an eight-step policy substantiation process. Circular economy also features in the framework implying recovery and re-use of resources. Indicators of successful implementation of conservation have also been most recently elaborated by GRAIN (2021) that supports farmers and social movements at the local level.

One of the best examples of collective action at the regional level is the Africa CSA Alliance (ACSAA). It operates through the NEPAD-iNGO Alliance on CSA—endorsed during the 31st African Union Summit (Malabo, June 2014). This serves the African Union Vision to have at least 25 million farm households more practicing CSA by 2025. It is a country-driven and regionally-integrated initiative to provide tools for action and enable partnerships. It emphasizes accountability and learning, monitoring and collating data/information on performance, and results and impact at the grassroots (FAO undated). The initiative is defined under two inter-linked components (Fig. 1).

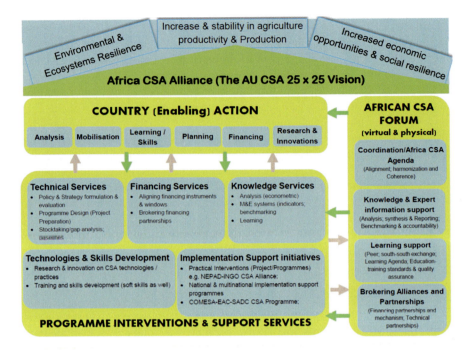

Fig. 1 Africa CSA vision 25 × 25—Africa's strategic approach for food security and nutrition in the face of climate change. *Source* NEPAD (2014: 9)

- Country action component, through *Country-Based Programs (CBP)*, with clear focus on national/community level capacity and enabling environment to accelerate and expand implementation. Countries are benefiting from *supporting services* provided by partners and facilitated by a NEPAD facilitation unit. According to their individual needs, countries select from a menu of supporting services: *(i) Analysis; (ii) Planning; (iii) Financing; (iv) Execution; (v) Tracking and assessing impact and learning;*
- Focus on fostering appropriate partnerships and alliances, knowledge support and learning including evidence-based support to program and policy design and review of individual (country, region, sector, etc.) performance against continental and global benchmarks. This includes: (i) Coordination; (ii) Knowledge exchange, expert information support and learning; (iii) Brokering alliance and partnerships; (iv) Communication and advocacy.

It will be obvious from the above presented seven strands that the governance of CSA continues to co-evolve at three levels. The *first* is about the intent of resilience, increasingly translated into a reality world over; albeit through short- and medium-term initiatives. Logically, the *second* is about the felt need to upscale successes through drastic transformations; not in an incremental manner. The *third* is about an ever-increasing spread and depth of socio-economic environmental and regulatory considerations at local, regional, and global scales. These are in response to calls for concerted action that may only get shriller with more intense and frequent onslaughts on climate. There is hope, especially with the integration of ESG elements in funding.

The analysis includes a special emphasis on evidence-based policy making and capacity building of communities to use appropriate decision support tools with the potential to enable a pole-vaulting into a climate-resilient future.

References

Abhilash SC, Alka R, Arti K, Narayan SR, Kavita K (2021) Climate-smart agriculture: an integrated approach for attaining agricultural sustainability. https://www.researchgate.net/publication/349030937_Climate-Smart_Agriculture_An_Integrated_Approach_for_Attaining_Agricultural_Sustainability

Acevedo M, Pixley K, Zinyengere N, Meng S, Tufan H, Cichy K, Bizikova L, Isaacs K, Ghezzi-Kopel K, Porciello J (2020) A scoping review of adoption of climate-resilient crops by small-scale producers in low- and middle-income countries. Nat Plants 6:1231–1241. https://www.nature.com/articles/s41477-020-00783-z.pdf

ADRC (2021) Climate Adaptation and Resilience (CARE) for South Asia, Project Brief Component-2. http://www.adpc.net/igo/category/ID1659/doc/2021-qVMx50-ADPC-CARE-Component_2_-_Brief_for_Web.pdf

Alvar-Beltrán J, Elbaroudi I, Gialletti A, Heureux A, Neretin L, Soldan R (2021) Climate Resilient Practices: Typology and guiding material for climate risk screening. FAO, Rome. http://www.fao.org/3/cb3991en/cb3991en.pdf

Anschell N, Salamanca A (2020) ASEAN guidance for climate-smart land use practices: a review. Deutsche Gesellschaft für Internationale Zusammenarbeit (GIZ), Jakarta. https://www.giz.de/en/downloads/2021-ASEAN%20Guidance%20for%20CSLU%20Practices-a%20Review.pdf

Arun K-C, Anjali P, Aggrwala Pramod K, Vardhan VV, Akhilesh Y (2019) Stakeholders prioritization of climate-smart agriculture interventions: evaluation of a framework. Agric Syst 174:23–31. https://reader.elsevier.com/reader/sd/pii/S0308521X18306085?token=ABFB71761D17EA46147BA990AA514795CEB55D266D1810814F21029E3F4DEE083EC47BC624F0E7A4A97926C3F2767B96&originRegion=eu-west-1&originCreation=20210620145931

Barasa PM, Botai CM, Botai JO, Mabhaudhi T (2021) A review of climate-smart agriculture research and applications in Africa. Agronomy 11:1255. https://doi.org/10.3390/agronomy11061255

Barry P, Richard L, Simon C, Naman G, Bahadur Aditya V (2018) Climate-resilient agriculture in South Asia: an analytical framework and insights from practice. http://www.acclimatise.uk.com/wp-content/uploads/2018/02/OPM_Agriculture_Pr2Final_WEB.pdf

Beaton C, Jain P, Govindan M, Garg V, Murali R, Roy D, Bassi A, Pallaske G (2019) Mapping policy for solar irrigation across the water-energy-food (WEF) nexus in india. GIZ IISD, GSI, TERI. https://www.teriin.org/sites/default/files/2020-01/2018ED24.pdf

Carter R, Ferdinand T, Chan C (2018) Transforming agriculture for climate resilience: a framework for systematic change. WRI. https://files.wri.org/d8/s3fs-public/transforming-agriculture-climate-resilience-framework-systemic-change_0.pdf

CCAFS, Biodiversity International and CIAT (2021) Short-term consultancy contract: sustainable finance originator climate smart agriculture in Brazil. https://ccafs.cgiar.org/sites/default/files/2021-03/SUSTAINABLE%20FINANCE%20ORIGINATOR%20CSA%20-%20Brazil.pdf

CFLI, EDFI and GIF (2021) Unlocking private climate finance in emerging markets -private sector considerations for policymakers. https://assets.bbhub.io/company/sites/55/2021/03/CFLI_Private-Sector-Considerations-for-Policymakers-April-2021.pdf

Christopher B, Purva J, Mini G, Vibhuti G, Rashmi M, Dimple R, Andrea B, Georg P (2019) Mapping policy for solar irrigation across the Water-Energy-Food (WEF) Nexus in India. GIZ IISD, GSI, TERI. https://www.teriin.org/sites/default/files/2020-01/2018ED24.pdf

CICERO and TERI (2019) Approaches to integrate climate change in agricultural finance case study analysis. https://www.teriin.org/projects/nfa/files/climate-finance-agr-report.pdf

Crouch L, Lapidus D, Beach R, Birur D, Moussavi M, Turner E (2017) Developing climate-smart agriculture policies: the role of economic modeling. RTI Press Publication No. OP-0034–1701. RTI Press, Research Triangle Park, NC. https://doi.org/10.3768/rtipress.2017.op.0034.1701, https://www.rti.org/rti-press-publication/climate-smart-agriculture/fulltext.pdf, https://www.rti.org/rti-press-publication/climate-smart-agriculture

Denton F, Wilbanks TJ, Abeysinghe AC, Burton I, Gao Q, Lemos MC, Masui T, O'Brien KL, Warner K (2014) Climate-resilient pathways: adaptation, mitigation, and sustainable development. In: Field CB, Barros VR, Dokken DJ, Mach KJ, Mastrandrea MD, Bilir TE, Chatterjee M, Ebi KL, Estrada YO, Genova RC, Girma B, Kissel ES, Levy AN, MacCracken S, Mastrandrea PR, White LL (eds) Climate change 2014: impacts, adaptation, and vulnerability. Part A: global and sectoral aspects. contribution of working Group II to the fifth assessment report of the intergovernmental panel on climate change. Cambridge University Press, Cambridge, United Kingdom and New York, NY, USA, pp 1101–1131. https://www.ipcc.ch/site/assets/uploads/2018/02/WGIIAR5-Chap20_FINAL.pdf

Department of Agriculture, Water and the Environment (undated) Roles and Responsibilities for Climate Change Adaptation in Australia. https://www.environment.gov.au/system/files/pages/2e55e020-e4d3-4e79-bf44-1846f720a8c8/files/coag-roles-respsonsibilities-climate-change-adaptation.pdf, https://www.environment.gov.au/climate-change/adaptation/planning-climate-change-nrm

FAO (undated) Regional CSA Alliances and platforms: Information sheet - The Africa CSA Alliance (ACSAA) and the NEPAD-iNGO Alliance on CSA. http://www.fao.org/3/bl862e/bl862e.pdf

FAO (2014) The water-energy-food nexus: a new approach in support of food security and sustainable agriculture. http://www.fao.org/3/bl496e/bl496e.pdf

FAO (2020) Agriculture and climate change. Law and governance in support of climate smart agriculture and international climate change goals. http://www.fao.org/3/cb1593en/CB1593EN.pdf

FAO (2021a) The impact of disasters and crises 2021 on agriculture and food security. http://www.fao.org/3/cb3673en/cb3673en.pdf

FAO (2021b) Scientific review of the impact of climate change on plant pests. A global challenge to prevent and mitigate plant pest risks in agriculture, forestry and ecosystems. http://www.fao.org/3/cb4769en/cb4769en.pdf

Food Farming and Countryside Commission (FFCC-UK) (2021) Farming for Change—Mapping a Route to 2030. https://ffcc.co.uk/assets/downloads/FFCC_Farming-for-Change_January21-FINAL.pdf

Global Green Growth Institute (GGGI) (2021) Compendium of Practices in Climate-Smart Agriculture and Solar Irrigation. https://gggi.org/site/assets/uploads/2021/03/Compendium-of-Practices-in-Climate-Smart-Agriculture-Web_Final.pdf

Govt of Madhya Pradesh (2017) Increasing adaptive capacity to climate change through developing climate-smart villages in select vulnerable districts of Madhya Pradesh. http://moef.gov.in/wp-content/uploads/2017/08/M.P.Detail-Project-Report_CSV-V-6.pdf

GRAIN (2021) Agroecology vs. Climate Chaos—Farmers leading the battle in Asia. https://grain.org/en/article/6632-agroecology-vs-climate-chaos-farmers-leading-the-battle-in-asia

Green Climate Fund (2020) Learning-Oriented Real-Time Impact Assessment (LORTA) Programme, Synthesis Report. https://ieu.greenclimate.fund/sites/default/files/evaluation/lorta-synthesis-report-2020-phase-iii-revised.pdf

Gupta Ni, Pradhan Sl, Jain A, Patel N (2021) Sustainable agriculture in india 2021-What We Know and How to Scale Up, Council on Energy, Environment and Water (CEEW), New Delhi. https://www.ceew.in/sites/default/files/CEEW-FOLU-Sustainable-Agriculture-in-India-2021-20Apr21.pdf

IPE Global (undated) Environment and climate change. https://www.ipeglobal.com/upload/spa_pdf/201012Img_201bb8_environmentandclimatechange_qcapproved.pdf

ICRIER-OECD (2018) Agricultural policy review of India. https://icrier.org/pdf/Agriculture-India-OECD-ICRIER.pdf

ICRISAT (2016) Building climate-smart villages—five approaches for helping farmers adapt to climate change. http://www.icrisat.org/wp-content/uploads/2016/11/Building-Climate-Smart-Villages.pdf

International Federation of Red Cross and Red Crescent Societies (IFRC) (2020) World Disasters Report 2020 Come Heat or High Water. IFRC, Geneva. https://media.ifrc.org/ifrc/wp-content/uploads/2021/01/20201113_WorldDisasters_6-INTRO.pdf

International Finance Corporation (IFC) (2017) Making agriculture climate-smart—a business perspective from South Asia, https://www.ifc.org/wps/wcm/connect/7132a965-db42-440f-ba9b-9ceee56e0e81/Making+Agriculture+Climate-Smart_FINAL.pdf?MOD=AJPERES&CVID=m086s0t

Josef S, Tubiello Francesco N (2009) Global food security under climate change. In: PNAS, vol 104, no 50, pp 19703–19708. http://www.pnas.org/cgi/doi/10.1073/pnas.0701976104

Kennedy M, Lalisa D, Judith N, Peter M (2021) Ecosystem-based adaptation practices as a nature-based solution to promote water-energy-food nexus balance. Sustainability 13:1142. https://doi.org/10.3390/su13031142

Kerr RB, Madsen S, Stüber M, Liebert J, Enloe S, Borghino N, Parros P, Mutyambai DM, Prudhon M, Wezel A (2021) Can agroecology improve food security and nutrition? A review. Glob Food Secur 29:100540. https://doi.org/10.1016/j.gfs.2021.100540

Maricclis A, Kevin P, Nkulumo Z, Sisi M, Hale T, Karen C, Livia B, Krista I, Kate G-K, Jaron P (2020) A scoping review of adoption of climate-resilient crops by small-scale producers in low- and middle-income countries. Nat Plants 6:1231–1241. https://www.nature.com/articles/s41477-020-00783-z.pdf

Mark R, Singh Santosh K (2021) Climate change—agriculture and policy in India, Report Number: IN2021–0070, USDA & GAIN. https://apps.fas.usda.gov/newgainapi/api/Report/DownloadReportByFileName?fileName=Climate%20Change%20-%20Agriculture%20and%20Policy%20in%20India_New%20Delhi_India_05-25-2021.pdf

Mbow C, Rosenzweig C, Barioni LG, Benton TG, Herrero M, Krishnapillai M, Liwenga E, Pradhan P, Rivera-Ferre MG, Sapkota T, Tubiello FN, Xu Y (2019) Food security. In: Shukla PR, Skea J, Calvo Buendia E, Masson-Delmotte V, Pörtner H-O, Roberts DC, Zhai P, Slade R, Connors S, van Diemen R, Ferrat M, Haughey E, Luz S, Neogi S, Pathak M, Petzold J, Portugal Pereira J, Vyas P, Huntley E, Kissick K, Belkacemi M, Malley J (eds) Climate change and land: an IPCC special report on climate change, desertification, land degradation, sustainable land management, food security, and greenhouse gas fluxes in terrestrial ecosystems. In press. https://www.ipcc.ch/site/assets/uploads/sites/4/2021/02/08_Chapter-5_3.pdf

Mikolajczyk S, Mikulcak F, Thompson A, Long I (2021) Unlocking smallholder finance for sustainable agriculture in Southeast Asia. Climate Focus and WWF. https://sustainablefinanceasia.org/wp-content/uploads/2021/03/WWF-2021-Unlocking-Smallholder-Finance-for-Sustainable-Agriculture.pdf

Miller M, Anderson M, Francis C, Kruger C, Barford C, Park J, McCown BH (2013) Critical research needs for successful food systems adaptation to climate change. J Agric Food Syst Commun Dev 3(4):161–175. https://www.foodsystemsjournal.org/index.php/fsj/article/view/206

Ministry of Environment, Forest, and Climate Change, Government of India (MoEFCC) (2021) India: third biennial update report to the united nations framework convention on climate change. https://unfccc.int/sites/default/files/resource/INDIA_%20BUR-3_20.02.2021_High.pdf

Mizik T (2021) Climate-smart agriculture on small-scale farms: a systematic literature review. Agronomy 11:1096. https://doi.org/10.3390/agronomy11061096

Muthee K, Duguma L, Nzyoka J, Minang P (2021) Ecosystem-based adaptation practices as a nature-based solution to promote water-energy-food nexus balance. Sustainability 13:1142. https://doi.org/10.3390/su13031142

Nelson GC, Rosegrant MW, Palazzo A, Gray I, Ingersoll C, Robertson R, Tokgoz S, Zhu T, Sulser TB, Ringler C, Msangi S, You L (2010) Food Security, farming, and climate change to 2050:

scenarios, results, policy options. IFPRI, Washington DC. http://ebrary.ifpri.org/utils/getfile/collection/p15738coll2/id/127066/filename/127277.pdf

NEPAD (2014), Africa CSA Vision 25 x 25—Africa's strategic approach for food security and nutrition in the face of climate change. https://www.nepad.org/file-download/download/public/15637.

Niti G, Shanal P, Abhishek J, Nahya P (2021) Sustainable agriculture in India 2021—What We Know and How to Scale Up, Council on Energy, Environment and Water (CEEW). New Delhi. https://www.ceew.in/sites/default/files/CEEW-FOLU-Sustainable-Agriculture-in-India-2021-20Apr21.pdf

OECD Global Forum on Agriculture (2021) Policies for a more resilient agro-food sector. https://www.oecd.org/agriculture/events/documents/agenda-oecd-global-forum-agriculture-2021.pdf

Porter JR, Xie L, Challinor AJ, Cochrane K, Howden SM, Iqbal MM, Lobell DB, Travasso MI (2014) Food security and food production systems. In: Field CB, Barros VR, Dokken DJ, Mach KJ, Mastrandrea MD, Bilir TE, Chatterjee M, Ebi KL, Estrada YO, Genova RC, Girma B, Kissel ES, Levy AN, MacCracken S, Mastrandrea PR, White LL (eds) Climate change 2014: impacts, adaptation, and vulnerability. Part A: global and sectoral aspects. contribution of working Group II to the fifth assessment report of the intergovernmental panel on climate change. Cambridge University Press, Cambridge, United Kingdom and New York, NY, USA, pp 485–533. https://www.ipcc.ch/site/assets/uploads/2018/02/WGIIAR5-Chap7_FINAL.pdf

Pound B, Lamboll R, Croxton S, Gupta N, Bahadur AV (2018) Climate-resilient agriculture in south asia: an analytical framework and insights from practice. http://www.acclimatise.uk.com/wpcontent/uploads/2018/02/OPM_Agriculture_Pr2Final_WEB.pdf

Rachel BK, Madsen S, Stüber M, Liebert J, Enloe S, Borghino N, Parros P, Mutyambai DM, Prudhon M, Wezel A (2021) Can agroecology improve food security and nutrition? a review. Global Food Secur 29:100540. https://doi.org/10.1016/j.gfs.2021.100540

Rani A, Kumari A, Singh RN, Kumari K (2021) Climate-smart agriculture: an integrated approach for attaining agricultural sustainability. https://www.researchgate.net/publication/349030937_ClimateSmart_Agriculture_An_Integrated_Approach_for_Attaining_Agricultural_Sustainability

Rasul G, Neupane N (2021) improving policy coordination across the water, energy, and food, sectors in South Asia: a framework. Front Sustain Food Syst. https://doi.org/10.3389/fsufs.2021.602475

Robert Z, Samuel P, Mathieu O, Bamidele O, Timothy T, Augustine A, Polly E, Mohammed S, Abdulai J (2016) Toward climate-smart agriculture in West Africa: a review of climate change impacts, adaptation strategies and policy developments for the livestock, fishery and crop production sectors. Agric Food Secur 5:26. https://agricultureandfoodsecurity.biomedcentral.com/track/pdf, https://doi.org/10.1186/s40066-016-0075-3.pdf

Rosman M, Singh SK (2021) Climate change - agriculture and policy in India, report number: IN2021-0070, USDA & GAIN. https://apps.fas.usda.gov/newgainapi/api/Report/DownloadReportByFileName?fileName=Climate%20Change%20-%20Agriculture%20and%20Policy%20in%20India_New%20Delhi_India_05-25-2021.pdf

Scherr Sara J, Seth S, Rachel F (2012) From climate-smart agriculture to climate-smart landscapes. Agric Food Secur 1:12. http://www.agricultureandfoodsecurity.com/content/1/1/12

Schmidhuber J, Tubiello FN (2007) Global food security under climate change. In: PNAS, vol 104, no 50, pp 19703–19708. https://www.pnas.org/cgi/doi/10.1073/pnas.0701976104

Stephanie W, Kirsten W, Margarethe S (2021) Options for multilateral initiatives to close the global 2030 climate ambition and action gap—policy field sustainable food systems. Umweltbundesamt. https://www.ecologic.eu/sites/default/files/publication/2021/2021-04-08_cc_13-2021_mitigation_options.pdf

Stephen K, Seth Asare O, Mensah Seth O, Ahmed A, Owusua Y, Kita M (2021) Are local development plans mainstreaming climate-smart agriculture? a mixed-content analysis of medium-term development plans in semi-arid Ghana. Socio-Ecol Practice Res 3:185–206. https://link.springer.com/content/pdf, https://doi.org/10.1007/s42532-021-00079-2.pdf

UNFCCC Secretariat (2021) Socioeconomic and food security dimensions of climate change in the agricultural sector. Subsidiary body for scientific and technological advice. Advance Version. FCCC/SB/2021/2. https://unfccc.int/sites/default/files/resource/sb2021_02_adv.pdf

UNDP (2020). Project document Swaziland: increasing farmer resilience to climate change—upscaling market oriented climate smart agriculture project. https://info.undp.org/docs/pdc/Documents/SWZ/SWZ%20Climate%20Smart%20Agriculture%20Project%20Document.pdf

UNESCO (2021) Skills development and climate change action plans Enhancing TVET's contribution. https://unevoc.unesco.org/pub/skills_development_and_climate_change_action_plans.pdf

UNFCCC (2016) Submission on indicators of adaptation and resilience at the national and/or local level or for specific sectors, from the International Center for Tropical Agriculture (CIAT) on behalf of the CGIAR Research Program on Climate Change, Agriculture and Food Security (CCAFS). https://unfccc.int/sites/default/files/resource/08_CIAT-CCAFS_%20Multi%20scale%20Smar_submission%20on%20Indicators.pdf

US Farmers and Ranchers in Action (2021) Transformative investment in climate-smart agriculture—unlocking the potential of our soils to help the U.S. achieve a net-zero economy. https://usfarmersandranchers.org/wp-content/uploads/2021/02/USFRA-Transformative-Investment-Report.pdf

USDA (2021) Climate-smart agriculture and forestry strategy: 90-Day Progress Report. https://www.usda.gov/sites/default/files/documents/climate-smart-ag-forestry-strategy-90-day-progress-report.pdf

van den Ende MA, Wardekker JA, Mees HLP, Hegger DLT, Vervoort JM (2021)Towards a climate-resilient future together: a toolbox with participatory foresight methods, tools and examples from climate and food governance. https://www.researchgate.net/publication/345733771

Vermeulen Sonja J, Campbell Bruce M, Ingram John SI (2012) Climate change and food systems. Ann Rev Environ Resour 37:195–222. https://www.annualreviews.org/doi/pdf, https://doi.org/10.1146/annurev-environ-020411-130608

Vidya V (2021) Evolving a legal framework for climate inclusive agriculture: exploring regulatory insights from foreign jurisdictions. RGNUL Stud Res Rev 7(2):51–84. http://rsrr.in/wp-content/uploads/2021/05/Vidya-Vijayaraghavan.pdf

Willingham Z (2021) Promoting climate-resilient agricultural and rural credit. https://www.americanprogress.org/issues/economy/reports/2021/01/14/494574/promoting-climate-resilient-agricultural-rural-credit/

Chapter 3
Rethinking the Agricultural Development Model in Post-COVID-19 Era Based on Scientific Knowledge: The Moroccan Case

Mohamed Taher Sraïri

Abstract The agricultural sector is a strategic economic activity in Morocco, still accounting for around 14% of GDP and employing 40% of the active population. With the emergence of the COVID-19 pandemic, it has been literally shaken up, with growing uncertainties about market opportunities and workers' availability, given the drop in consumers' incomes and the requested social distancing. However, citizens have rapidly acknowledged the importance of agriculture, as it ensured a regular food supply during the lockdown at relatively affordable prices. The pandemic, which has emerged in a particularly dry year in Morocco, has provided an opportunity to revise the constraints facing the agricultural sector, particularly water stress and work remuneration. It has also drawn attention to food sovereignty due to consumers' awareness of the significant share of staple food imported. It is true that these imports represent an amount of virtual water which allow over-passing the domestic water shortage; however, the food trade balance is still in deficit despite the role of agriculture in the export of high value commodities as emphasized by public authorities. In addition, recent studies have shown that agricultural exports rely on growing amounts of groundwater, in areas with an arid to semi-arid climate. This has happened despite the significant state subsidies awarded to farmers to convert old furrow irrigation means to drip irrigation systems. On-farm investigations have demonstrated as well that drip irrigation has mainly resulted in an expansion of the area with irrigated cash crops. This situation has been considered as the opposite of the goal sought by the Moroccan agricultural strategy, which stated that irrigated fruits and vegetables would guarantee higher economic water productivity than rain-fed crops. Altogether, these rapid trends have simply ignored the important added value of rainfall. It seems, therefore, crucial to recognize that the post-COVID-19 agriculture should be different from what it has become. Indeed, there is a need for a paradigm shift where rain-fed agriculture has to get at the top of the political agenda, with significant financial allocation. Moreover, this will have to be coupled with a change in farmers and consumers' awareness about the positive effects of short circuits, to decrease

M. T. Sraïri (✉)
Department of Animal Production and Biotechnology, Head of the School of Agricultural Sciences, Hassan II Agronomy and Veterinary Medicine Institute, Rabat, Morocco
e-mail: mt.srairi@iav.ac.ma

© The Author(s), under exclusive license to Springer Nature Switzerland AG 2022
M. Behnassi et al. (eds.), *Food Security and Climate-Smart Food Systems*,
https://doi.org/10.1007/978-3-030-92738-7_3

fossil energy inputs, as well as to promote a low external inputs' agriculture with an insight on the sustainability of diets. In this chapter, we discuss these issues as it appears that it is not only food production which is at stake while talking about post-COVID-19 farming. This has to encompass wide topics, such as environment preservation, rural development, sustainable food systems, etc. This is a compulsory vision based on scientific knowledge, to ensure social inclusiveness and sustainable use of limited resources, to reduce rural exodus, and to guarantee the attractiveness of farming activities to young generations.

Keywords Agriculture · Food sovereignty · Income · Scientific knowledge · Water productivity · Work

1 Introduction

In Morocco, to govern is to rain
Théodore Steeg (1868-1950)

In Morocco, since the Independence (1956), the agricultural sector has always been considered as a key development priority, given the weight of rural affairs in governance and political issues (Leveau 1972). Successive governments have devoted major interest and financial means to enhance the performance of the agricultural sector, particularly through the development of irrigation. In fact, since the colonization era (Préfol 1986) and under the reign of King Hassan II (1961–1999), a national effort of building large-scale dams has been undertaken (Funnel and Binns 1989) within the framework of the so-called 'Million ha irrigation plan'. This was a mandatory orientation, as the arid and semi-arid climate characterizing most areas of the country—more than 80% of its surface, with average annual rainfall levels below 400 mm, and with a very long and hot summer season—implies the use of irrigation to secure the harvest of vital crops, such as fruits, vegetables, sugar beet, maize and lucerne (to feed dairy herds), etc.

By the year 2008, public authorities launched an ambitious agricultural strategy named the 'Green Morocco Plan'(GMP) which aimed to accelerate the rhythm of irrigation adoption, by subsidizing at levels up to 80% and even 100% (in the smallholder farms, with less than 5 ha) of the investments needed to dig wells and boreholes, as well as drip irrigation equipment. This plan also relied on an innovative approach based on the liberal dogma, by allowing private entrepreneurs to get access to state-owned land, through public private partnerships, and thus encouraging large investment projects in farms earlier managed by state-owned enterprises (Mahdi 2014), which often showed poor performance. These measures have ushered a new dynamic of agricultural intensification, through the mobilization of important financial means, estimated to have reached around 20 billion US $, although it did not benefit fairly to all the operators in the sector; women being particularly excluded (Montanari and Bergh 2019). Although it has allowed improving the output of several commodities, such as fruits, vegetables and poultry meat, the GMP has however not succeeded in securing a steady increase in vital products, particularly cereal grains, dates, milk, etc. For the specific case of the dairy sector, Morocco has witnessed a milk boycott during 2017, which has revealed the setbacks in the chain's governance, hindering all the efforts made to increase its value and milk consumption average levels, in order to improve animal source food intake. Morocco remains, therefore, still importing huge quantities of staple food products: for example, more than 4 million metric tons of soft wheat annually (almost 120 kg *per capita* per year).

With the emergence of the COVID-19 pandemic, some important players in the exports of cereal grains have announced that they could adopt protectionist measures and limit the volumes injected in global markets (Espita et al. 2020). Consequently, local voices in Morocco have tried to inform the public opinion on the risks of these new developments. They have also insisted on the idea that former free trade-oriented policies should be questioned, given the sensitive issues represented by food sovereignty and its consequences on domestic affairs. Moreover, the COVID-19 pandemic has also implied a surge of cases in some agricultural related activities, particularly in berries' handling workshops, mainly export oriented.

Given that the COVID-19 pandemic has emerged in a context marked by the end of the GMP and the launch of a new strategy called the 'Green Generation 2020–2030', which aims to consolidate the GMP's spirit and overcome its shortcomings, it seems inevitable that an analysis of the possible developments of Moroccan agriculture has to be undertaken. This has to be achieved by questioning the ability of scientific knowledge to efficiently influencing relevant decision-making processes (Sager et al. 2020), given that significant criticism has been targeting the GMP's methodology (Akesbi 2015) since its implementation. In fact, this plan almost ignored the important existing literature on the risks linked to agricultural intensification under arid and semi-arid conditions as well as the diversity in farms' structures and strategies (Faysse 2015). Altogether, these ideas impose an objective thinking on the future paths that may be followed by the Moroccan agricultural sector. This is the main purpose of this chapter which tries to analyse the potential developments of agriculture sector in a post COVID-19 context based on empirical evidence and assessments of global agriculture and food systems (Dernini and Berry 2015), as well as domestic data sources (Office of Change 2020) to establish baselines for comparison with the case of Morocco.

2 The Importance of the Agricultural Sector in Morocco

During the three-month lockdown, the importance of agriculture has become obvious: food stockpiles rapidly constituted by households have gained a value higher than the one of prestigious societies' shares in stock markets. The population has, therefore, learnt about the crucial roles of agricultural sector, mainly the regular and steady supply of vital food commodities, job creation, territories' maintenance, etc. In fact, with the interruption of the services in large cities, numerous workers of rural origin were forced to go back to their villages, where they tried to recover a modest income, through farming activities. During this lockdown, the vital contribution of farmers' work, which ensures food supply, has been acknowledged by the public.

Such developments are in total accordance with the thesis of the American anthropologist David Graeber who insists on the emergence in modern societies of hundreds of trivial jobs (or 'bullshit jobs'), whereas the most important activities needed to ensure the well-being (food, education, health, etc.) have become less considered as they do not guarantee decent wages (Graeber 2018). This comes as a key finding which is now well-established; indeed, analysing the remunerations in the agricultural sector demonstrates that even though it employs 40% of the workforce (3.7 million people), it only ensures 13–14% of the GDP (Haut Commissariat au Plan 2015). This means that the wages are limited and not steady, given the seasonality of farming activities, mainly in rain-fed systems. And even in large-scale irrigation schemes, as well as in areas with complementary irrigation (i.e., regions which have turned to the groundwater economy with private wells), a working day will only allow a limited income, generally less than 10 US $ (Sraïri et al. 2013; Sraïri and Ghabiyel 2017). This is often due to the heavy workload induced by livestock rearing (routine daily work needed to feed the animals, milk the dairy stock, etc.) (Cournut et al. 2018), but which sometimes is the only source of income for smallholder farmers who do not own sufficient assets (capital, land, and water) to practice cash crops. Indeed, it is worth mentioning that smallholder farms (i.e., with less than 5 ha of arable land) still represent more than 80% of the total number of farms. Therefore, with such a structural constraint, livestock has to be considered as the 'wealth of the poor' (Duteurtre and Faye 2009). This is not the case for intensive poultry production, which has faced a severe downturn with the outbreak of the COVID-19 pandemic, given that fast food consumption has plummeted, inducing a significant drop in eggs and meat demand as well as a sharp decrease in farm gate prices (Kouame 2020). This has been worsened by the fact that all ceremonies—such as weddings and pilgrimages which traditionally constitute events where poultry products are massively used—have been halted.

Finally, with regard to the work in agricultural sector, most of farmers—who do not enjoy social protection—are less motivated to remain active in farming activity forever. This certainly explains the significant number of persons, particularly farmers' sons who prefer migration to urban centers or abroad, seeking alternative sources of livelihood. This dynamic is also exacerbated by other factors, especially water scarcity.

3 Water Stress and Its Consequences for Water Productivity Evaluation

Water stress and its impacts on agriculture sector in Morocco have become an inevitable issue for debate. This is mainly due to the arid and semi-arid nature of the country, as the average annual rainfall levels do not exceed 400 mm in more than

85% of its total area (Fig. 1). This means an acute water stress, since the renewable annual water volume availability *per capita* has become below 730 m^3 (as it was above 2,500 m^3 in the 1960s), given the population growth (less than 11,6 million in 1960, above 35,9 million in 2020) and the decrease in rainfall levels (Fig. 2).

Since agriculture utilizes more than 85% of the total volumes in Morocco in a context of water shortage, a serious threat to its development is represented by water scarcity. This was already pointed out by previous research, which found that developing agriculture in a country facing also limited land availability could be compared to 'mirages' (Swearingen 1986). Analysing water uses by the agricultural sector under such constraints implies a series of precautions, given that this resource is not uniform, as it has several sources: green water (rainfall); blue water (irrigation, whether from surface sources—dams and springs—as well as groundwater—wells and boreholes—and even unconventional sources such as treated wastewater and desalinated seawater); and finally virtual water (which means the water volumes used in the countries of origin of food imported in another location). In fact, the latter source of water has been recognized as a significant mean to overpass scarcity

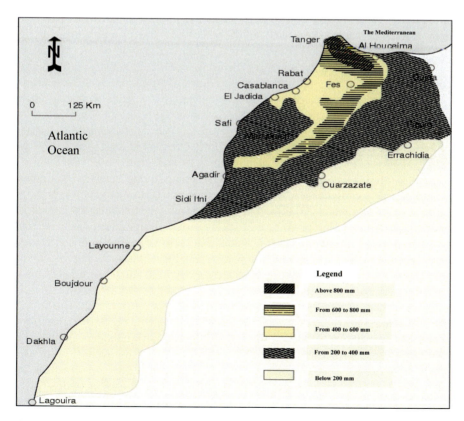

Fig. 1 Regional average rainfall levels in Morocco

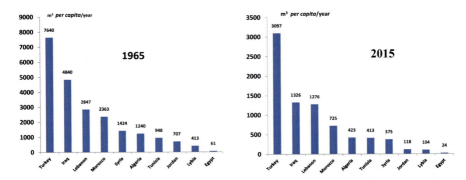

Fig. 2 Renewable water availability *per capita* per year in the Middle East and North Africa (MENA) countries. *Source* The World Bank (2017)

in water-stressed regions like the MENA, where water stress is acute (Antonelli and Tamea 2014).

In addition, climate change scenarios (an increase of 2 °C in average temperatures and less rainfall; −15%, with more frequent extreme climate events like flash floods and long periods of drought) imply that the water stress is going to get amplified, exacerbating social and economic vulnerability of people whose livelihoods are mainly linked to farming activities (Schilling et al. 2020). This might be even more felt in rain-fed systems with no possibilities of irrigation, as these continue to represent more than 80% of the total arable land. This is really worrying for the future of the agricultural sector, given that irrigation has been considered since the colonization era as the unique vector of development; this approach seems reaching its utmost limits. Since the launching of the GMP, its dynamic was to reach a climax and the public authorities have allocated significant budgets to ensure the adoption of irrigation means (from wells and boreholes digging to drip irrigation equipment installation). As a consequence, the area converted from furrow irrigation to drip irrigation has almost reached 500,000 ha (around a third of the total irrigation surface), ensuring, according to the promoters of such a program, significant water savings. However, on-farm follow-ups have revealed that these savings are almost theoretical (Batchelor et al. 2014), whereas in reality, drip irrigation spreading has mainly resulted in an increase of the surface irrigated per farm, and the substitution of traditional rain-fed crops (cereals, leguminous pulses, autumn fodder like oats and barley) by high water-consuming cash crops (orchards, vegetables, maize, etc.). The follow-ups have also revealed the important means of 'bricolage' adopted by farmers and technicians to allow the spreading of such an innovation (Benouniche et al. 2014) in systems where formal training is almost lacking. As a consequence, water uses have soared, mainly from private sources (i.e. wells and boreholes), implying the emergence of the groundwater depletion phenomenon (Molle and Tanouti, 2017).

This situation is particularly obvious in the most arid areas of the country (i.e. its southern and eastern zones). Moreover, such financial rationality applied to the agricultural sector has also induced that the rain-fed crops—which used to be its actual pillar and a barometer that is still used to assess its annual performance by spring (i.e. the expected output of cereal grains which in fact influences the whole economic growth of the country)—are increasingly considered as elements of the past. At the contrary, irrigated crops (mainly orchards) have reached an emblem of high investment agriculture, enabling farmers who practice them to get social consideration, given the skills they necessitate and their contribution to the country's export potential. This dichotomy between irrigation and rain-fed agriculture has created dangerous drifts, which have become unacceptable. For instance, some ideas began to defend that irrigated crops systematically perform better than rain-fed ones, as they allow higher yields (without taking into consideration the low level of dry matter their products contain…) and better profitability. But these ideas do not consider that the irrigation expansion has amplified water withdrawal from aquifers, generating groundwater depletion (El Moustaine et al. 2014). Moreover, it has also resulted in output surpluses for certain kinds of fruits and vegetables. Therefore, it has not always allowed reaching the goal which was sought, i.e. an increase in water economic productivity (US $ per cubic meter of water).

There are many examples which confirm these findings. For instance, the difficulties to sell citrus fruits, due to important volumes ensured by newly-planted domains, which benefited from public subsidies (irrigation, seeds, etc.) in addition to the limited competitiveness of the Moroccan product on global markets (as some other countries have lower prices), have even definitely pushed certain farmers to pull up trees. This has also been precipitated in some regions, like the Souss Massa (south-west of the country), due to severe water shortages as well as the increase of irrigation water salinity levels, implying that farmers simply abandon their trees. Moreover, water reserves in dams have collapsed due to several successive years of extreme drought meaning that the public authorities have informed farmers they cannot rely anymore on surface water to irrigate their plots; the remaining volumes in dams being reserved to cities' supply. A similar scenario has also been witnessed in the oasis regions, where the expansion in desert borders of irrigation areas from groundwater sources has quite exclusively been devoted to the cultivation of watermelon and date palms, creating a dichotomy between traditional oases and 'modern' expansion areas (Hamamouche et al. 2018). Logically, the same phenomena of output surplus and water resources depletion have occurred, aggravated by the very arid climate of these regions (less than 100 mm per year). Therefore, the steady supply of water to urban centers has become hampered, as the level of the aquifer has deepened. In addition, watermelon profitability is not definitely guaranteed, as it can suffer from a drop in the local demand or an abundant output, which is sometimes not absorbed by the export markets.

These facts do not however prevent those who benefited from public subsidies to exploit groundwater to lobby for further support, particularly for solar energy pumping, which will amplify water withdrawals. This can only generate social exclusion, given that those who cannot handle additional investments for digging deeper boreholes have already withdrawn from the groundwater economy (Ameur et al. 2017). The expansion of drip irrigation area has been promoted to the status of a doxa for all operators in agricultural sector, from farmers to technicians and researchers, implying that all the efforts converge toward more production with additional water, even from unconventional sources (i.e. treated wastewater and desalinated seawater). Very few voices now dare expressing different ideas from this single thought, like for example the improvement of rain-fed crops' yields, particularly cereals which prove adapted to the specific climate of North-African centuries.

In fact, the output of cereal grains did not improve enough with the launching of the GMP, despite some support measures, like the promotion of the cereal seed system (Bishaw et al. 2019). The existing data show that the annual cereal production in Morocco has remained largely determined by rainfall levels and their distribution, implying an important inter-annual variability (Fig. 3). In fact, the GMP almost added nothing to the existing patterns of production, as it only reinforced the substitution of barley by soft wheat as the leading species, in terms of the area sown, given the important demand for that particular cereal in consumption habits. No significant improvement in the average yield could be achieved and their levels remain very far from the average global (3 tons/ha) and even African (2.3 tons/ha) levels. Despite the efforts to generalize the use of fertilizers, at a time where Morocco is one of the leading countries in rock phosphate production (Lyon et al. 2020), low cereal yields also reflect the poor water management in soils. They also reveal the weak extension systems, implying that for this strategic sector, long-term interventions to spread conservation agriculture adoption are urgently needed, as it has been shown in a similar North African context (Tunisia) (Bahri et al. 2019).

The single thought legitimizing the continued depletion of water from arid areas is in fact based on the theory of comparative advantages, where Morocco is assigned to export more and more fruits and vegetables by mobilizing a cheap workforce at a time where he has to import a large share of the staple food used (cereal grains, pulses, vegetal oils, sugar, feedstuffs for livestock, cheese, etc.) (Table 1). Likewise, this single thought neglects all the collateral damages inflicted to the environment,

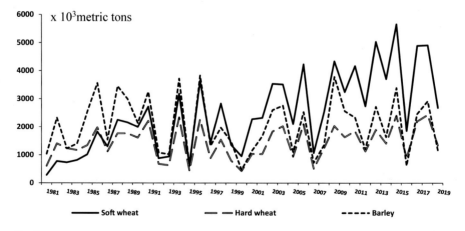

Fig. 3 Inter annual variability of the output of cereal crops (1981–2019). *Source* Adapted from ONICL (2020)

Table 1 Main imports of food commodities in Morocco, 2018 and 2019 ($\times 10^6$ US $)

Commodity	Imports in 2019	Imports in 2018
Wheat	923,2	912,4
Oil meal	524,1	491,1
Maize	523,5	464,7
Sugar	395,7	350,6
Tea	223,5	206,9
Dates	162,3	114,3
Cheese	95,1	103,0

Source Adapted from theOffice of Change (2020)

such as desertification and degradation of many fragile ecosystems, and promotes an agricultural development model whose limitations have been totally unveiled by the COVID-19 pandemic.

4 Rethinking an Innovative Agricultural Development Model

The sudden outbreak of the COVID-19 pandemic has had an appalling impact on the Moroccan economy, as it has prompted a sharp recession, the first of its kind since 1995 (World Bank 2020). This has meant a drop of about 6% in the GDP due to less production and services as well as a decline in tourism incomes (more than 250,000 jobs lost in this sector alone), which will generate an overall deficit whose value reaches around 8% of GDP. The multiple costs (economic, financial, health, etc.) of the pandemic have called for innovative thoughts related to a more balanced world, where the human well-being should be the top priority rather than profit, underlining the importance of implementing efficient education systems and sustainable consumption patterns as well as solidarity. These ideas have to nurture in order to help conceiving alternative ways of living, allowing future generations to evolve in a still liveable planet, in spite of the numerous challenges linked to climate change and the growing pressure on rare resources because of the demographic growth and rapid urbanization (Acuto 2020).

Therefore, the COVID-19 pandemic constitutes an opportunity to be ceased by the Moroccan agricultural sector. It offers an occasion to realize an objective and in-depth assessment of the GMP and the identification of alternative pathways for a sustainable development. The most urgent decision to be taken should aim at reviewing the previous priorities of intense water withdrawals. Indeed, improving water productivity in the agricultural sector requires considering that "a crop needs more than a drop" (Kuper et al. 2009). It has to be accompanied by a necessary assessment of the sources of the resources used whether 'green water', 'blue water' or 'grey water' (Hoekstra and Mekonnen 2012). As a consequence, public policies have to be assessed carefully by using the latest scientific knowledge, in order to give the priority to 'green water' (Rockström et al. 2009). This has to be adopted wherever rainfall levels allow profitable farming systems to develop, meaning that rain-fed crops and activities have to be effectively encouraged: cereals, pulses and livestock using rangelands and fodder resources without irrigation.

In the same line, Sraïri et al. (2009a) have demonstrated in the Tadla irrigation scheme that dual purpose herds (both milk and meat at the same time) require important volumes of water: respectively 1.7 and 9.1 m^3 of water per kg of milk and live weight. The same authors also found in the Saïss plain (i.e. a rain-fed area with cereals and pulses, but which turned recently to the groundwater economy, by planting orchards), that the dairy stock within smallholder farms uses rainfall as the main source of water to get milk and live weight. In addition, livestock also uses significant amounts of virtual water, like imported grains and even bread leftovers. These findings mean that the livestock chains do not cause any harm to the groundwater table in areas with more than 500 mm of annual rainfall levels (Sraïri et al. 2016). At the contrary, onions and orchards, which have developed significantly in the area because of the public incentives levied by GMP, contribute to the groundwater depletion phenomenon. In fact, rain-fed crops, particularly cereals and legume pulses, not only ensure food supply for humans, but they also produce various by-products such as straws and stubble, which are strategic to rear livestock (Magnan et al. 2012). In addition, livestock plays a crucial role in farming systems' resilience as it allows diversifying the sources of incomes, and it represents an insurance to face climate crises as well as economic disturbances (Ryschawy et al. 2013). It often represents the unique source of revenues in smallholder farms, where the existing assets (land, capital and even water) are not sufficient to ensure livelihoods. Livestock will however mobilize important volumes of routine work, with limited remunerations (Sraïri et al. 2018).

Moreover, the governance of supply chains implying thousands of smallholder farms impose a bottom-up organization, where stakeholders (farmers, processors, retailers, consumers' organizations, etc.) have negotiating space to discuss urgent hot topics related to the chain: quality assessment and remuneration, technical support for farmers, value chain distribution, etc. (Sraïri et al. 2009b; Sraïri et al. 2011; Ourabah et al. 2017). Unfortunately, that was not respected in GMP's goals regarding dairy chain. The efforts converged towards encouraging mega dairy farms which benefitted from a consequent share of subsidies (heifers' imports, milking devices, drip irrigation, etc.), but without taking into account the evolution of smallholder systems, whose constraints (especially limited farm gate milk price, several setbacks linked to insufficient feed availability, and unbalanced dietary rations) remained quite unchanged (Sraïri et al. 2015). Altogether, these findings have finally resulted in the emergence of the phenomenon of the 'dairy boycott' which was witnessed during spring 2017 for a period of six months. Even if it targeted only one dairy processor (the leading one in the market which used to collect around 50% of the volumes, i.e. 'Centrale Danone'), it soon affected all the chain, as milk collection circuits were all severely disrupted, given the significant drop of consumption levels. As a consequence, farmers could not easily market their daily output. It is estimated, according to FAOSTAT (2020), that Morocco has lost around 25% of its annual milk production from 2017 to 2018, and this means that all efforts undertaken within the framework of GMP for this chain, particularly huge amounts of public money, have been simply wasted (Fig. 4).

The difficulties of implementing a sustainable dairy chain remind everyone of the constraints which have to be taken into consideration when it comes to resources management (Sraïri et al. 2019). Hence, the importance to promote a sound governance of the whole agricultural sector is obvious, by designing policies through inclusive approaches. As the water constraint is surely the most biting, it has to be acknowledged that adding value to rainfall must be the top priority in the future agenda. As a consequence, improving cereals' average yield has to be sought by all means, not by additional irrigation or by sowing more surfaces devoted to these crops in irrigation schemes. At the opposite, improving the cereal output has to be achieved in the rain-fed areas, mainly in favourable years with enough and regular rainfall levels. This has to be entailed by support efforts as well as research programs which should allow the reinforcement of synergies between rain-fed crops and livestock, in order to lay the foundations of a sustainable agricultural development on the long run (Szymczak et al. 2020). This also has to be done in low external input agricultural systems with a minimum use of irrigation water, fossil energy, fertilizers and pesticides, through the integral recycling of biomass residues as well as manure, in what is currently known as the 'circular economy', where livestock plays an essential role (Peyraud et al. 2019). This concept is far from the pillars sustaining the recent development of poultry products' chains, either chicken meat or laying hens, which have allowed Morocco to reach by 2020 average annual consumption levels of 22 kg of poultry meat and 185 eggs *per capita* (FISA 2020). In fact, such an output from poultry facilities is almost entirely based on imported inputs, particularly maize grain (2.2 million metric tons per year) and soya meal, implying a significant carbon footprint for these products (Sraïri 2011).

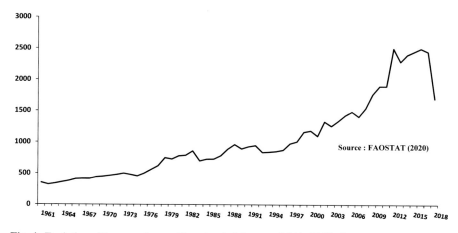

Fig. 4 Evolution of the annual raw milk output in Morocco (1961–2019). *Source* FAOSTAT (2020)

In addition to farming systems design in favour of integrated crop-livestock activities, adding value to the irrigation water is also mandatory, as it remains limited in many chains (Schyns and Hoekstra 2014). This might also need to be accompanied by the promotion of priority proximity markets. All these concepts represent opposite directions to the philosophy of GMP which has mainly promoted the model of very large specialized farms, with the domination of monoculture with a decoupling between livestock and crops: either orchards or dairy herds fed with maize silage irrigated from groundwater. Such models are nowadays criticized because of their environmental and even economic vulnerabilities (Garrett et al. 2020). In fact, in Morocco, it has appeared that large farms specialized in one species (citrus, olives, etc.), which benefitted from important public incentives to plant trees and to dig boreholes as well as to equip themselves with drip irrigation means, were particularly vulnerable to any hazard related to heath (the emergence of a new pathogen), economy (the decrease in prices because of a surplus output or because consumers lost their incomes) or even climate change (extreme events like drought, flood, hails or freezing).

Given all these rapid developments and taking into account the increased food prices volatility in global markets, it has to be acknowledged that the key concept of 'food sovereignty' has to be considered again, as Morocco imports around 200 kg *per capita* per year of cereal grains (mainly soft and hard wheat, but also maize for poultry and ruminants). Recognizing the specificities of food and agriculture means a massive investment in local production systems while avoiding an overdependence on global markets. Moreover, it also ensures thinking about the sustainability of future's food systems in Morocco, a country which used to adhere strongly to the precepts of the Mediterranean diet, but, because of the shifting dietary patterns, it is starting to suffer from significant outbreaks of overweight and obesity: respectively 46.8 and 16.4% of the total population according to the World Health Organization (WHO 2011). In fact, adding value to rainfall for food production not only will decrease fossil fuel used for pumping irrigation water as important volumes of subsidized butane are consumed for that purpose (Doukkali and Lejars 2015), but it will also ensure preserving diets with an important contribution of locally grown staple food of vegetal origin, one of the pillars of sustainable food systems (Willett et al. 2019). Finally, promoting the local production and short circuits will also allow decreasing the important amounts of fossil fuel needed to export and import hundreds of thousands of tons of raw food commodities, thus participating to global mitigation efforts.

5 Conclusion

One of the major teachings of the COVID-19 pandemic for Morocco has certainly been the increased interest devoted to agriculture and food systems by the public opinion. Citizens have been forced to acknowledge the vital role of agriculture and farmers' labour. The continuity of agricultural activities during the three-month lockdown, despite health risks and economic uncertainties, has allowed securing a steady food supply. However, the average consumer has discovered during this period the huge amount of imported staple food (cereal grains, pulses, feed for animals, etc.), a fact which can only persuade him of the current situation of food dependency towards global markets. These imports currently represent around 130 US $ per capita and per year, and have to be considered as imports of virtual water, certainly allowing to find a solution to the structural water stress felt all over the country. These imports have also contributed to supply food to numerous households, given their relatively affordable prices, particularly in the case of families having lost their sources of income and becoming totally dependent on public furlough funding. However, what would have happened if the prices have significantly increased in global markets?

The pandemic having triggered in a very dry year, its effects on the domestic economy will be exacerbated and felt on the long term. This should not constitute an excuse to get back to past reflexes where the expansion of irrigation and the extreme mobilization of water resources (even unconventional such as desalinated seawater and treated sewage water) was the norm. On the contrary, it is urgent to engage a responsible debate on water uses in the agricultural sector at all scales, even if it requires the revision of the ambitions, while accepting that the scientific knowledge has to be associated to policy making, as it was the case during the COVID-19 pandemic management and efforts to avoid its spread. This is particularly of concern since numerous voices have alerted on the risks of a growing water stress which would disturb the supply of this vital resource to important urban centres.

Altogether, these ideas imply a paradigm shift towards the valorisation of rainfall water, by building the capacity of operators in the agricultural sector, particularly through the promotion of good management practices: the choice of seeds, crop fertilization and protection, sufficient and balanced dietary rations for animals, management of manure and liquid wastes, etc. These represent numerous challenges which imply reconsidering the roles of agricultural training, in its various levels (from farmers to technicians and conception engineers). In order to achieve the ambitions of the 'Green Generation' 2020–2030 strategy and allow an inclusive growth, which entails decent wages for the majority of workers in the sector, while ensuring a sustainable use of the needed resources (capital, land and water), a bottom-up approach with inclusive goals has to be implemented. Finally, the ongoing COVID-19 pandemic and the acute economic recession it is inducing should be considered as an opportunity to get rid of past counterproductive ideas and practices. It is more than urgent to see realities as they are and admit that given the numerous structural constraints and hazards impacting the agricultural sector in Morocco (water stress, prices' volatility, emerging diseases, etc.), it is totally illusive to continue to promote it as the locomotive of the economy; investments in effective education systems being by far at the top of the priorities for a new development model.

Acknowledgements The author wishes to thank many crop-livestock farmers who allowed him to pursue long-term farms' follow up, making possible to understand the complexities of issues related to resources' sustainable and profitable use, particularly water and work. Last but not least, the author also thanks his students who assisted in the realization of research works in various areas of Morocco (the irrigation schemes, the oases, the rain-fed plains, etc.), ensuring the collection of reliable data about crop-livestock integration and the challenges it will have to face in the near future.

References

Acuto M (2020) COVID-19: Lessons for an urban(izing) world. One Earth 2:317–319. https://doi.org/10.1016/j.oneear.2020.04.004

Akesbi N (2015) Qui fait la politique agricole au Maroc ? Ou quand l'expert se substitue au chercheur. Annales de l'INRA Tunisie 88:104–126

Ameur F, Kuper M, Lejars C, Dugué P (2017) Prosper, survive or exit: contrasted fortunes of farmers in the groundwater economy in the Saiss plain (Morocco). Agric Water Manag 191:207–2017

Antonelli M, Tamea S (2014) Food-water security and virtual water trade in the Middle East and North Africa. Int J Water Resour Dev 31:326–342. https://doi.org/10.1080/07900627.2015.1030496

Bahri H, Annabi M, M'hamed HC, Frija A (2019) Assessing the long-term impact of conservation agriculture on wheat-based systems in Tunisia using APSIM simulations under a climate change context. Sci Total Environ 692:1223–1233

Batchelor C, Reddy VR, Linstead C, Dhar M, Roy S, May R (2014) Do water-saving technologies improve environmental flows? J Hydrol 518:140–149

Benouniche M, Zwarteveen M, Kuper M (2014) *Bricolage* as an innovation: opening the black box of drip irrigation systems. Irrig Drain 63:651–658. https://doi.org/10.1002/ird.1854

Bishaw Z, Yigezu YA, Niane A, Telleria RJ, Najjar D (eds) (2019) Political economy of the wheat sector in Morocco: seed systems, varietal adoption, and impacts. International Center for Agricultural Research in the Dry Areas, Beirut, Lebanon, p 300

Cournut S, Chauvat S, Correa P, Dos Santos Filho JC, Dieguez F, Hostiou N, Khahn Pham D, Servière G, Sraïri MT, Turlot A, Dedieu B (2018) Analyzing work organization by the work assessment method: a meta-analysis. Agron Sustain Dev 38:58. https://doi.org/10.1007/s13593-018-0534-2

Dernini S, Berry EM (2015) Mediterranean diet: from a healthy diet to a sustainable dietary pattern. Frontiers Nutr https://doi.org/10.3389/fnut.2015.00015

Duteurtre G, Faye B (2009) L'élevage, richesse des pauvres. Versailles, France, EditionsQuæ

Doukkali MR, Lejars C (2015) Energy cost of irrigation policy in Morocco: a social accounting matrix assessment. Int J Water Resour 31:422–435. https://doi.org/10.1080/07900627.2015.103696

El Moustaine R, Chahlaoui A, Bengoumi D, Rour E-H (2014) Effects of anthropogenic factors on groundwater ecosystem in Meknes area (Morocco). J Mater Environ Sci 5:2086–2091

Espita A, Rocha N, Ruta M (2020) COVID-19 and food protectionism: the Impact of the pandemic and export restrictions on world food markets. World Bank Policy Research Working Paper No. 9253, p 30. https://papers.ssrn.com/sol3/papers.cfm?abstract_id=3605887

FAOSTAT (2020) Cattle milk annual output in Morocco (1961–2018). Available from: http://www.fao.org/faostat/fr/#data/QL

Faysse N (2015) The rationale of the Green Morocco Plan: missing links between goals and implementation. J North Afr Stud 20:622–634. https://doi.org/10.1080/13629387.2015.1053112

FISA (Fédération Interprofessionnelle du Secteur Avicole) (2020) Statistiques du secteur avicole. Available at: https://www.fisamaroc.org.ma/index.php?option=com_content&view=article&id17&Itemid=53

Funnel DC, Binns JA (1989) Irrigation and rural development in Morocco. Land Use Policy 6:43–52

Garrett RD, Ryschawy J, Bell LW, Cortner O, Ferreira J, Garik AVN, Gil JDB, Klerkx L, Moraine M, Peterson CA, dos Reis JC, Valentim JF (2020) Drivers of decoupling and recoupling of crop and livestock systems at farm and territorial scales. Ecol Soc 25:24. https://doi.org/10.5751/ES-11412-250124

Graeber D (2018) Bullshit Jobs. Editions Les Liens qui Libèrent, Paris, p 416

Hamamouche MF, Kuper M, Amichi H, Lejars C, Ghodbani T (2018) New reading of Saharan agricultural transformation: continuities of ancient oases and their extension. World Dev 107:210–223

Haut Commissariat au Plan (2015) The national survey on household consumption and expenditure in Morocco (in French). http://www.hcp.ma/Introduction-de-Monsieur-Ahmed-LAHLIMI-ALAMI-Haut-Commissaire-au-Plan-a-la-presentation-des-resultats-de-L-enquete_a1819.html

Hoekstra AY, Mekonnen MM (2012) The water footprint of humanity. Proc Natl Acad Sci (PNAS). 109:3232–3237. https://doi.org/10.1073/pnas.1109936109

Kouame JM (2020) Aviculture : la filière perd des plumes. L'Économiste, édition n°5 830. 27 Août 2020. www.leconomiste.com/article/1066363-aviculture-la-filiere-perd-des-plumes

Kuper M, Bouarfa S, Errahj M, Faysse N, Hammani A, Hartani S (2009) A crop needs more than a drop: towards a new praxis in irrigation management in North Africa. Irrig Drain 58:S231–S239

Leveau R (1972) Le fellah marocain, défenseur du trône. Éditions la Découverte, Paris, p 275

Lyon C, Cordell D, Jacobs B, Martin-Ortega J, Marshall R, Camargo-Valero MA, Sherry E (2020) Five pillars for stakeholder analyses in sustainability transformations: the global case of phosphorus. Environ Sci Policy 107:80–89

Magnan N, Larson DM, Taylor JE (2012) Stuck on stubble? The non-market value of agricultural by-products for diversified farmers in Morocco. Am J Agr Econ 94:1055–1069

Mahdi M (2014) Devenir du foncier agricole au Maroc : un cas d'accaparement des terres. New Medit. 13(4):2–10

Molle F, Tanouti O (2017) Squaring the circle: Agriculture intensification versus water conservation in Morocco. Agric Water Manag 192:170–179. https://doi.org/10.1016/j.agwat.2017.07.009

Montanari B, Bergh SI (2019) A gendered analysis of the income generating activities under the Green Morocco Plan: who profits? Hum Ecol 47:409–417. https://doi.org/10.1007/s10745-019-00086-8

Office of Change (2020) Results of foreign exchanges until December 2019 [In French]. Available at: www.oc.gov.ma/fr/actualites/communique-resultats-des-echanges-exterieurs-a-fin-decembre-2019

Ourabah Haddad N, Ton G, Sraïri MT, Bijman J (2017) Organisational challenges of Moroccan dairy cooperatives and the institutional environment. Int J Food Syst Dyn 8:236–249. http://centmapress.ilb.uni-bonn.de/ojs/index.php/fsd/article/view/835/720

ONICL (Interprofessional Office Of Cereals and Pulses) (2020) The output of the main cereal grains' crops (soft and wheat, barley and maize) (In French). Available from: www.onicl.org.ma

Peyraud J-L, Aubin J, Barbier M, Baumont R, Berri C, Bidanel J-P, Citti C, Cotinot C, Ducrot C, Dupraz P, Faverdin P, Friggens N, Houot S, Nozières-Petit M-O, Rogel-Gaillard C, Santé-Lhoutellier V (2019) Science for tomorrow's livestock farming: a forward thinking conducted at INRA. INRA Prod Anim 32 :323–338. https://doi.org/10.20870/productions-animales.2019.32.2.2591

Préfol P (1986) Prodige de l'irrigation au Maroc. Le développement exemplaire du Tadla. 1936–1985. Les Nouvelles Editions Latines. Paris, p 266

Rockström J, Falkenmark M, Karlberg L, Hoff H, Rost S, Gerten D (2009) Future water availability for global food production: the potential of green water for increasing resilience to global change. Water Resour Res 45:W00A12

Ryschawy J, Choisis N, Choisis JP, Gibon A (2013) Paths to last in mixed crop-livestock farming: lessons from an assessment of farm trajectories of change. Animal 7:673–681

Sager F, Mavrot C, Hinterleitner M, Kaufmann D, Grosjean M, Stocker TF (2020) A six-point checklist for utilization-focused scientific policy advice. Climate Policy. In press https://doi.org/10.1080/14693062.2020.1757399

Schilling J, Hertig E, Tramblay Y, Scheffran J (2020) Climate change vulnerability, water resources and social implication in North Africa. Reg Environ Change 20:15. https://doi.org/10.1007/s10113-020-01597-7

Schyns JF, Hoekstra AY (2014) The added value of water footprint assessment for national water policy: a case study for Morocco. PLoS ONE. https://doi.org/10.1371/journal.pone.0099705

Sraïri MT (2011) Le développement de l'élevage au Maroc : succès relatifs, dépendance alimentaire. Le Courrier de l'Environnement de l'INRA. 60 :91–101. http://www7.inra.fr/dpenv/pdf/C60TaherSrairi.pdf

Sraïri MT, Ghabiyel Y (2017) Coping with the work constraints in crop-livestock farming systems. Ann Agric Sci 62:23–32. http://www.sciencedirect.com/science/article/pii/S0570178317300015

Sraïri MT, Chatellier V, Corniaux C, Faye B, Aubron C, Hostiou N, Safa A, Bouhallab S, Lortal S (2019) Durabilité du développement laitier : réflexions autour de quelques cas dans différentes parties du monde. INRA Prod Anim 32:339–358. https://doi.org/10.20870/productions-animales.2019.32.3.2561

Sraïri MT, Bahri S, Ghabiyel Y (2018) Work management as a means to adapt to constraints in farming systems: a case study from two regions in Morocco. Cahiers Agricultures. http://www.cahiersagricultures.fr/articles/cagri/pdf/2017/01/cagri160177.pdf

Sraïri MT, Benjelloun R, Karrou M, Ates S, Kuper M (2016) Biophysical and economic water productivity of dual purpose cattle farming. Animal 10 :283–291. http://journals.cambridge.org/action/displayAbstract?fromPage=online&aid=10082209&fulltextType=RA&fileId=S1751731115002360

Sraïri MT, Sannito Y, Tourrand J-F (2015) Investigating the setbacks in conventional dairy farms by the follow-up of their potential and effective milk yields. Iran J Appl Anim Sci 5:255–264

Sraïri MT, Bahri S, Kuper M (2013) Le travail et sa contribution aux stratégies d'adaptation de petites exploitations agricoles familiales mixtes d'élevage bovin/polyculture au Maroc. Biotechnologies, Agronomie, Société et Environnement 17 :463–474. http://www.pressesagro.be/base/text/v17n3/463.pdf

Sraïri MT, El Jaouhari M, Saydi A, Kuper M, Le Gal P-Y (2011) Supporting small scale dairy farmers increasing their milk production: evidence from Morocco. Trop Anim Health Prod 43:4–49

Sraïri MT, Rjafallah M, Kuper M, Le Gal P-Y (2009) Water productivity of dual purpose herds (milk and meat) production in a Moroccan large-scale irrigated scheme. Irrig Drain 58:S334–S345

Sraïri MT, Benhouda H, Kuper M, Le Gal PY (2009b) Effect of cattle management practices on raw milk quality on farms in a two stage dairy chain. Trop Anim Health Prod 41:259–272. https://doi.org/10.1007/s11250-008-9183-9. http://link.springer.com/article/

Swearingen WD (1986) Moroccan mirages: agrarian dreams and deceptions 1912–1986. Editions I.B. Tauris, London, p 254

Szymczak LS, Carvalho PCDF, Lurette A, Moreas AD, Nunes PADA, Martins AP, Moulin CH (2020) System diversification and grazing management as resilience-enhancing agricultural practices: the case of crop-livestock integration. Agric Syst 184

World Bank (2020) Morocco Economic Monitor, p 44. Available at: https://documents.worldbank.org/en/publication/documents-reports/documentdetail

World Bank (2017) Beyond scarcity: water security in the Middle East and North Africa. MENA Development Report. The World Bank Group. Washington. Available at: https://www.worldbank.org/en/topic/water/publication/beyond-scarcity-water-security-in-the-middle-east-and-north-africa

World Health Organization (WHO) (2011) Non communicable diseases country profiles 2011. Global report. Available from: http://www.who.int/nmh/publications/ncd-profiles2011/en/index.html

Willett W, Rockström J, Loken B, Springmann M, Lang T, Vermeulen S, Ganett T, Tilman D et al (2019) Food in the anthropocene: the *EAT*-Lancet commission on healthy diets from sustainable food systems. Lancet 393:447–492. https://doi.org/10.1016/S0140-6736(18)31788-4

Chapter 4
Impacts of Climate Change on Agricultural Labour Force and Food Security in Pakistan: The Importance of Climate-Smart Agriculture

Syed Ghazanfar Abbas, Mirza Barjees Baig, and Gary S. Straquadine

Abstract Pakistan is the sixth most populous country in the world with a population of 220 million according to recent figures. With an annual population growth rate of around 2%, it is projected to be the fifth most populous country in the world by 2050. Urban areas of Pakistan and its entire region of the subcontinent are expected to be under severe risk of socioeconomic losses due to climate change by 2050. Increased and intense heat waves are expected to extend the limits of human endurance and survivability. A Press Release of the United Nations dated 8th October, 2019 after its 74th session, suggested that countries with lower per capita GDP have lesser financial means to adapt quickly to climate change. Innovative research is, therefore, needed to overcome any likely disaster. Review of the literature suggests that hot and humid countries like Pakistan are expected to experience a significant increase in temperature and humidity by 2050. Field workers' efficiency is likely to be affected due to higher temperatures and fatigue. A United Nations Development Program' (UNDP) report (2016) reconfirms this scenario stating a possible 10% loss in effective working hours in heat-exposed regions. The Intergovernmental Panel on Climate Change (IPCC) Fifth Assessment Report (AR5) for the Asia region reported that climate change threats in agriculture-dependent economies, such as Pakistan, come from their distinct geography, demographic trends, and socioeconomic factors. However, the greater factor could be the lack of adaptive capacities. The climate change projections of the AR5 for South Asia show that warming is likely to be above the global mean. There is no doubt that climate change will impact glacial melting rate and precipitation patterns, particularly affecting the timing and strength of monsoon rainfall. Consequently, this will significantly influence the productivity and efficiency of

S. G. Abbas (✉)
Ex-Senior Director, Farm Mechanization, Pakistan Agricultural Research Council (PARC), Islamabad, Pakistan

M. B. Baig
Prince Sultan Institute for Environmental, Water and Desert Research, King Saud University, Riyadh, Saudi Arabia
e-mail: mbbaig@ksu.edu.sa

G. S. Straquadine
Career and Technical Education, Utah State University (Eastern), Logan, UT, USA
e-mail: gary.straquadine@usu.edu

© The Author(s), under exclusive license to Springer Nature Switzerland AG 2022
M. Behnassi et al. (eds.), *Food Security and Climate-Smart Food Systems*,
https://doi.org/10.1007/978-3-030-92738-7_4

water-dependent sectors—such as agriculture and energy (IPCC 2014). Therefore, the impact of climate change is forcing a change in habits and habitats. This chapter aims at informing decision makers around the world about those dynamics, with emphasis on Pakistan. The goal is to better assess, adapt and bring issues before the policy makers so that the risks associated with the climate change are promptly addressed.

Keywords Climate change · Agricultural labor · Smart agriculture techniques · Use of computers in agriculture

1 Introduction

Pakistan is located between longitude 61 and 75° East and latitude 24 and 37° North. The country is surrounded by Iran to the west, Afghanistan to the north-west, Arabian Sea to the south, China to the north-east and India to the south-east. Due to its topography, it has four seasons and around the year it carries a wide difference in temperatures. Pakistan has a cool, dry winter from December through February and a hot and dry spring from March to May. There is a summer rainy season, with southwest monsoon period, from June to September. Usually, the retreating monsoon period of October and November helps in water storage just before the main wheat crop is sown from late November until early January in various parts of the country. However, as in other parts of the world, due to global climatic changes this weather pattern has become more unpredictable. Hailstorms in March, when wheat crop is maturely standing, inflicts damages in some areas. In the last decade, heavy rains at the wheat harvest (May and June) have also been very damaging for farmers.

In the year 2020, with a population of above 220 million, Pakistan's economy is heavily dependent on its agriculture. The total area of Pakistan is 79.6 million hectares, of which 70 percent is arid to semiarid. About 50.88 million ha (or 63.9%) are rangelands and only 22 million ha (or 27.6%) are cultivated lands. Agricultural research has played a significant role in producing enough food for Pakistan's populations and the country is considered self-sufficient in wheat, rice, sugarcane, and maize. To reemphasize the positive role of agriculture growth in Pakistan, it can be gauged by quoting that in the mid-1960s, combined population of East and West Pakistan was 90 million (50 million in east Pakistan and 40 million in west Pakistan). Whereas now current population of Pakistan is 220 million. During this time, Pakistan has lost vast agricultural areas due to urban and infrastructural development. Therefore, attaining self-sufficiency in staple food items is a great achievement given these circumstances, so the role of agricultural research cannot be underplayed. There are many factors that contribute to food availability and security in any high populated country like Pakistan. This chapter summarizes only the resultant effects of trends in agricultural labour force related to climate change, as well as the technological developments in the last few decades.

2 Agricultural Labour Force

Agricultural farm power sources components include: human power in the shape of agricultural labour force, animal power, and machinery (and electricity for tube well, etc.). Out of these, human and animal power is directly affected by climate change.

Abbas (1999) compiled data for the Food and Agriculture Organization (FAO) and reported agricultural labour force projections for the year 2015/2030. The following strategy was considered when making projections with regards to human labour:

- Estimated the current total population by relying on the available demographic data.
- Segregated the data into: urban, rural, agricultural population, and non-agricultural population.
- Assessed the efficiencies of agricultural labour force (Box 1). For example, Dabbits (1993) has quoted that the average human capacity is often estimated as 0.075 kw.
- Assessed the current human labour involvement in producing food.
- Estimated future food demand.
- Calculated the labour force necessary to meet the estimated food demand.

Box 1 Estimated Agricultural Labour Force Efficiency
WHO indicates that after eight hours of work, an adult of good health, between 20 and 39 years old, who can eat sufficiently, produces daily intake between 10.9 and 14.2 MJ (2,600–3,400 kcal) of energy. For productive work, there is then available:
Total energy value 15.50 MJ
Rest (sleeping) 2.09 MJ
Leisure time 4.64 MJ (6.73 MJ)
Productive labour 8.77 MJ (2.43 kwh)

Animal power and machinery do not fall in the purview of this chapter; thus they are not elaborated further.

The climate and eating habits have changed over the period and, therefore, revisiting any such data is insightful. Due to excessive temperatures, especially in the summer, there is likelihood that working efficiency as well as total number of working hours will reduce. Therefore, improved versions of mechanical power and related farm machinery shall play an important role in the overall efficiency and, ultimately, total agricultural production of future. Table 1 shows that the share of agriculture in total employment reached a peak in 2010–2011 at 43.5%, but fell to 37.4% in 2017–2018. The share of agricultural workers who are males has generally trended down, reaching 29.6% at the end of 2017–2018. The female share of agricultural workers is usually about two times the male share.

Table 1 Share of agriculture in total employment in Pakistan (%)

Indicators	2006–2007	2007–2008	2008–2009	2009–2010	2010–2011	2012–2013	2013–2014	2014–2015	2017–2018
Both sexes	42.0	42.8	43.3	43.4	43.5	42.2	42.2	41.0	37.4
Male	35.0	35.2	35.7	35.2	34.9	33.1	33.2	32.0	29.6
Female	71.4	73.8	72.7	73.9	74.2	74.9	72.9	71.8	66.1

Source Pakistan Employment Trends (2018)

Kjellstrom et al. (2009) claim that global climate change could impair the health and productivity of millions of active people. They suspect that in most regions, climate change will decrease labour productivity. They estimate that by the 2080s, the greatest absolute losses in labour capacity (in the range 11% to 27%) are seen in Southeast Asia, Central America, and the Caribbean. They suggest that workers may either have to work longer hours, or more workers may be required to achieve the same output. Based on these scenarios, in the next section, we will discuss the increase in food demands, and the impact of reduced labour due to climate change.

3 Growing Population and Food Security

According to the United Nations' projections, global population will reach 9.8 billion people in 2050. Natural disasters such as earthquakes, floods, typhoons, hurricanes, and more recently the spread of COVID-19 pandemic, inflict serious damage to economies, but have little effects on population growth. In this context, we assume that the current pandemic will not change the trend of population growth worldwide that have been quoted at various stages in this chapter.

According to the United Nations Population Fund (UNFPA) with the current growth rate of 2% in Pakistan, population is expected to cross the 250 million mark by 2030 and reaches above 300 million by 2050. Thus, it is likely that this population growth will outstrip developmental gains and continue to adversely affect the economy, environment, health, education, and quality of the life in the country (Fig. 1).

Bokhari et al. (2017) warned that any change in local weather conditions due to climate change poses threats to the agricultural productivity in Pakistan. They pointed

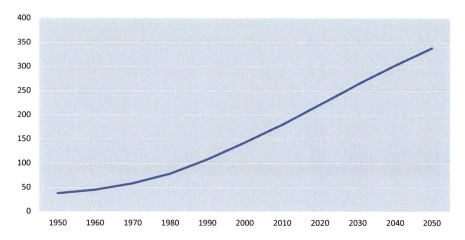

Fig. 1 Pakistan population trend (1950–2050). *Source* World Population Review—Live

out that extreme meteorological events, including high temperatures or heavy rains, are expected to increase as a result of global warming that may raise concern over how future climate change will impact natural and human systems. They quoted that the 2014 Global Climate Risk Index have placed Pakistan at 12th place in terms of the impacts of weather-related events. According to them, the summer and winter precipitation over the last 40 years have increased in northern areas of Pakistan and mainly along the foot-hills of the Himalayas. In another research, Baig et al. (2020) have quoted Eckstein et al. (2018) citing 2019 Global Climate Risk Index indicating Pakistan is expected to be among the ten countries most affected by climate change. It further cautions citing the World Wildlife Fund (WWF) report (2012) that average temperatures are expected to rise faster than most other locations, reaching 4 degrees by the year 2100. That is the reason that we now more frequently hear about untimely and frequent floods or landslides in these locations.

As per Agricultural Statistics of Pakistan (2017–2018), agriculture sector has recorded a remarkable growth of 3.81% (Table 2). The statistics reveal that sugarcane and rice surpassed their production targets of 2017–2018, while cotton crop production also managed to exceed its previous year's production. Wheat and maize crops' production remained subdued. In additional to these crops, increased production of fodder, vegetables, and fruits was also reported.

As already mentioned, wheat is one of the four main agricultural crops in Pakistan and is mostly consumed as a staple food. Almost all farmers (around 80%) grow wheat covering nearly 40 percent of the country's total cultivated land (GoP 2018). Due to favourable weather conditions and timely rains throughout the growing period, the general crop condition has been reported very good in recent years. The country's average yield has now increased to nearly 3.1 tons per hectare with total production in the vicinity of 24–26 million tons, which is more than their current total consumption per year. Every year before the harvest, the Government announces a support price to encourage farmers to plant. In March 2020, the support price was Rs. 1,400 per 40 kg which is Rs. 50 more than the previous year's support price. Table 3 and Fig. 2 show the exact area under wheat cultivation across Pakistan for the 2018–2019 crop.

The Global Agricultural Information Network Report (2019) reveals that two-thirds of the country's water for irrigation are sourced from snow and glacier melts, with the balance supplied by seasonal monsoon rains. Therefore, climate change is very important for agriculture in general, and for the wheat crop in particular.

After wheat, rice is Pakistan's second largest staple food crop. About 10 percent of Pakistan's total agricultural area is under rice cultivation during the summer (PACRA 2018). Rice growing areas of Pakistan are broadly classified into four zones as depicted in Table 4.

Pakistan is a leading producer and exporter of Basmati and IRRI rice and those exports are a major source of foreign currencies. Rice production in Pakistan has always been 2–2.5 times more than the total national consumption. Yields have steadily increased over the past decade (PACRA 2018). As rice crop is heavily dependent on irrigation water, any adverse climatic change resulting in reduced water supply could markedly affect its production. Thus, Pakistan should be ready with modified rice growing techniques that require less water. The introduction of direct

Table 2 Agriculture growth percentages (2011–2012 to 2017–2018)

Sector	2011–2012	2012–2013	2013–2014	2014–2015	2015–2016	2016–2017	2018–2018
Agriculture	3.62	2.68	2.50	2.13	0.15	2.07	3.81
Crops	3.22	1.53	2.64	0.16	−5.27	0.91	3.83
Important crops	7.87	0.17	7.22	−1.62	−5.86	2.18	3.57
Other crops	−7.52	5.58	−5.71	2.51	0.40	−2.66	3.33
Cotton ginning	13.83	−2.90	−1.33	7.24	−22.12	5.58	8.72
Livestock	3.99	3.45	2.48	3.99	3.36	2.99	3.76
Forestry	1.79	6.58	1.88	−12.45	14.31	−2.37	7.17
Fishing	3.77	0.65	0.98	5.75	3.25	1.23	1.63

Source Pakistan Bureau of Statistics

Table 3 Wheat area by province, 2018–2019

Province	Area (Million Hectares)	Percentage of total area
Punjab	6.50	74
Sindh	1.16	13.2
Khyber Pakhtunkhwa	0.75	8.5
Baluchistan	0.38	4.3
Total	8.79	100

Source Global Agricultural Information Network Report No. 1907 dated 29th March, 2019

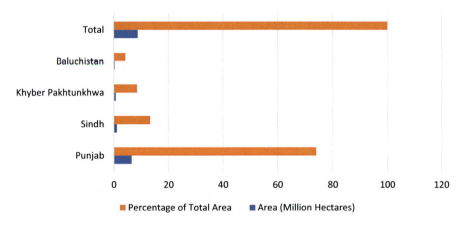

Fig. 2 Area under wheat crop in Pakistan (2018–2019)

Table 4 Rice growing zone of Pakistan

Zone I	Northern high mountainous areas of Khyber Pakhtunkhwa (Swat and Khagan). Sub-humid climate and average rainfall of 750–1000 mm
Zone II	Lies between the Ravi and Chenab Rivers in the central Punjab. Sub-humid, sub-tropical climate with average rainfall of 400-700 mm. This is the famous premium zone where Basmati rice is grown
Zone III	West bank of Indus river in upper Sindh and Balochistan. Larkana, Jacobabad (Sindh), Nasirabad and Jaffarabad (Balochistan). High temperature and sub-tropical climate with average rainfall of 100 mm. It is best suited for long grain rice
Zone IV	Indus delta basin in Lower Sindh (Badin and Thatta Districts). Climate is arid tropical and is suited for coarse varieties

Source Global Agricultural Information Network Report No. 1907 dated 29th March, 2019

rice seeding, as done in Italy, could be a good strategy. Agricultural innovations and smart agricultural techniques will need to offset any reductions in irrigation water.

Climate change may also affect adversely the prospect of achieving food security, not only in Pakistan but also in many parts of the world. Enhancing agricultural production will not be easy because the gap between actual and potential yields has narrowed in recent years due to traditional agricultural research and development, such as proper use of chemical fertilizers (Abdul Rehman et al. 2015). Most climate models indicate that the agricultural potential of developing countries may be more adversely affected than the world average because of excessive urbanization and conversion of agricultural lands into infrastructural or new housing development. In addition to such risks, the impacts of climate change may potentially alter the precipitation and evapotranspiration patterns, hence affecting renewable water that is essential not only for human beings but also for agricultural activities, forestry, and livestock.

Aamir et al. (2020) have reconfirmed that agriculture is one of the most climate-sensitive sectors in an economy. Climate change impacts temperature, precipitation, and soil radiation, which are factors directly associated with crop growth at various stages. Rising temperature, uneven distribution of precipitation, floods, droughts, and other climatic disasters have also affected human life in addition to socio-economic sectors of the world's over-populated regions, including Pakistan. Table 5 and Fig. 3 depict scenarios for the world economy in 2050 considering variables like GDP,

Table 5 Baseline development (percent change)

Variable (Growth Rate)	2011–2020	2020–2030	2030–2040	2040–2050
GDP	55.7	41.07	29.11	22.56
Population	21.13	17.45	15.10	12.99
Capital	44.29	61.39	33.12	28.30
Skilled labour	68.32	42.02	27.51	23.09
Unskilled labour	25.68	24.23	18.5	16.16

Source Aamir et al. (2020)

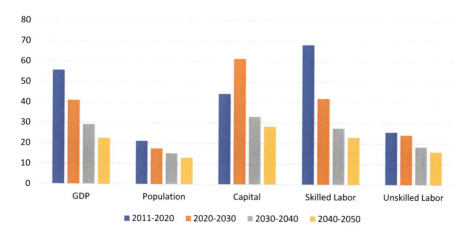

Fig. 3 Scenario of the world economy in 2050 with variables like GDP, population, capital, skilled and unskilled labour

population, capital, and skilled and unskilled labour. The important thing to note is the decline in available skilled and unskilled labour with each passing decade.

4 Agricultural Production by 2050: A Case Study of Pakistan

Hunter et al. (2017) have projected that food demand in 2050 will rise as the world population reaches above 9 billion and the greater wealth drives per capita consumption. Due to its population explosion, Pakistan may need some supplementary crops, fruit, and vegetables besides the above-mentioned wheat and rice crops, in order to feed its population. Table 6 gives current wheat and rice production levels, consumptions levels, and the estimated future demands for 2030, 2040, and 2050.

Innovative agricultural techniques will have to be adopted for enhancing production of vegetables and other crops. Production of good quality fodder crops for livestock will also need special attention. Economists have studied the agricultural land availability and water and yield constraints for future years to estimate food crop needs until 2030 and 2050. The International Food Policy Institute (IFPRI) stressed that production needs to increase between 25 and 70% to meet food demands of 2050 (IFPRI 2010). Therefore, there is an urgent need to move towards mechanized agriculture and the use of smart-agricultural techniques to promote sustainability and ensure food security in Pakistan.

Based on the above projected demands for wheat and rice, we need to know the status of agricultural labour force and farm power that will be needed to achieve such production levels. Agricultural labour data can be deduced from the population data and its projections for the coming years until 2050. Similarly, the farm power data can also be estimated from the tractor population that is available in various agricultural machinery census. Table 7 shows total population, total agricultural population, and agricultural labour force estimates until 2050.

Climate change can greatly affect the performance of the agricultural labour force. Olsen (2009) described how international labour standards could be made more relevant to climate change when the labour efficiency will decrease due to adverse weather conditions. The report assessed various studies and observations of

Table 6 Current wheat and rice production, consumptions and the future requirements (2030/2050)

	2019–2020	2019–2020	2030	2040	2050
Crop	Production	Consumption	Projected Demand*	Projected Demand[a]	Projected Demand[a]
Wheat	25.6	25.4	30.5[b]	37.0	46.0
Rice	7.5	3.3	3.9	4.8	6.0

[a] Estimated based on expected population
Source [b] Supply and Demand for Cereals in Pakistan, 2010–2030

Table 7 Projection on Pakistan's total, agricultural, and agricultural labor force (2030/2050)

	2000	2010	2020	2030	2040	2050
Total population	156	180	220	245	277	300
Labour population	58	76	93	103	116	125
Agricultural population	27	35	42	46	51	55

the global climate system. The IPCC has provided evidence regarding the serious risks of continued global warming for the labour force, which includes exertion and fatigue. The report further revealed that over one billion people were employed in agriculture during the year 2008.

Ali et al. (2017) have considered climate change as a global environmental threat to all economic sectors—specifically agriculture. Recently, Pakistan has faced extreme weather events like untimely and heavy rainfall, hailstorms, and floods in mountainous areas whose impacts have damaged major crops and farm properties. It is suspected that much of these disasters are induced by climate change. These weather patterns are not likely to change soon, so there is a need to adopt agricultural practices that can more readily deal with these problems. Adaptation strategies can be very effective, but many farmers do not use them. Adaptation to the negative impacts of climate variability and change is essential to improve the food security of Pakistan and to protect rural households. Climate-Smart Agriculture (CSA) is one way forward. Experts claim that despite adverse effects of global warming, unexpected rains, floods, hailstorms, high temperatures, and lower efficiency of agricultural labour force, food security of rapidly growing population can be ensured using CSA techniques and practices.

5 Examples of CSA Practices

Agricultural productivity, income, and food security have been negatively affected by climate change impacts. In this regard, CSA offers attractive opportunities for strengthening the agriculture sector and improving food security for the future generations while adapting to climate change. Therefore, there is a need to start exploring and adopting changes in agricultural practices in order to manage existing problems and to cease opportunities while meeting the needs of an ever-growing population. World Agriculture Towards 2030/2050 in the 2012 revision edition through Nikos Alexandratos and Jelle Bruinsma in a publication "Global Perspective Studies Team FAO Agricultural Development Economics Division" have projected estimates based on food consumption pattern and population growth, which show that agriculture production needs to increase by 65% by 2050. The International Centre for Tropical Agriculture (CIAT) and the World Bank have identified measures that can be adopted in this perspective ranging from innovative technological practices like laser land leveling and solar powered irrigation systems, to management changes like crop diversification, proper cropping patterns, and optimized planting dates. Investing in research to develop high-yielding heat resistant, drought tolerant, and pest resistant crop varieties and livestock breeds is especially critical. Nagargade et al. (2017) have outlined the following benefits of CSA practices and techniques:

- *Increased and better-quality production*: optimized crop treatment such as accurate planting, watering, pesticide application and harvesting directly affects production rates. Also, it helps in analysing production quality with regards to insecticide and pesticide use.
- *Lowered operational costs*: automating processes in planting, plant protection, and harvesting can reduce resource consumption, human error, and overall cost.
- *Water conservation*: weather predictions and soil moisture sensors apply water only when it is needed.
- *Real-time data and production insight*: farmers can visualize production levels, soil moisture, sunlight intensity more remotely and in real time to accelerate decision-making process.
- *Increased use of applications for marketing of agricultural produce.*
- *Accurate farm and field evaluation*: accurately tracking production rates by field over time allows for detailed input application and prediction of crop yields.
- *Improved livestock farming*: sensors and machines can be used to detect reproduction and health events earlier in animals. Geofencing location tracking can also improve livestock monitoring and management.
- *Reduced environmental issues*: all conservation efforts, such as water usage and increased production per land unit, directly affect the environmental footprint positively.
- *Remote monitoring*: local and commercial farmers can monitor multiple fields in multiple locations around the globe using internet connection. Decisions can be made in real-time and from anywhere.
- *Equipment monitoring*: farming equipment can be monitored and maintained according to production rates, labour effectiveness and failure prediction.

The below benefits can be directly related to agricultural practices that need special attention because of higher expected temperatures:

- *Efficient Irrigation Management:* Due to limited groundwater resources and other environmental changes, farmers need to move towards high efficiency irrigation systems.
- *Organic Practices:* The production of organic fruit and vegetable reduces dependency on chemical fertilizers and herbicides. This will increase soil health and its capacity to absorb carbon. Many agricultural technologies and practices such as minimum tillage, alternative methods of crop establishment, nutrient and irrigation management, and residue management can improve crop yields, increase nutrient and water-use efficiency, and reduce greenhouse gas (GHG) emissions from agricultural activities. In addition, conservation agricultural techniques have proven to reduce labour needs in the context of harsher working conditions. Abbas (1997) stated that energy used for tillage accounts for approximately 11% of total

Table 8 Energy, cost and time factors in minimum and conventional cultivation

	Energy GJ/ha	Fuel L/ha	Labour used Hours/ha
Ploughing	2.26	67.4	N.A
Conservation tillage	1.02	18.5	0.97
Conventional tillage	2.63	71.0	2.24

direct farm energy use, while fertilizer application accounts 0.7%, and pesticide application accounted for 0.8%. Thus, a reduction or elimination in tillage would bring substantial energy savings. Table 8 shows differences in fuel consumption, total energy and labour used under different tillage systems. There is massive savings of energy, fuel, and labour when adopting conservation tillage over the conventional tillage practices.

Besides many other aspects, if following aspects of smart agriculture are introduced widely in developing countries, like Pakistan, maximum and cost-effective benefits shall be achieved:

- *IT agricultural network for all agricultural inputs* such as tractors, farm equipment, quality seed, fertilizer, and availability of rental farm machinery. The internet of things (IoT) will allow better communication among buyers and sellers to improve utilization of these inputs. In Pakistan, the tractor and farm equipment rental business are an unorganized activity. Tractors are usually only rented out to nearby locations. However, since 1985, with the introduction of Combined Harvesters (CH), 'contract farming' has spread widely to overcome the underutilization of this very costly machine. Further, it encourages more farmers to invest in these machines because they can be rented to others. With ever decreasing farm sizes, it has become essential that 'Machinery Pools' are made available for easy access to the farming community. In addition, Selvaraju et al. stated that the provision of needs-based climate information to farmers can support management of the most important agriculture resources (land, water, and genetic resources). They have further claimed that within the resource-management context, climate must be recognised as a resource, assessed in quantitative terms by climate and agriculture services, and communicated to, and properly managed by farmers. The likely outcomes and impact of adopting such IT innovations will:
    – encourage tractor / combine harvesters' contractors and farm equipment dealers;
    – encourage laser-land leveling contractors;
    – introduce full packages of 'transfer of technology (TOT)' by companies, e.g., rice transplantation technology—nursery, transplanting, weeding, harvesting;
    – encourage legislative cover on farm equipment rental service; and
    – encourage the establishment of farm equipment operational training institutes.
- *Credit* to farmers can be facilitated by linking various financial institutions to assess and choose the best possible option. Agricultural credit is an important

and delicate issue for small farmers in Pakistan. Agricultural loans by different commercial and private institutions are available; these institutions have their own rules and regulations for disbursement of loans. Small farmers in many cases get trapped by local lenders because they don't have information about other credit options. By using current IT advancements, all loan schemes can be pooled in a way that small farmers can make better decisions and avoid expensive loan terms. The likely outcomes and impacts of such measures consist of: encouraging small land holding farmers and their families to look at various options before applying for an agricultural loan; taking benefit of existing public and private financial programs that are announced from time to time; and allowing the farmer to keep the loan record readily accessible.

- *Progressive tagging of livestock* starting with cattle and moving towards camels and other animals. Pakistan's livestock sector is quite large and spreads all over the country. However, most livestock are present in the remote areas where tracking is an issue. Animal tracking data helps understand how individuals and populations move within local areas, migrate across provinces, or cross over country boundaries. If such data is made available by using current IT packages or software, environmental challenges such as climate and land-use change, biodiversity loss, invasive species, and the spread of infectious diseases can be addressed more easily. The likely outcomes and impacts consist of the following: tagging animals have many benefits for farmers and authorities for managing their livestock, performing disease control and traceability, and animal migration; facilitating the quick management of problems at specific locations for any specific disease; and the immediate identification of any animal becomes possible and animal census becomes easier.

6 Conclusions

Pakistan is a climate-sensitive country since the mean temperature has increased by 0.5 degree Celsius in 30 years. By 2050, Pakistan's mean temperature is expected to rise by 1.4–3.7 °C. Wheat and rice, as staple food crops, along with maize, sugarcane and gram/pulses will suffer the most important negative impacts on yields. The 2050 wheat projected area is expected to decline by 2.5% under climate change as compared to no climate change. The same will be true for other agricultural food and livestock crops.

In short, climate change is likely to disrupt food availability, reduce access to food, and affect its quality. Crop yields will fall and the production costs will increase due to various factors including projected increases in temperatures, changes in precipitation patterns, changes in extreme weather events, and reductions in water availability. Harsher working conditions in the field due to higher temperatures will affect performance of agricultural labour force and may lead to seeking more efficient farm machineries to offset their inefficiency. The business-as-usual approach will not efficiently work, especially in Pakistan as a developing country. Therefore, farmers must be encouraged to adopt improved agricultural techniques. CSA is a form of agriculture which uses modern technologies to sustainably increase the quantity and quality of agricultural outputs. Farmers have more access to GPS, soil scanning, data management, and faster Internet to get reap from CSA's benefits. These improved agricultural practices have the potential to help achieve food security through the satisfaction of growing food demands that are estimated to increase by at least 70% from the current agricultural production levels (FAO Agricultural Development Economics Division 2012).

Therefore, environmental and climatic changes and reduced land area due to population pressure and urbanization need to be addressed as soon as possible. The agricultural labour force is migrating to urban areas due to higher wages and the preferred lifestyle in cities. There is a strong need for more mechanized agricultural production techniques. These needs are more important for a populous country like Pakistan; therefore, it is essential that the policy makers must realize and promote changes as discussed in this chapter if food security has to be improved and ensured. Some of the recommendations involve the application of CSA through Internet. It is suggested to adopt conservation tillage and marketing of agricultural produce using various applications. Similarly, concept of shared farm equipment can increase the feasibility of equipment purchases and improve the productivity of small farms. The adoption of improved CSA techniques can also improve farmers' decisions on crop choice, planting time, plant protection, and harvest. It can also improve time management, reduce water and chemical use, and produce healthier crops with higher yields.

The adoption of smart agricultural techniques include: application of Internet of Things (IoT) in agriculture for weather and market forecasting; use of appropriate renewable energy technologies; improved agricultural production systems that are less laborious for agricultural workers; generation of farm power from the windmills and solar panels that can be used for water supply and storage; and improved and efficient user-friendly farm equipment.

The above recommendations, if properly considered by all stakeholders—including farmers, agricultural marketing and financial institutions, agricultural machinery rental contractors, and agricultural policy makers—may help maintain a good food security system in a changing climate context.

References

Aamir M, Tahir A, Khurshid N, Ahmed M, Boughanmi H (2020) Economic effects of climate change-induced loss of agricultural production by 2050: a case study of Pakistan

Ali S, Liu Y, Ishaq M, Shah T, Abdullah AI, Ud Din I (2017) Climate change and its impact on the yield of major food crops: evidence from Pakistan. Food 24, 6(6):39. https://doi.org/10.3390/foods6060039.

Abbas SG (1997) Development of a computer based decision support system for introducing no-till technique. PhD Thesis, Massey University, New Zealand

Abbas SG (1999) End of the assignment report as FAO consultant on global farm power assessment (2015/2030)

Baig MB, Burgess PJ, Fike JH (2020). Agroforestry for healthy ecosystems: constraints, improvement strategies and extension in Pakistan, Agroforestry Systems, available online at: https://www.researchgate.net/publication/338392832_Agroforestry_for_healthy_ecosystems_constraints_improvement_strategies_and_extension_in_Pakistan

Bokhari SAAG, Rasul AC, Hoogenboom RG, Ahmad A (2017) The past and future changes in climate of the rice-wheat cropping zone in Punjab, Pakistan. Pakistan J Meteorol 13(26)

Dabbits HJ (1993). Human and animal power crop production, past experiences and outstanding problems. In: Human and draught animal power in crop production workshop proceedings. Harare, Zimbabwe

Eckstein D, Hutfil ML, Winges M (2018) Global climate risk index 2019. Who suffers most from extreme weather events? Weather-related loss events in 2017 and 1998 to 2017

FAO Agricultural Development Economics Division (2012) World Agriculture Towards 2030/2050, edited by Nikos Alexandratos and Jelle Bruinsma (Global Agriculture towards 2050, 2009). http://www.fao.org/fileadmin/templates/wsfs/docs/Issues_papers/HLEF2050_Global_Agriculture.pdf

Government of Pakistan (GoP) (2018) Ministry of Finance: Economic Survey of Pakistan. Agriculture. Chap. 2. Available at: http://www.finance.gov.pk/survey/chapters_19/2-Agriculture.pdf

Global Agricultural Information Network Report No. 1907 dated 29th March, 2019

Hunter MC, Smith RG, Schipanski ME, Atwood LW, Mortensen DA (2017) Agriculture in 2050: recalibrating targets for sustainable intensification. BioScience 67(4)

Intergovernmental Panel on Climate Change (IPCC) Fifth Assessment Report (AR5) (2014). https://www.ipcc.ch/report/ar5/syr/

International Food Policy Research Institute (IFPRI) (2010) Food security, farming, and climate change to 2050: scenarios, results, policy options. ISBN 978–0–89629–186–7

Kjellstrom T, Kovats RS, Lloyd SJ, Holt T, Richard SJT (2009) The direct impact of climate change on regional labor productivity. Arch Environ Occup Health 64(4)

Nagargade M, Tyagi V, Singh MK (2017) Climate smart agriculture: an option for changing climatic situation, plant engineering, Snježana Jurić, IntechOpen. https://doi.org/10.5772/intechopen.69971. Available from: https://www.intechopen.com/books/plant-engineering/climate-smart-agriculture-an-option-for-changing-climatic-situation

Olsen L (2009) The employments effects of climate change and climate change responses: a role for international labor standards? by global union research network

Pakistan Credit Rating Agency (PACRA) (2018). http://pacra.com.pk/uploads/doc_report/Rice%20Sector%20Study-Dec18-Final.pdf

Pakistan Employment Trends (2018) Government of Pakistan, Ministry of Statistics, Pakistan Bureau of Statistics

Rehman A, Jingdong L, Shahzad B, Chandio AA, Hussain I, Nabi G, Iqbal MS (2015) Economic perspectives of major field crops of Pakistan: an empirical study. Pac Sci Rev 1(3):145–158

The International Centre for Tropical Agriculture (CIAT) and the World Bank (2017) Climate-smart agriculture in Pakistan. CSA Country Profiles for Asia Series. International Center for Tropical Agriculture (CIAT).,The World Bank. Washington, D.C., p 28

UNDP Report (2016) Climate change and labor: impacts of heat at workplace. Climate and Labour Issue Paper. https://www.ilo.org/wcmsp5/groups/public/---dgreports/---dcomm/documents/genericdocument/wcms_476051.pdf

World Population Review—Live. https://worldpopulationreview.com/countries/pakistan-population/

World Wildlife Fund (2012) WWF Report https://wwfeu.awsassets.panda.org/downloads/lpr_living_planet_report_2012.pdf

Chapter 5
Impacts of Climate Change on Horticultural Crop Production in Sri Lanka and the Potential of Climate-Smart Agriculture in Enhancing Food Security and Resilience

W. M. Wishwajith W. Kandegama, Rathnayake Mudiyanselage Praba Jenin Rathnayake, Mirza Barjees Baig, and Mohamed Behnassi

Abstract Sri Lanka is experiencing dramatic climatic changes in the last decades, which are severely affecting continuous crop production and distracting farmers' livelihood and food supply chain. Hence, there is a growing concern about climate change and its effect on food security and that majority of crops cultivation depends on the monsoon pattern. The annual production merely fulfils the country's food demand; therefore, food security is a continuous high priority. Among all cultivations, horticultural crops are being reduced in terms of yield and nutritional quality to a greater extent due to the perishable nature of harvest. By reference to relevant literature, this chapter evaluates the importance of shifting conventional cropping systems to a climate-smart agriculture (CSA) system since it allows to achieve simultaneously many objectives such as combating climate change impacts on horticultural crops and achieving sustainable food security and resilience. The research, conducted by reviewing the available literature on the country's current adaptation efforts, have identified many issues such as: a decline in crop productivity and livelihood of rural farmers; postharvest losses and destruction of supply chain; and a high

---

W. M. W. W. Kandegama (✉) · R. M. P. J. Rathnayake
Department of Horticulture and Landscape Gardening, Faculty of Agriculture and Plantation Management, Wayamba University of Sri Lanka, Makandura, Gonawila, Sri Lanka
e-mail: wishwajith@wyb.ac.lk

M. B. Baig
Prince Sultan Institute for Environmental, Water and Desert Research, King Saud University, Riyadh, Saudi Arabia
e-mail: mbbaig@ksu.edu

M. Behnassi
International Politics of Environment and Human Security, College of Law, Economics and Social Sciences, Ibn Zohr University of Agadir, Agadir, Morocco
e-mail: m.behnassi@uiz.ac.ma

Center for Environment, Human Security and Governance (CERES), Agadir, Morocco

© The Author(s), under exclusive license to Springer Nature Switzerland AG 2022
M. Behnassi et al. (eds.), *Food Security and Climate-Smart Food Systems*,
https://doi.org/10.1007/978-3-030-92738-7_5

investment in food export. According to findings, horticultural crops could be used in CSA to address climate change and achieve food security in Sri Lanka due to their potential in increase productivity, improving resilience, and reducing carbon emissions. Furthermore, CSA helps fill knowledge gaps and serve as a guide for climate-smart investments and development initiatives. However, to fully benefit from such a framework, further research is needed to determine the magnitude of climate change impacts on horticultural crop productivity and how such impacts vary depending on the agricultural geographical zone, variety, and customer demand.

Keywords Climate change · Horticultural crops · Food security · Climate-smart agriculture · Sri Lanka

1 Introduction

The rapidly changing climatic patterns of Sri Lanka have led to high fluctuations in temperature and precipitation levels throughout the day. This situation has created uncertainty in making decisions regarding cropping systems and practices. Fruits and vegetables are considered vitally important horticultural crops with a high demand from all world markets. With an increased awareness on healthy nutrition, consumers aspire to include fruits and vegetables into their daily diet. Food and Agriculture Organization (FAO) and World Health Organization (WHO) recommended to incorporate 400 g of fruits and vegetables in a daily dietary intake; excluding potatoes or other starchy vegetables (FAO 2017).

Growers in Sri Lanka often focus on rainfed irrigation systems in their productions, relying on natural resources and low usage of intensive cropping methods. Therefore, crop production by local farmers is irregular throughout the year. Maintaining a continuous production is an obstacle with such systems since changes in rainfall patterns have affected farmers in receiving expected yield. In situations of unexpected yields, farmers face the problem of meeting the buyers' price of middlemen. The prices fluctuate and the involvement of middlemen from farm to fork is high, resulting in a lot of postharvest losses annually. As a result, the prices fluctuate highly between consumers and farmers, resulting in undetermined profits and losses.

Moreover, significant impacts of climate change have affected farmers, resulting in making poor decisions regarding the type of crop to be cultivated. In such a context, farmers lack guidance from local institutions regarding the adaptation of their cultivations towards sustainability and resilience, and this has created losses for the country. In this perspective, climate-smart agriculture (CSA) emerges as an efficient approach to formulate responses such mitigating the effects of climate change and improving farmers' livelihood and productivity (Rainforest Alliance 2020). It also allows the integration of sustainable agriculture methods to avoid the negative effects of single crop cultivation methods. Therefore, integrated cropping systems will be environmentally friendly and allow farmers to generate income from multiple practices. They provide local farmers with a wide range of opportunities as full-time growers. Huge economic returns of CSA can improve the productivity of lands in the long term. Moreover, improvements in growers' security help retain farmers in the country. Also, the lack of profits and high cost of inputs cause a decline in agricultural production; therefore, awareness tools and pricing strategies can improve the situation and uplift farmers.

This research focuses on the impacts of climate change on horticultural crops and growers and explores the potential of CSA in enhancing both agricultural systems' climate resilience and food security.

2 Agricultural Geographical Zones in Sri Lanka

Climatic conditions in Sri Lanka are limited to wet and dry seasons. Alternative changes in seasons are rapid in the past decades with variations of climatic conditions. Initially, the geographical distribution classified 15 agro-ecological zones, which were provided with recommendations for specific crop cultivations (Panabokke 1996). With climatic changes and inaccuracies of the predicted cultivation, the altitudinal positions of the land were revised to 45 agro-ecological zones. Such zones were classified according to rainfall, soil, forestry, vegetation, and land use (Fig. 1).

Based on *annual rainfall*, the distribution of agro-ecological zones into regions includes: the Dry zone (D) <1750 mm; the Intermediate zone (I) 1750–2500 mm; and the Wet zone (W) >2500 mm. Based on *elevation*, it includes: the Upcountry (U)>900 m; the Mid-country (M) 300–900 m; and the Low-country (L) <300 m. The numbers depict the degree of rainfall and the level of moisture and evaporation in an area (1–5). The letters are denoted by the degree of wetness decrease from (a-f), from sub regions determined by rainfall distribution (Chithranayana and Punyawardena 2008). Based on agro-ecological regions of Sri Lanka (Fig. 1), the integrated effect of climate is considered to be uniform, providing suitable conditions for crop production as recommended. The agro-ecological regions are categorized based on climatic zones and the rainfall of the regions shows fluctuation; therefore, such agro-ecological regions have increased subdivisions. The histogram representation shows an annual rainfall variation from January to December, which gives a better explanation to the farmers on the ideal cultivation periods. Farmers from all regions of Sri Lanka rely on rainfall as the water source and depend on the peak season to improve the vegetative growth of plants. Intermediate zone upcountry onwards, the agro-ecological regions have been revised. There is a high variation of rainfall, and certain regions tend to receive less rainfall compared to early categorization.

Recommended agricultural crops are mentioned in the description along with terrain and major soil groups. Soil classification of the regions are abbreviated: i.e. LHG (Low humic gley); and RBL (Reddish brown latosolic). Wet zone and Intermediate zone prominently contain red yellow podsolic (RYP) and RBL soil; the grounds are suitable for export agricultural crops, forest plantations, fruit crops, mixed home gardens, rubber, paddy, tea, vegetables, and rainfed upland crops are recommended to be grown in dry zone low-country.

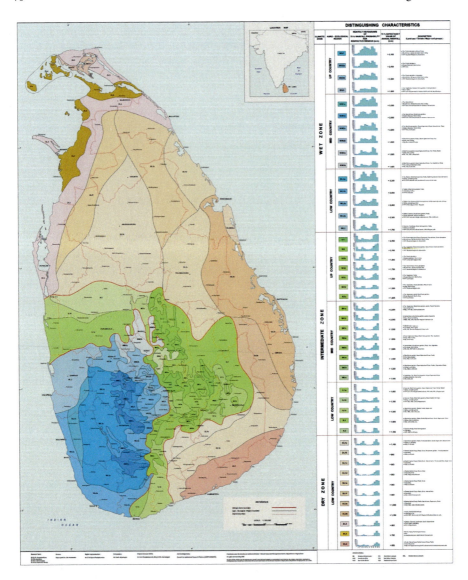

Fig. 1 Agro-ecological regions in Sri Lanka

Horticultural crops have a wide diversity of cultivation practices in regions. Fruit crop cultivation has not been defined in the agro-ecological regions due to social practices of cultivating fruit crops as mixed home gardens. Apart from the cultivation, land is utilized in forest plantations to create high economic value.

3 Crop Cultivation Patterns in Sri Lanka

The extent of 2.3 million hectares is available as agricultural lands, of which around 80% is under smallholder growers, especially of rice, fruits, vegetables, and other crops. The horticultural crop production has been adapted to the rainfed cropping system for generations. The suitability of crops to different areas are provided in the ecological map for farmers' guidance. The production of major staple crops is extensive in Sri Lanka. Farmers from generations to generations tend to cultivate rice and have maintained rituals in farm practices of rice cultivation (Withanachchi et al. 2014).

In the past, horticultural produce received less attention due to low demand and interest by consumers. Smallholder farmers cultivate variety of vegetables. Exotic vegetables cultivated in the upcountry have high economic value and demand; even though the productivity is different from time to time, there is a consistent demand for consumption, e.g. carrots, leeks, cabbage, cauliflower, broccoli etc. Low-country vegetables grown according to chena cultivation methods—like chilli, brinjal, okra, gourds and cucumber—are widely present in different dry and intermediate zone regions.

Most of the fruits were grown in home gardens with minimum effort on extensive cultivation. Among the fruit crops, papaya, banana, mango, and pineapple have high economic importance and are stable in the local consumption market. The availability of other fruits is inconsistent and seasonal. Papaya, banana, mango, and pineapple are grown mainly in wet zone and zones. There are many indigenous varieties among the fruits that have no export value, such as *karthacolomban* variety of mango and *ambul, kolikuttu, and anamalu* varieties of banana. Under-utilized fruits have received attention from locals due to awareness programs to enhance cultivation. Properties of phytochemicals in such under-utilized fruits have been also identified. There is an emerging trend for cultivation of under-utilized fruits, eg. *Anoda* and Guava. Certain fruit crops have been introduced to the country for large scale exportation industries, eg. *Cavendish* banana (Udari et al. 2019).

Prevailing cropping patterns are increasingly impacted, which prevent from obtaining maximum yields. Certain crops are highly seasonal but with high economic value in season, eg. *rambutan* and *mangosteen*. High availability of crops to consume maintains fluctuating economic value in the country. Consumers' preferences are not uniform and production remains unpredictable (Senauer et al. 1986; Jayawardene et al. 2013; Wijesekere 2015).

3.1 Seasonal Cropping

Time intervals of obtaining harvest vary in crops: perennial crops have a long-life span; annual crops have one-year life span; and biennials have two years life span. Most of the annual and biennial crops are cultivated with the onset of monsoon rains for sufficient water supply in rainfed cultivations. Perennials retain for a long time and, depending on the economically important harvest, the plants obtained harvests. Economically valuable fruit forming perennials have seasonal fruiting, and the seasonal production of fruits affects the availability of fruits.

In Sri Lanka, the majority of seasonal fruits are available from May to July, and trees and crops undergo flowering in February and March, when the north-east monsoon rains have stopped: e.g., for Rambutan, Mangosteen, Java apple, Durian, etc. the period of obtaining the optimum harvest is called season, and obtaining a low harvest is called off-season. Increasing high demand and supply in season and off-season of fruits leads to high prices. Most of the harvest is obtained from home gardens due to extensive trees' growth; large-scale cultivations of seasonal fruits are not available. Yet home gardens filled with seasonal fruits are sold with the involvement of collectors and middlemen. Certain seasonal crops tend to provide a great profit with high economic value and storability. Seasonal production of cashew occurs with large-scale plantations or sometimes cultivations intercropped with coconut lands. There is a high demand for cashew nut, which has long shelf life, and the cashew apple is discarded due to high perishability. There is a considerable demand for cashew for the flavour and extensive processing involved in consumption. Annual and biennial crops have the ability to obtain the desired harvest in recommended times, e.g. leafy vegetables produce quick yields compared to carrot cultivation. Meanwhile, perennial crops with annual harvests annually tend to have a peak production interval. Vernalization facilitates fruit development during the transition from hot and wet climates (Malaviarachchi et al. 2019; Bandula and Nath 2020).

3.2 Water Availability for Crop Cultivation

The rapid fluctuations of climate in 6.56 million hectares of Sri Lanka have created opportunities for growers to cultivate a wide range of crops in different geographical zones. Therefore, fresh produce in the local market is harvested from different ecosystems. There is an extensive network of free water resources that provides adequate water quantities to most cultivated areas. Meantime rainfall patterns have multiple origins from conventional, monsoon, and depression winds. Winds from Bay of Bengal impact the weather variation in the north-east regions and south-west regions. Seasonal changes in monsoon depict the precipitation levels at different time intervals, e.g. May to September influences the first inter monsoon known as the *Yala* season, which is preferred by most farmers to undertake seeding in their lands. Meanwhile, water is abundant in wet zone regions throughout the year. Yet, intermediate zone and dry zones have abundant water supply in *Maha* season due to monsoon rains (National Agency for Public Private Partnership 2018).

Therefore, growers have adopted rainfed cultivation systems over decades but subsequent climatic changes have shifted their outcomes. The Insufficiency of rainwater created a concern among growers, which encourages them to look for irrigation methods based on prevailing water resources. However, irrigation bears additional expenses for growers who shifted from rainfed cultivation. Due to the high cost of production, irrigation systems are available in regions with less rainfall, while rainfed cultivation prevails in regions with high rainfall.

3.2.1 Irrigated Cultivation

The over-utilization of land has incurred dramatic changes to the environment and climate. In the meantime, unpredictable variations in the environment and climate have impacted the crop cultivations while improving irrigation methods. In dry zone environment, drip irrigation systems are extensively applied to large-scale cultivation of fruit crops to obtain high productivity. Since dry zone regions have arable land for cultivation, fruit crops in such regions need to be irrigated. Sprinkler irrigation is commonly used as a practice in vegetable cultivation. Vegetables with physiological adaptation of short or modified stems—like pineapple production and exotic vegetable productions like beet cultivation, carrot, and cabbage—are effective in obtaining a good harvest with such an irrigation technique. These crops have better water retention capacity with short stems and less evaporation (Gunaratne et al. 2021). Meantime, drip irrigation became popular in dry zone regions. As a result of water conservation for agricultural lands, the government's initiatives to develop irrigation schemes have increased arable land extent to 483,000 hectares. Similarly, institutional and infrastructural development programs are conducted on crop diversification (Aluwihare 1991; Karunaratne and Pathmarajah 2002). The improvement of the water-resource use allowed farmers to minimize the requirement for potential crops. This adaptation helped manage cultivations in the dry zone for a certain period of time. Depending on the maintenance facilities of irrigation systems, farmers have now adapted to include traditional methods like mulching to improve soil conditions.

Paddy cultivation prevails as a staple crop despite large water requirements; therefore, irrigated and rain-fed cultivation systems are available. Due to extensive irrigated paddy fields, crop diversification has been introduced to rice fields. Major irrigation schemes involve cultivating non-rice crops—like banana and sugarcane—in well-drained soils and rice on poorly-drained soils. This helps prevent leaching soil nutrients to underground water sources. In minor irrigation schemes, the groundwater table is used as other water sources decline in dry periods. The wells are associated with semi-perennial crops like banana, papaya, lime or vegetable with rice cultivations. In the intermediate zone, paddy lands in the dry spell are utilized for cultivation of vegetables and onion, which allow farmers to generate a good income (Jayawardane and Weerasena 2001) (Table 1).

Crop diversification projects are carried on vegetable production in the mid-country intermediate zone (IM regions), and banana production in *Udawalawe* River irrigation scheme. Crop diversification patterns have less potential in the wet zone of Sri Lanka. The cultivation of vegetables, fruits, and other crops allows water retention in topsoil—e.g. banana cultivations can be successfully grown in the dry zone

Table 1 Relief of vegetation recommended in agro-ecological regions for crop diversification with paddy lands

Agro Ecological zones	Vegetation
Up country Intermediate Zone	Vegetables: Tomato, Beans, Carrot, Cabbage, Beetroot and Leeks Other Crops: Potato
Mid country Intermediate Zone	Vegetables: Tomato, Cabbage, Beans, Shallots and Capsicum Other Crops: Onion, Tobacco
Low Country	Vegetables: Cucumber, Gourds, Long Beans, Okra and Capsicum Fruits: Melon
Low Country Wet Zone	Root and Tuber Crops: Coleus and Sweet Potato Vegetables: Long Bean, Bitter Gourd and Okra Fruits: Banana
Upcountry	Other Crops: Cowpea, Onion, Potato, Chilli Vegetables: Gherkin
Upcountry Intermediate Zone	Other Crops: Potato, Onion and Chilli
Dry and Intermediate Zones	Vegetables: Gherkin
Wet Zone	Fruits: Banana and Papaya

Source Jayawardane and Weerasena (2001)

with an inadequate supply of water for rice cultivation since banana is a water loving crop it retains moisture necessary (Marambe et al. 1996).

3.2.2 Rainfed Cultivation

The impact of rainfall has brought up species' survival and easiness of crop cultivations followed by a rainfed system. The annual rainfall distribution in Sri Lanka has allowed growers to adapt to the growing conditions of farms depending on geographical locations. Recommendations by the Department of Agriculture (DOA), Sri Lanka hold to specific Agro-Ecological Regions depending on climatic conditions. Climatic hazards such as extreme rainfall patterns, extreme temperature, floods and droughts have severely impacted cropping systems. Moreover, Sri Lanka as an island is highly vulnerable to changes in sea levels, which impact the coastal regions of the country (Climate Change Secretariat, Ministry of Mahaweli Development and Environment 2016). The uncertainty in obtaining quantitative yields, influenced by climate change, has affected growers in adjusting their market values. Extreme climatic and environmental conditions, combined with poor handling, storage, and transportation practices, cause significant postharvest losses. The rise in temperature decreases the shelf life of fruits and vegetables and negatively affects highly perishable foods.

The integration of traditional farming methods allows farmers to practice rainfed systems, especially in the wet zone with sufficient rainfalls. Dry spells are abundant in dry and intermediate zones; therefore, rainfed cultivation practices are disappearing from cultivations (Amarasingha et al. 2015).

4 Food Security in Sri Lanka

Global pricing strategies have been implemented in order to meet the demands of a free market environment. Individuals tend to purchase food based on its monetary value, with the vast majority of consumers preferring to purchase low-cost items. The prices of products are essentially based on the production cost and transport but such prices tend also to be higher for healthy product. Therefore, consumers fail to maintain a healthy diet by being price conscious about what they consume. Public policies concerning a healthy diet are lacking; instead, they are more focused on profit-generating farming activities. For example, coconut oil manufacturing has a low price for the kernel made from spoiled kernels, and fresh kernel-based coconut oil has a different price. With the increased demand for coconut oil in the export market, consumers face high prices of high-quality coconut oil produced for exportation.

Controlling food prices by employing food pricing strategies—like the control over subsidies and taxes—help improve population diets. Government initiatives can be adopted to develop nutrition-sensitive agriculture strategies by providing intensives to consumers. For example, rural people living in the estate sector are vegetarian and do not have refrigerators; therefore, these people consume pulses as a source of proteins due to low price and storability (FAO 2018).

Personalized diet patterns specific to gender, age, physical activity level, and physiological state have an impact on health vitality and substantiality (WHO 2015). Food consumption patterns are adjusted in a manner based on receiving nutrients appropriately. Therefore, adjusting consumption patterns has a great impact on food demand control. It is important to include less than 30% of total energy intake from fats and less than 10% of total energy intake from free sugars (Hooper et al. 2015). But communities have adapted to consuming high amounts of carbohydrates due to the attention given to staple foods. High levels of cholesterol and sugars lead to high Low-Density Lipoprotein (LDL) cholesterol diseases, diabetes, and other non-communicable diseases (NCDs). The consumption of rice with other carbohydrates leads to low demands for fruits and vegetables. Diet patterns in Sri Lanka lead to the prevalence of undernutrition and micronutrient deficiencies in rural community, and a rise in over-nutrition population leading to a higher rate of NCDs. The vulnerable population is poor and highly affected by food prices. This creates food insecurity among people with limited supply of essentials at different times of the year. Food and nutrition insecurity prevails from food shortage and food price surging, which cause rural households to skip meals, especially breakfast, and reduce quantities of consumption (Ranasinghe et al. 2015). As observed by WHO (2015), most people consume too much sodium through salt (an average of 9–12 g of salt per day) and less consumption of potassium (less than 3.5 g). Adults are recommended to consume at least 400 g of fruits and vegetables per day. This results in a high blood pressure and increased risks of heart diseases and stroke. Reduced salt intake to less than 5 g per day could reduce 1.7 million deaths per year (Mozaffarian et al. 2014).

Coherence planning of national policies and investment plans promotes healthy diet and protects public health. Incentives should be adopted to enhance the productivity of fresh fruits and vegetables and minimize the production of processed foods containing saturated fats, trans-fats, free sugars, salt, and sodium. Food products should be reformulated to reduce the contents of saturated fats, trans-fats, free sugars, salt, and sodium (WHO 2015). Consumers should be encouraged by recommending foods, beverages, regulatory instruments, and serving portions. Consumers' awareness programs and healthy dietary practices in schools—such as culinary skills—would enhance good eating habits. Point of sale information, nutrition labelling, and standardized nutrition content would also facilitate consumers' understanding on foods. Government subsidies are provided on essential items, which are energy-dense food like dhal, rice, sugar, big onion, potato, canned fish, milk powder, and chicken meat as recognized by the Consumer Affairs Authority in 2014, thus making their pricing at very low costs. Comparatively, low energy-dense foods like fruits and vegetables are priced very high. This has affected the nutritional outcome of the country in a very significant way.

The Department of Census and Statistics has observed in 2016 that a large population of Sri Lanka remains food insecure and malnourished. Among this high-risk food insecure population, 15.1% of children under five-years old are age stunted, and 45.3% of women at the reproductive stage are overweight or obese (FAO 2019).

The food insecurity crisis occurs along with a reduced production in the country. As in Fig. 2, local communities involved in agriculture have decreased gradually from 12.69% (2009 GDP of agriculture) to 7.42% (2019 GDP of agriculture). At present, government policies provide fertilizer subsidies which enhance the use of chemical fertilizers by most farmers. Moreover, the import of healthy foods like nuts

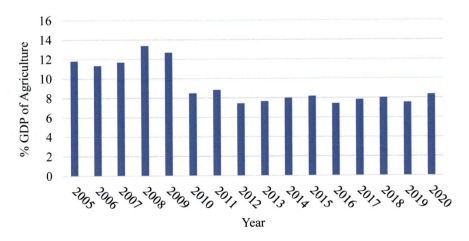

Fig. 2 GDP of agriculture in Sri Lanka (2005–2019). Source https://data.worldbank.org/indicator

and fruits is imposed by a differential cess tax of 25–30%, which is not imposed on important high energy-dense foods (FAO 2019).

Small-scale farmers have a limited access to markets; therefore, they lack insurances or loan programs to invest in the increase of their production. This creates difficulty for such farmers in improving their productivity.

5 Horticultural Crop Cultivation

Crops assigned to the horticultural industry includes: vine fruits, nuts, vegetables, aromatic and medicinal plants, ornamental plants and cut flower-based crops, and crops used in fruit orchards or landscape purposes. Horticultural crops have a high economic value due to its abundant use and intensive cultivation to meet customers' demands. There is a wide range variation in individual species of horticultural crops such as annuals, perennials, bushes, trees, roots, tubers, pot plants, cut flowers, turfs, shrubs, foliage and grasses (International Society for Horticultural Science 2021).

A wide range of horticultural crops is grown, adapted to the country's dry and wet weather conditions. The growth of plants is affected by growth conditions received compared to the optimum environment. Fruits and vegetables are of high economic value compared to ornamental plants in Sri Lanka. Food-based crops have a massive impact of being essential in meeting human dietary requirements. It is estimated that approximately 176,000 hectares of land are used for the production of fruits and vegetables (FAO 2018). The Department of Census and Statistics recorded in 2013 around 710,000 metric tons of vegetable production and around 540,000 metric tons of fruit production. Tropical crops and temperate crops grow well in the upcountry hilly areas of the wet zone, which is supported by the abundance of climatic variations. In Sri Lanka, a diverse range of fruits and vegetables, including indigenous crops, is available, each with a high economic value in the export market and a high level of importance in the local market.

Based on consumption patterns of fruits and vegetables, the amount of food intake is considerably low. Per capita fruit consumption in Sri Lanka is 5.0 kg/person/year, whereas in developed countries it is 45.0 kg/person/year (Food Balance Sheet 2019).

Horticultural produces are introduced to local and export markets, characterized by a high demand. Growers tend to grow crop varieties that provide the maximum output and reduce the growth of varieties with low output. As a result of such a selective cultivation, wild varieties of certain crops, which carry important genetic traits, have been naturally endangered in ecosystems: e.g., farmers tend to grow fruit crops with high sweetness (Mauritius pineapple and Red lady papaya) compared to fruits with low sweetness values.

5.1 Crop Varieties and Distribution

Most of the fruits and vegetables available in the market are cultivated crops of recommended varieties. Specific to various environmental conditions, crops and yields are recommended for different regions. There are many fruit crops available in the local market that are not recognized as recommended varieties by (DOA) due to their indigenous nature. Indigenous crop varieties show less demand in the local community and have abundant nutrients and important genetic properties to health and environment: e.g., *kohu amba*, a type of mango that demands the vigorous growth of tree, and it is considered as a potential root stock for budding. Meanwhile, the mango variety *kohu amba* is less preferred to consume due to the high fiber content in the fruit pulp.

Certain varieties are introduced from foreign market to comply with the quality standards for exportation. Meanwhile, most of the indigenous mango and banana varieties cannot be exported due to a non-compliance with quality standards: e.g., physiological maturity cannot be observed in most local mango varieties by the physical appearance in *Karthakolomban, willard, vellai colomban, gira amba* and other indigenous mango varieties. Moreover, banana varieties—like *Ambon, ambul, kolikuttu* etc.—cannot be identified from color break stage, which is not physiological maturity, and they are highly susceptible for diseases. Therefore, among banana and mango varieties, only Cavendish banana and Tom EJC mango are exported from Sri Lanka (Vidanagama and Piyathilaka 2011).

Recommended varieties for vegetable cultivation are listed by DOA depending on different growth conditions and regions. Thus, varieties are not interpreted in local market for consumers: e.g., TA2 and MI are DOA recommended varieties of snake gourd with pod lengths of 1 m and 0.5 m correspondingly. Yet, snake gourd available in the market are not sorted depending on the pod length.

5.2 Demand for Horticulture Crops

Fixed pricing strategies are not maintained in Sri Lanka. Therefore, the demand for horticultural crops is adopted in the local market with monetary value. Purchase is made on lowest cost and high perishable foods in the local market are provided with no grading. Therefore, products with different physiological maturity are available for consumers.

Annually, 1.1% of the local vegetable production is exported to other parts of the world (Fig. 3), whilst the cost of production per unit of product is high in Sri Lanka compared to other countries, which has resulted in low quantities of production and exportations. Most of the exportation includes indigenous species like bitter gourd, snake gourd, brinjal and pungent chili varieties. Exotic vegetables like carrot, leek, cabbage, and bean have high demand in the local market and priced higher than

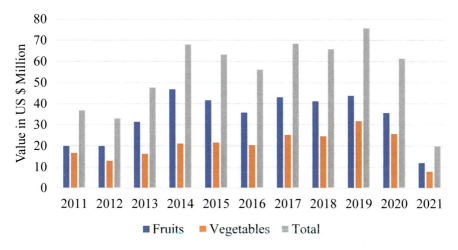

Fig. 3 Export performance of fruits and vegetables sector (2011–2021). *Source* Export Development Board (2021)

low country vegetables: e.g., long beans (LKR 120/kg) are cheaper than green beans (LKR 180/kg).

Indigenous crops have high economic value in the export market for their abundant nutrients and fiber and unique flavours: e.g., indigenous yams such as *innala (Lecranthus sp.), kiriala (Xanthasoma sagittifolium), kohila (Lasia spinosa)* and *nelum ala (Nymphea lotus)* and vegetables such as breadfruit, young jackfruit, and drumsticks. Certain fruit crops are organic by natural vegetation in default that creates a great space for exportation; high demand exists for pineapple, mangosteen, ripe jackfruit, avocado, rambutan, starfruit, and soursop.

The fluctuating demand for horticultural crops, especially during the prevailing COVID-19 pandemic situation, has a significant impact. The prices of vegetables quadrupled and the supply of upcountry vegetables to the Colombo region was halted during the onset of the pandemic.

5.3 Value-Added Production

The product development is emerging and receives popularity around the country. Value addition helps minimize postharvest losses by reducing the wastage of perishables. Most of the fruits and vegetables are highly perishable due to thin membrane. As a solution, the supermarket channels have adapted to create fruit plates with a minimum processing in refrigerators for a direct consumption (Reitemeier et al. 2021). The value-added production has increased in recent years (Fig. 4).

Fruits are incorporated in beverage, jam, cordial, confectionaries, cakes, biscuits, and sweetened local products with various brand names. People enjoy purchasing

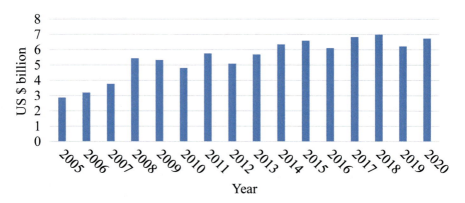

Fig. 4 Increasing demand for value-added production in Sri Lanka. *Source* World Bank (2021)

branded and unbranded products to explore various flavors. The local market is abundant with pickles, chutney, sauce, and pastes made of different fruits and vegetables: e.g., Mango Chutney, Malay Pickle, Tomato Sauce, and Garcinia Paste etc. There are small-scale exportations of horticultural produce in the forms of dehydrated, semi-processed, canned, pickles, and beverage forms. Chunks of fruits are incorporated as flavorings to different products made from tea, coconut, and rice. Snacks with spices are popular among locals and have a growing interest among other countries.

Indigenous value-added products made from various indigenous fruits or under-utilized fruits are locally available for purchase from street vendors: e.g., fruit pieces, fruit pickles, fruit blends, and spiced snacks. Such products have very low prices and will be solely dependent on vendor's choice of price.

Value-added products have less demand in the local market with a low supply, and most consumers are not interested in such products with low quality. This is mainly due to the low assurance of product hygiene during production and the abundance of food-borne diseases. Industries complying with Sri Lankan quality standards for value-added production from fruits and vegetables are limited. This is because the cost of obtaining standards is high and uniform production of products are low due to resource limitations, especially in obtaining uniform supply of desired fruits or vegetables.

6 Impacts of Climate Change and Environmental Degradation on Agriculture in Sri Lanka

Climate change-induced disasters and changes in the environment have severely affected the agriculture sector in Sri Lanka. The following section highlights the key climatic and environmental drivers and their related implications for the agriculture sector.

6.1 Environmental Factors and Related Practices Affecting Agriculture

The deterioration of soil quality with excessive agronomic practices causes farmers to leave their cultivation grounds as bare lands: e.g., paddy lands are left for years without farming. Most farmers do not cultivate in *yala* season due to low availability of water. Comparatively, most of the horticultural crops in annuals show constant production throughout the year with less management practices. Perennials have low agronomic practices, most of the trees are pruned with dry spells to bloom in wetter climates. Development programs are currently initiated to improve the productivity and climate resilience of small-holder businesses. Adaptation to good agricultural practices will enhance the scope of attention given to crops.

Screening lands for agricultural purposes where forests or wildlands are cleared for the purpose of cultivation and urbanization have impacted biodiversity. The government issues certain land for cultivation purposes called 'permitted lands' facilitated for rural livelihood (CSIAP 2018). Apart from private lands, humans have been attributed to 'unpermitted lands', which are currently used for banned cultivation such as cannabis. For clearing of lands, growers often control weeds by the burning method, which is a major factor behind forest fires in dry lands.

Most of the drylands for cultivation are cleared from trees for uniform integrity of the cultivation. Wetlands are cleared as well from trees for uniform plots along the contour plains. Loosening of soil results in soil erosion in upcountry that drastically drives landslides occurring in hilly areas.

Marshy lands in coastal areas are cleared and the dredging to create sufficient ground level for construction purposes. Houses and industries, which are developed closed to coastal areas, are a factor of quick contamination of water sources a damage of aquatic plants through water pollution. There is also less attention drawn towards industrialization and water pollution, which have increased salinity level of water. Therefore, most coastal habitats generate accumulated pollutants in water basins that deteriorate water quality. As a result, habitats are destroying wetland ecosystems: e.g., *Muthurajawela* wetlands in a natural ecosystem for freshwater biodiversity. Human exploitation has created significant losses, thus reducing the extent of ecosystems. Less attention provided by government authorities and citizens have disrupted such ecosystems, thus causing drastic losses in freshwater resources.

Excessive exploitation of natural ecosystems by humans, with less attention to plants and wildlife, has a huge impact in Sri Lanka. The exploitation of trees from rainforests for economic purposes, coupled with the disinterest of authorities, is abundant. Also, conservation measures are not implemented effectively with the involvement of key stakeholders.

The majority of postharvest losses concern horticultural produce due to its highly perishable nature. Most of the horticultural produce are sold at weekly markets by different regional sellers. The stalls are temporarily fixed and the main aim of sellers is to sell their products within the day. Therefore, post-harvest losses are high due to improper handling, and the municipal council collects the crop waste to process compost. Most of the municipal sites are free-range dumping sites and the composting emits biogas into the open air. Biogases are toxic when released to the atmosphere in large quantities and can be used as a biofuel if trapped. Treatment plants are not well upgraded in Sri Lanka; therefore, the biogas produce is not trapped in a large-scale manner. Instead, biogases are emitted in the air with improper treatment practices.

Local farmers apply large quantities of pesticides above the recommended levels to protect crops from pest and diseases. This results in a retention of residues in the crop and topsoil that leach to groundwater. As a significant effect, reservoirs in regions of Anuradhapura have heavy metal accumulation in water that resulted in kidney malfunctions of people who used the water for drinking purpose. Higher government subsidies to the purchase of fertilizers have led to the overuse of inorganic fertilizers. Chemical compounds in fertilizers adversely affect soil and human health. In 2014, the import of inorganic fertilizer expanded to 765,000 tons and 8,200 tons of pesticides for 70% paddy cultivations in agricultural systems (CSIAP 2018).

6.2 Climate-Induced Disasters and Their Impacts on Agriculture

Agriculture sector is suffering from the adverse effects of climatic changes and their induced extreme events, especially landslides, erosion of soil and coastal regions, salinity intrusion into soils and aquifers, tornado type winds, and extreme droughts and winds.

Climatic hot spot areas prevail in eleven administrative districts, which get severely impacted by climatic changes (World Bank; CIAT 2015). Several districts—*Ampara, Kurunegala, Puttalam, Polonnaruwa, Hambantota, Anuradhapura, Mullaitivu, Kilinochchi, Trincomalee, Batticaloa and Moneragala*—are prevalent climatic hotspots. The extent of cultivation and attainment of high yield are a barrier to the rural economy in a changing climate. Climatic changes are rapid and unpredictable with the rising change in global temperature, therefore yields from cropping systems have varied rapidly. Most farmers rely on rainfall to water their crops; however, the prevalence of climatic changes affects rain fed cropping systems, thus generating high costs for farmers in providing plant growth conditions.

Every year, there is a continuous rise of temperature: 0.01–0.03 °C per year in the planet Earth. These changing climate conditions cause wetter rainy seasons and drier dry seasons in Sri Lanka. As detected by climatic predictions, the annual rainfall will increase from 57 to 121 mm in the wet zone and the drought conditions will increase in the intermediate zone. Meanwhile, temperature increments in the dry and intermediate zones will increase from 1 to 1.2 °C (World Bank; CIAT 2015).

The sensitivity to time, amount, and intensity of sunlight is high in horticultural crops. Crops tend to receive high intensity of sunlight in *yala* season compared to *maha* season. The growth of horticultural crops widely depends on the level and intensity of sunlight; plants prefer to trap.

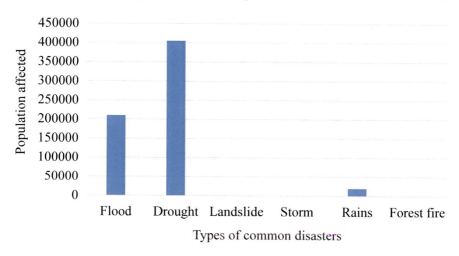

Fig. 5 Agricultural losses due to Natural Disasters from 1974–2008. *Source* UNDP, Sri Lanka

Categorically, short-day plants require less than 12 h of sunlight whereas long-day plants require more than 12 h of sunlight, and day neutral plants grow irrespective of the day length. Short-day plants such as soybean, rice, and many ornamental plants show quick flowering. Long-day plants such as potatoes, carrot, radish and beet take time to flower and require light in the short-term night to induce flowering. Cucumber and cabbage are day neutral plants (Esham and Garfoth 2013b).

Extreme weather conditions increase the scarcity of resources and result in a lower productivity of staple crops. Natural disasters—mainly flood, cyclones, high winds, droughts, and landslides—account for approximately two percent of total annual losses from disasters. Unsustainable agricultural practices exacerbated the prevailing natural disasters and make them more disastrous (FAO 2021). As a result, continuous climatic variations are abundant in the past five years, shifting the ecological patterns of cultivation (Fig. 5). The increased variability of climate has significantly affected smallholder farmers.

Floods are very common with annual rainfall depending on the distribution and longevity of rain. Rainfall patterns vary from region to region causing floods in regions with low-sea levels and areas closer to rivers. Floods have caused significant losses of harvest in lowlands, result in landslides that can critically damage a cropping system, and can cause disease development in young plants. Due to excess levels of water, plants get damaged by damping-off, stem rot, and root rot. High moisture levels in the immediate environment provide a comfortable environment for pests to thrive, enhancing diseases to increase with potential vectors. Farmers rely on rain for seedling development but floods can damage young seedlings on water. Moreover, floods result in landslides in hilly areas, damaging cultivations and homelands. Soil water increases and loosens the soil in bare lands, causing erosion. Therefore, cultivations in hilly regions are adopted to contour lines in order to avoid alterations in contours resulted from erosion. Damages from erosion can severely affect ecosystems.

Currently, droughts have a significant effect on livelihoods in drylands and their impacts result in a lower productivity of staple crops. Indeed, extended droughts cause large losses in crop production with early signs of crop wilting. Long-term droughts affect growers in seedling development and in maintaining productivity. Extreme rise in temperature levels and a decrease in rainfall have a direct impact on cultivation of staple food crops due to sensitivity in changes of air temperature. Droughts incur the highest economic damage compared to other natural disasters. It is expected to increase the extended dry spells creating large losses in crop production (Burchfield and Gilligan 2016).

Water reserves have dried up in regions with extensive droughts and irrigation canals are not effective for certain regions as well. The accumulation of heavy metals in freshwater sources by agricultural chemicals has led to intoxication in certain areas: e.g., cyanotoxins in water reserves at Anuradhapura due to the accumulation of heavy metals run off from agrochemicals utilized by farmers (Ruwanpathirana et al. 2019).

Sea-level rises in coastal regions of Sri Lanka reduce the habitants in such areas. Ground waters are contaminated by sea-water intrusion and salinization increases salt content of freshwater sources. The cost of refining seawater for human activities is high; comparatively it is easier to purify groundwater and freshwater sources. As a result, the water availability is reduced in the country. Inward intrusion of salt water affects the breeding sites of various amphibians: e.g., turtles hatch on the sea shore and obtain warm environment to nurture. Breeding grounds contaminated with salt water can reduce fertile strips of land, thus limiting species diversity.

Changing rainfall patterns and unpredicted periods of high rainfall are expected to strain the capacity of irrigation systems and increase the risk of landslides in some areas (Koresawa et al. 2018). The impacts of natural disasters will increase the cost of production in the efforts made to optimize environmental conditions.

6.3 Impacts of Climate Change on Crop Production in Sri Lanka

Intensive cultivation practices are adopted for most of the crop products in Sri Lanka. The adverse impacts of climate change affect all traditional paddy channels, coconuts, fish, and most fruits and vegetables. Frequent weather changes affect crop management practices in all cultivations, resulting in undetermined productivity levels. As a result, profits earned by farmers are altered with intense climatic variations. Therefore, production quantities of all crops have sine wave variation (peaks and off peaks). There are serious effects on underutilized fruits grown in intermediate and dry zones from climatic variations (Ratnayake et al. 2019).

Natural disasters have caused significant losses in production of all kinds of crops in Sri Lanka. The change in crop production has fluctuated the price of crops within the market, thus affecting consumers purchase capacity compared to their concern on quality. Consumers are often price conscious when buying crops rather than being quality conscious. The impact from production to transport and distribution is immense, based on management practices adopted in the ecosystem. The excessive and unsustainable use of natural resources has affected the sustainability and productivity of agriculture sector, especially due to the degradation of soil and the increase of water salinity. Indeed, the adverse usage of chemical fertilizers and pesticides has affected the microbiota and quality of soil; excessive soil applicants are always washed off with rains and floods due to low retention in soil and accumulated in water reserves.

Maintaining optimum growth conditions at critical stages of development is essential. The adverse impacts of climate change lead to mutation and premature developments (premature flowering). The harvestable portion of the crop is not uniform with extreme size variations, thus reducing consumers' preferences.

Local markets do not grade products based on sizes, and most of the consumers prefer purchasing moderate-sized harvests for most of the crops. Consumers are free to choose the products with required sizes and pay the same amount. Consumer bias sorting of produce results in a selected quantity of the produce being unacceptable. Such deselected produce cannot be sold with flexible pricing strategies; therefore, the produce is discarded. High postharvest and instability of food prices lead to food insecurity.

6.3.1 Main Export Crops

Plantation crops are closely monitored for production variations to minimize risks of low productivity. There is a great effect of productivity to maintain export quality and quantity for high prices. Tea production varies in flavors and tannin content depending on the agro-ecological region. Such variations of tea production have been recognized and categorized to seven major principal regions—*Nuwara Eliya, Dimbulla, Kandy, Uda Pussellawa, Uva, Sabaragamuwa*, and *Ruhuna*. Extreme rainfall affects the plantations because the tea gardens are established in contours of hilly areas and landslides occur with soil erosion in tea lands. The vigorous growth of buds depends on moisture; thus, the lack of sufficient moisture produces dormant buds that have a low economic value. A reduction of monthly rainfall by 100 mm could reduce productivity by 30-80 kg/ha/month of tea (Wijeratne et al. 2007).

Rain guards are recommended for plantations to lower the loss of tapping interferences from rains. Rain hinders rubber latex tapping as it can lead to panel diseases with high moisture on trees. Rain collects water on the canopy of the tress seeping to the trunk. The mosses on the trunk absorb water and retain the trunk wet for some time. Extreme rainfall reduces the number of days available for rubber tapping (Nugawela 2008).

Coconut cultivations show sensitivity to rainfall and dry spells in dry regions. Changes in atmospheric temperature and annual rainfall are the key factors influencing the variability of coconut cultivation (Esham and Garfoth 2013b). The production of coconut has decreased recently, thus increasing prices drastically. The increase in temperature in recent years has caused the spread of pest and disease incidence. Therefore, researches are working on the introduction of new varieties for cultivation in different regions which are adapted to droughts and strong winds. In the first eight months of 2020, the coconut output has decreased by 12% and the price of a coconut has reached LKR 120 compared to only LKR 10 cents in 1948 (Wijewardene 2020).

6.3.2 Other Export Crops

The export development board pays attention to crops cultivated other than plantation crops with a stable market position. Most commonly, spices generate high value due to the domestic authentic flavors. However, most of the export crops are produced in low quantities due to a lack of knowledge and non-establishment of a market for raw material purchases. Moreover, there is also a lack of sufficient processing plants for certain export agricultural crops such as cinnamon, pepper, cardamom, clove, nutmeg, vanilla, coffee, cocoa, *Garcinia*, and turmeric.

Due to high pricing, adulteration of crops has been abundant. This has led to the disintegration of export market such as the case of black pepper adulterated with papaya seeds. The production chain is not established as well to encourage local farmers: e.g., there are no collection points for vanilla pods. In-house processing is expensive for local farmers and requires access to potential processing plants that would purchase their harvest. Climate conditions are critical for proper drying and rainfall leads to fungal development on drying harvest, which results in a low quality.

Due to a lack of bulk production, most of the spices are purchased from native countries at much lower price. The COVID-19 crisis resulted in halting the import of spices and in increasing the prices of local spices in the market: e.g., the prices of local turmeric increased from LKR 750 to LKR 4500 per kg and such a dramatic increase affects the ability of local consumers to eat turmeric (Sirinivasan 2020).

Most the farmers who cultivate export agricultural crops practice the intercropping method with plantation crops. The plant prefers shady conditions for growth, and processing requires extremely dry and hot conditions to obtain the desired quality. Without sufficient levels of moisture, most of the crops generate a low harvest. The availability of improved varieties is lower, which draws the attention of researchers to develop drought-resistant varieties. Measures to expand the cultivation of such valuable export agricultural crops are also required.

6.3.3 Paddy Cultivation

From seed sowing to harvest processing, annual rainfall should fluctuate from high to low precipitation in order to obtain optimum yield. Therefore, the climate has an important role in paddy cultivation due to a high sensitivity to water and photoperiod. Due to such a high sensitivity, in addition to inter-generational cultural ethics, farmers are reluctant to diversify their production (Esham and Garforth 2013a). Paddy cultivations in wet zones are rain-fed, and fields in dry and intermediate zones are irrigated and rain-fed. Most irrigated conditions are river basin-oriented complex water management system from natural water reserves. Therefore, the paddy cultivations are severely impacted by drought conditions. Ambient temperatures above 33 °C for 60–90 min would result in high temperature injuries to rice plants which are at the flowering stage as the flowers dry up and shed. In 2014, severe drought conditions in dry zones result in 40% reduction of the paddy harvest (FAO 2021).

Improper irrigation techniques and poor drainage have affected the salinity level of agricultural lands. As an impact, the crops suffer from bad soil health. This reduces the crop productivity and challenges the food security. Available projections show that, by 2050, the majority of the rice growing areas will face water stress conditions due to less rainfall in the major seasons (Esham et al. 2017). Rice varieties are developed for drought tolerance to resist the water shortage. Water stress conditions result in a reduction of panicle development and filling of panicles.

The rice processing is highly affected by climate patterns during the handling, storage, and distribution practices. Since rainfall patterns affect threshing and drying, it is important to have dry weather conditions for better storage life. The lack of proper storage conditions for paddy leads to high postharvest losses. The Paddy Marketing Board (PMB) does not provide sufficient storage facilities for the total production, leading to immense postharvest losses.

To face all these challenges affecting paddy cultivation, it is increasingly important to provide optimum growth conditions to meet the self-sufficiency of rice in Sri Lanka.

6.3.4 Horticultural Crops

Horticultural crop production requires abundant water supplies and farmers invest on various irrigation methods in the cropping systems such as hose irrigation, sprinklers, drippers, and land preparation techniques. Most of the horticultural crops are highly perishable due to their physiological state. An increase in temperature results in an increased respiration rate of fruits and vegetables: e.g., climacteric fruits ripen quickly after harvesting. The shelf life of the produce is gradually reduced, resulting in a low retention time of the desired physiological maturity state.

Flower bud initiation of perennial fruit crops induces from a prolong dry spell; these conditions help prevent premature wilting of flowers. In such a critical period, if it rains the premature flowers fall off the quantity of flower subjecting to maturation is reduced. This condition has become a common cause for reduction of fruit harvest in certain periods of time. Also, the variations in climate result in a delayed and early flowering of perennial crops. Rains at the onset of flowering result in dropping of flowers and reducing fruits as in mango trees.

Under-utilized fruit crops are receiving the attention of local consumers due to a high availability of phytochemicals; therefore, such crops are considered to have a high medicinal importance. Indigenous fruit species—like Ber and Tamarind—have waxy leaves, sunken stomatas, and thick cuticle to minimize water requirement. Meanwhile, fruits like *Kirala* have very low postharvest losses leading to less handling practices. Pineapple is a crop of commercial importance, which has a high adaptation potential to different climatic conditions and water-use efficiency. This makes pineapple suitable for most areas of cultivation. Intercropping is practiced in coconut plantations and this makes water available for coconut palms. Cultivations of papaya are adapted to drip irrigation methods to survive drought conditions. However, papaya is more prone to diseases in extreme wet and dry conditions; therefore, optimum climatic conditions are necessary. For banana, climatic requirements vary from climate to climate: i.e., *Ambul* banana cultivations require more water than *Ambon* banana. Low-soil moisture results in an internal browning disorder of banana and *Kolikuttu* banana requires a dry spell for flowering. Furthermore, surveys are carried out to study the behavioral changes of fruit crops from flowering to fruiting (Renuka and Edirimanna 2019).

The economical maturity of vegetables varies from young shoots to ripen fruits; the perishability of crops depends mainly on the level of maturity and outer coating. Climatic conditions from harvest to consumer effect the postharvest life of the fruits. Exotic crops grown in hilly areas require cold transport conditions to maintain the quality of the harvest. However, the cold chain transport is not used in local market to handle horticultural produce due to the high cost of transportation; yet this is important and widely used in export industries.

7 Climate-Smart Agriculture as an Adaptation Strategy

Climate-smart agriculture (CSA) is an ambitious approach to improve the integration of agriculture development and climate responsiveness while achieving food security and development goals (World Bank; CIAT 2015). Such an approach transforms and orients agricultural production systems and food value chains to support sustainable development and food security (FAO 2019). In addition, CSA aims to increase sustainable and decarbonized agriculture with high productivity and enhanced resilience of agro-systems (IICA 2017). The framework defines the planning of tradeoffs and synergies between productivity, adaptation, and mitigation.

Centuries back, resource-poor farmers used intelligently genetic diversity and integrate crops and livestock in ways adapted to unique environmental conditions. Therefore, the present state of environment and climate has created a strong foundation for sustainable agriculture requirements.

Cultivations in Sri Lanka accompany 42% of the total land area, of which 80% belongs to smallholder farms. Therefore, home gardens contribute to household-level food security in most villages. Whilst most of the cultivations are practiced extensively for mono-cropping, vegetables and high value crops are practiced both extensively and intensively.

Certain unacceptable farm management practices critically affect the sustainability of agriculture. To reverse such trends, farmers require guiding information and alternative methods through institutions and recommendations. CSA provisions on farm management practices that correlate to sustainable agriculture should be developed. Recommended CSA practices for drought and prolonged dry season cultivation include the use of cover crops and mulch to conserve the water availability. Sudden heavy rains generating floods are currently recurrent due to climatic changes. To avoid surface run off and direct damage, trenching and cover crops help reduce the direct effects on cultivated crops.

The implementation of traditional farming practices and modern agricultural methods that protect the environment helps conserve resources. Farmers should be provided with extensive knowledge on multiple cropping methods that encourage crop diversification (World Bank; CIAT 2015). Such a program helps farmers improve the productivity of their crops, thus enhancing their livelihood. Prolonged cultivations of crops may enhance climate resilience and neutrality given that this strategy allows the control of greenhouse gas emissions.

Excessive usage of synthetic fertilizers to paddy cultivations implies costly import of inputs. Improving fertilizer management practices, combining the use of organic and inorganic fertilizers, developing farmer's training on nutrient management, allowing access to low-cost raw materials to help improve the productivity, and enhancing public awareness on environment protection are recommended measures to be included in public policies. Stakeholders commonly tend to limit the cropping patterns to economic viable varieties and methods. However, they should be more conscious about the role of CSA in enhancing agricultural sustainability.

7.1 Introduction of Climate-Resilient Varieties

Complementary programs organized by DOA improve farmers' awareness on crop germplasm collection and systematic crop comparison. Farmers in different agro-ecological regions can improve access to new climate-adapted genetic material (World Bank; CIAT 2015). Practical exposure to farmers through workshops is needed to adapt to complex agricultural ecosystems.

Adapting cropping systems to conserve crop variability helps improve genetic sustainability. The vulnerability of traits has increased over generation with over-exploitation. Conserving the genetic diversity and crop cultivations, diversified with indigenous crops, improves ecosystems and growing conditions in most regions. Stress-tolerant varieties have increased adaptation to climate risks such as drought, flood, heat, and temperature stresses. Indigenous crop varieties have a high resistance to different pests and diseases. Through research and development, improved varieties help determine tolerance for abiotic stresses. There are commercial varieties of rice, tea, and maize with tolerance to submergence, drought, flood, heat, and temperature.

Planting cycles should be strategically changed, with minimum use of raw materials, to suite the changing climatic and soil conditions. Systematic crop comparison, respective to different agro-ecological regions, gives a better idea to farmers in selecting crops.

7.1.1 Water Management

The conservation of limited water resources is essential to all agro-ecological regions. Shared cultivations of crops and drought-resistant varieties, with low and high requirements of water, allow the optimization of the water use. Using management practices, low tillage, rainwater harvesting, and micro-irrigation methods help growers in intermediate and dry zones.

Maize cultivations with the onset of rains reduce losses from changing water patterns due to the high availability in the seedling stage. Selective cultivation of maize in agro-ecological regions, with long rainy seasons, can reduce the crop vulnerability to changing water and climatic patterns.

7.1.2 Rainwater Harvesting

Adaptation of farms' operations to harvest, storage, later use of rainwater for domestic and farming purposes, soil conservation and environmental management are recommended practices (LRWHF 2020). For agricultural requirements, land surfaces and rock catchments are used. Groundwater recharging is among the methods to store water in soil for later use.

Currently, rice cultivations in the dry zone adapted to seasonal planting have established rainwater harvesting from village tanks to efficiently use and manage the resource. Dry-sowing methods of paddy minimize the water use and conserve soil moisture when combined with zero or minimum tillage practices.

7.1.3 Micro-Irrigation

Micro-irrigation is an improved technological method, which can be applied using sprinklers and drip systems. The system development in farms is costly for smallholder farmers, yet the cost of water application can be mitigated. Sprinkler irrigation methods are adopted among farmers compared to drip irrigation systems due to the high-water delivery for low-country vegetable cultivations; although most of the farmers adopt irrigation using hose pipe instead of micro-irrigation techniques (Udagedara and Sugirtharan 2018).

7.1.4 Low-Tillage Practices

The preparation of soil for seed placements are minimized in low tillage; thereby, practices like pulverization, cutting or movement of soil are avoided. The minimization of tillage practices eliminates soil erosion and increases organic matter retention and water infiltration to soil. This results in previous crops retaining as a mulch in the soil to the new crops. The retention of previous crops can lead to disease development; therefore, crop rotation can be practiced to avoid any infestation. Meantime, cash crops and cover crops alternatively planted give sufficient resources for growth, nutrition, and development.

Minimum tillage practices contribute to the improvement of soil structure, soil fertility, soil water-holding capacity, and carbon sequestration which reduces the threat of global warming by enhancing soil quality.

7.2 Soil Management

Land and soil degradation is a critical result of climate change—along with overexploitation and inappropriate management—through changes in natural soil conditions. Therefore, minimizing human involvement and promoting sustainable soil and land management practices help optimize soil conditions. Moreover, the management of resources such as water, nutrients, crops, livestock, and associated biodiversity helps reduce climate change negative impacts and enhance soil biodiversity with nitrogen fixing bacteria and burrowing animals. Reducing tillering practices will retain the soil air structure and infiltration of water, thus reducing wind and water erosion. The sustainability of soil moisture can be enhanced by mulching and use of cover crops; cover crops are a living mulch in gardens that controls weeds and retains nutrient (FAO 2013).

The low use of nitrogen fertilizers reduces the nitrous oxide emissions into the environment. Therefore, tea growers are adopting organic fertilizers that can enhance soil quality, water retention, and soil functions, thus increasing the potential to overcome climatic shocks. Composting practices, known as manure harvesting using organic residues of plant and animal matter, provide nutrients and better soil conditions. Soil texture may be improved with volume, microbiota, air, and moisture. Manure is produced commercially to enhance organic export-oriented cultivations. Therefore, annual fruits and vegetables cultivation practices should be using manure to enhance organic production.

The use of salt-tolerant varieties helps survive water-stress conditions and different soil environments with high intoxication from pollutants. Lands overexploited using chemical fertilizers and pesticides are not cultivated due to a low productivity and heavy metal accumulation. Therefore, resistant and tolerant varieties can help overcome the burden of poor soil conditions (Pradheeban et al. 2014).

Contour planting is practiced in hilly areas to minimize soil erosion and better root growth. This facilitates down the flow of water in the planting lands. Contour planting, which is highly practiced for plantation crops like tea and exotic vegetable cultivations, facilitates crop rotation. In Nuwara Eliya region, vegetables cultivation is practiced using mixed cropping in rows alternating lettuce, beets, knolkhol, leeks, and carrots. This allows a better use of land and retainment of soil moisture.

Soil conditions can be improved by providing altering temperature to nursery seedlings; this results in a resistance to pests and diseases at seedling stage, especially damping off. Most commonly, nursery seedlings of fruits and vegetables are protected with a mulch. Mulch forms a barrier for pest and disease infestations and allows plants to survive the initial stages of growth. To improve the nutrient retention of soil and combat erosion, 60% of tea growers have adopted mulching and thatching. The conservation of soil moisture can be enhanced by the use of residual moisture and weed control. Moreover, coconut cultivations may use live mulches—such as leguminous crops—to reduce the requirements of nitrogen fertilizer of crops and enrich the soil fertility (Fig. 6).

Mulching improves the soil structure by water infiltration and retention. This increases carbon storage in soils, which improves the soil fertility, with reduced use of synthetic fertilizer.

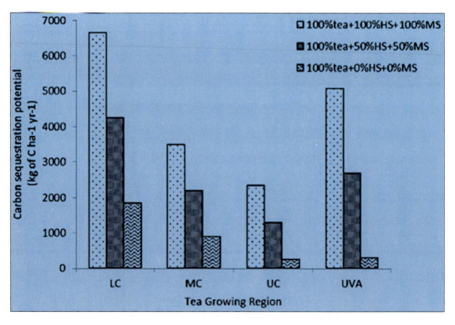

Fig. 6 Comparison of carbon sequestration potential of tea lands among different tea growing regions with different densities of shade trees relating to different levels of compliance with TRI recommendations. HS: High Shade; MS: Medium Shade; LC: Low Country; MC: Mid Country; UC: Up Country. *Source* Wijeratne et al. (2014)

7.3 Crop Diversification

The cultivation of several crops on a single piece of land results in developing symbiotic associations between crops, which help improve the harvest, sustain ecosystems, and enhance the longevity of cultivation methods: i.e., plantation crops like coconut are intercropped with fruiting cover crops like pineapple, papaya, and banana that provide optimum sunlight conditions for the maximum growth of those cover crops.

Small-scale farmers have adopted changing planting cycles and strategies due to high costs of mono-cropping associated with more inputs and high labor-intensive cultivation techniques. Diversification of planting cycles help improve the soil-use efficiency, resulting in an improved use of minerals in the soil this limits the leaching off of nutrients from the soil.

7.4 Animal-Integrated Cropping Systems

Established lands with livestock integration helps develop complex food chains that cycle materials within the ecosystem. The animal manure from livestock extensively helps produce organic manure to fertilize cropping systems. Compost can be created using plant debris and animal manure given its high economic value in the local market. Landlords traditionally followed these methods of cropping systems to sustain the production of all foods in home gardens. However, animal-integrated cropping systems have been introduced in a few farming systems to control weeds.

Rearing and conservation of indigenous cattle are highly important to resist diseases and heat stresses. The integration of adaptive breeds would enhance productivity: e.g., breeds like CPRS chickens and *kottukachchiya* goats are not adopted widely. As a significant effect the sufficiency of milk production in Sri Lanka is limited to 50% in 2015 (Marambe et al. 2015). The cultivation of fodder grasses allows free range farming for rearing animals with a low susceptibility to diseases by adapting to the environment. In addition, waste management is effective with the integration of animals, due to the high degradability of nutrients that enhances the composting process or biogas production.

7.5 Alternative Farming Methods

Approaches that integrate renewable sources of energy improve efficiency by reducing labour. Tools and equipment create smart-energy usage systems and large-scale cultivations incorporate solar energy to generate power. Wind farms may be established in areas with strong winds to generate electricity. Effluent water treatment plants may be established as well to treat sewage and industrial waste; therefore, polluted water can be purified in the premises and used for drinking purposes.

The establishment of biogas extraction units to generate energy allows efficient recycling of farm waste. Biochar, utilized as a manure for vegetables, provides soil micro-organisms, enhances moisture retention in soil, and tolerates dry spells.

Aquaponics allow year-around production of high value vegetables along with fish, thus benefitting from declining traditional farming methods. The challenges of extreme weather can be avoided by continuous retention of water.

The adaptation to modern technologies allows farmers to modify their agronomic practices with precision agriculture: i.e., sensors allow farmers to identify the health status of crops and monitor the effects of temperature. Few large-scale producers use technology in monitoring their nurseries and greenhouses to meet exportation standards. Therefore, industries involved in precision agriculture are under research-level projects.

8 Conclusion and Recommendations

Global food requirements are increasing and agricultural crops have a significant role in providing food and restoring ecosystems. Human activities have damaged biodiversity through degradation and overuse and the potential exists for restoring the lost nature. CSA provides hope for the enhancement of sustainable agriculture. Most of the horticultural farming practices followed in Sri Lanka use inorganic chemicals. Methods to adopt environmentally friendly management practices require much attention by growers due to the prevailing health risks. NDCs are rapidly increasing in the globe and mostly due to inappropriate dietary habits. Provisioning to better production of fruits and vegetables can help enhance food security, resilience, and rural livelihood.

Few educated farmers minimally adopt the promising CSA practices. The needed knowledge and investments by supporting institutions can help uplift smallholder farmers in this direction. This enhances the production of nutrient-rich foods which meet consumers' needs and preferences. Adopting diversified crop cultivations—through a shift from mono-cropping to shared cultivation methods—can help enhance climate resilience and reverse the adverse effects of natural disasters. Potential natural rectifications should meet the increasing demand for crops with proper strategies. Therefore, the adoption of CSA practices and technologies will help achieve food security and enhance the sustainability of agriculture.

References

Amarasingha RPRK, Suriyagoda LDB, Marambe B, Gaydon DS, Galagedara LW, Punyawardena R et al (2015) Simulation of crop and water productivity for rice (Oryza sativa L.) using APSIM under diverse agro climatic conditions and water management techniques in Sri Lanka. Agric Water Manag 160:132–143

Aluwihare PB (1991) Irrigation investment trends in Sri Lanka: New construction and beyond. IWMI 56–60
Bandula A, Nath TK (2020) Underutilized Crops in the Agricultural Farms of Southeastern Sri Lanka: farmers' knowledge, preference, and contribution to household economy. Econ Bot 74(2):126–139
Burchfield E, Gilligan J (2016). Agricultural adaptation to drought in the Sri Lankan dry zone. Appl Geography 77:92–100
Chithranayana RD, Punyawardena BVR (2008) Identification of drought prone agro-ecological regions in Sri Lanka. J Natl Sci Found Sri Lanka 36(2):117–123
Climate Change Secretariat, Ministry of Mahaweli Development and Environment (2016) National adaptation plan for climate change impacts in Sri Lanka, pp 26–32
Climate Smart Irrigated Agriculture Project (CSIAP), Environmental Assessment & Management Framework (2018). Ministry of Agriculture, Sri Lanka
Climate Smart Irrigated Agriculture Project (CSIAP) Pest Management Plan, Adopted from Agriculture Sector Modernization Project (2018) Government of Sri Lanka
Department of Census and Statistics (2016) Demographic and health survey. Government of Sri Lanka, Colombo
Esham M, Garforth C (2013a) Agricultural adaptation to climate change: insights from a farming community in Sri Lanka. Mitig Adapt Strat Glob Change 18(5):535–549
Esham M, Garforth C (2013b) Climate change and agricultural adaptation in Sri Lanka: a review. Clim Dev 5(1):66–76. ISSN 1756–5537. https://doi.org/10.1080/17565529.2012.762333
Esham M, Jacobs B, Rosairo HSR, Siddighi BB (2017) Climate change and food security: a Sri Lankan perspective. Environ Dev Sustain. https://doi.org/10.1007/s10668-017-9945-5
FAO (2013) Module 1A: introducing climate smart agriculture? Why Climate-Smart Agriculture, forestry and fisheries in the Climate Smart Agriculture Sourcebook. Available at http://www.fao.org/climate-smart-agriculture-sourcebook/concept/module-a1-int roducing-csa/a1-overview/en/. Accessed 27 February 2021
FAO (2017) Fruit and vegetables for health initiative, Licence: 6807EN/1/02.17
FAO (2018) country gender assessment of agriculture and the rural sector in Sri Lanka. Colombo. 80 pp. Licence: CC BY-NC-SA 3.0 IGO
FAO (2019) Policy brief food prices: an effective way of shfiting diets towards healthier habits in Sri Lanka, Colombo. Licence: CC BY-NC-SA 3.0 IGO
FAO (2021) Assessing risk in times of climate change and COVID-19: City region food system of Colombo. Sri Lanka, Rome. https://doi.org/10.4060/cb2897en
Food Balance Sheet (2019) Department of census and statistics, ministry of economic reforms and public distribution, pp 8–20
Gunaratne LHP, Hemachandra KS, Kumudumali YMK, Manawasinghe NKGKR, Sathischandra HGAS, Soorasena JM et al (2021) Comparative assessment of vegetable crop performances and ecological indicators during transition from conventional to ecological agriculture. Asian Res J Agric 1–9
Hooper L, Abdelhamid A, Bunn D, Brown T, Summerbell CD, Skeaff CM (2015) Effects of total fat intake on body weight. Cochrane Database Syst Rev (8)
Inter-American Institute of Cooperation for Agriculture (IICA) (2017) Climate smart agriculture in the Eastern Caribbean States: a compendium of stories from farmers by IICA is published under license Creative Commons. Available at http://creativecommons.org/licenses/by-sa/3.0/igo/. Accessed 15 March 2021
International Society for Horticultural Science (2021) Defining Horticulture. Available at https://www.ishs.org/defining-horticulture. Accessed 02 March 2021
Jayawardane SSBDG, Weerasena LA (2001) Crop diversification in Sri Lanka. Food and Agriculture Organization of the United Nations Regional Office for Asia and the Pacific Bangkok, Thailand
Jayawardena R, Byrne NM, Soares MJ, Katulanda P, Hills AP (2013) Food consumption of Sri Lankan adults: an appraisal of serving characteristics. Public Health Nutrition 16(4):653–658

Karunaratne ADM, Pathmarajah S (2002) Groundwater development through introduction of agrowells and micro-irrigation in Sri Lanka. Use of Groundwater for Agriculture in Sri Lanka

Koresawa A, Popuri S, Talpahewa C (2018) State of Sri Lankan cities report, p. 253; 250

Lanka Rain Water Harvesting Forum by Stredge Tech Solutions (LRWHF) (2020), What is rainwater harvesting? http://www.lankarainwater.org/what-is-rain-water-harvesting/. Accessed 4 March 2021

Malaviarachchi MAPWK, Karunathilake WAK, Perera RACJ, Nijamudeen MS (2019) Threats and related research towards adaptation of other field crops to climate change in the dry zone of Sri Lanka. In: Marambe B (ed) Proceedings of the workshop on present status of research activities on climate change adaptations. Sri Lanka Council for Agricultural Research Policy, Colombo. pp 21–29

Marambe B, Silva P, Weerahewa J, Pushpakumara G, Punyawardena R, Pallawala R (2015) Enabling policies for agricultural adaptations to climate change in Sri Lanka. Handbook of Climate Change Adaptation Springer-Verlag, Berlin Heidelberg

Marambe B, Sangakkara UR, Azharul Haq K (1996) Crop diversification strategies for minor irrigation schemes. In: Proceedings of the workshop organized by the irrigation research management unit, irrigation department and the Sri Lanka National Program, International Irrigation Management Institute, pp 10–19

National Agency for Public Private Partnership (2018) Environmental Assessment & Management Framework (EAMF). In: Framework Development and Infrastructure Financing to Support Public Private Partnerships, pp 20–24

Mozaffarian D, Fahimi S, Singh GM, Micha R, Khatibzadeh S, Engell RE, Powles J (2014) Global sodium consumption and death from cardiovascular causes. N Engl J Med 371(7):624–634

Nugawela A (2008) Do not allow rain to interfere rubber tapping and lower your income levels. Daily News. http://archives.dailynews.lk/2008/03/18/fea03.asp, Retrieved March 11, 2021

Panabokke CR (1996) Soils and agro-ecological environments of Sri Lanka. Colombo, Sri Lanka: NARESA. xvi, p 220

Pradheeban L, Nissanka NAASP, Suriyagoda LDB (2014) Clustering of rice (Oryza sativa L.) varieties cultivated in Jaffna District of Sri Lanka based on salt tolerance during germination and seedling stages. Trop Agric Res 25(3):358–375

Rainforest Alliance (2020) What is climate-smart agriculture? Available at https://www.rainforest-alliance.org/articles/what-is-climate-smart-agriculture. Accessed on March 29, 2021.

Ranasinghe P, Pigera ASAD, Ishara MH, Jayasekara LMDT, Jayawardena R, Katulanda P (2015) Knowledge and perceptions about diet and physical activity among Sri Lankan adults with diabetes mellitus: a qualitative study. BMC Public Health 15(1):1–10

Ratnayake SS, Kumar L, Kariyawasam CS (2019) Neglected and underutilized fruit species in Sri Lanka: prioritisation and understanding the potential distibution under climate change. Agronomy 10:34. https://doi.org/10.3390/agronomy10010034

Reitemeier M, Aheeyar M, Drechsel P (2021) Perceptions of food waste reduction in Sri Lanka's commercial capital Colombo. Sustainability 13(2):838

Renuka KA, Edirimanna ERSP (2019) Present status of research on fruit crops for climate change adaptation and future needs of Sri Lanka. In: Marambe B (ed) Proceedings of the workshop on present status of research activities on climate change adaptations. Sri Lanka Council for Agricultural Research Policy, Colombo, pp 43–47

Ruwanpathirana T, Senanayake S, Gunawardana N, Munasinghe A, Ginige S, Gamage D, Amarasekara J, Lokuketagoda B, Chulasiri P, Amunugama S, Palihawadana P, Caplin B, Pearce N (2019) Prevalence and risk factors for impaired kidney function in the district of Anuradhapura, Sri Lanka: a cross-sectional population-representative survey in those at risk of chronic kidney disease of unknown aetiology. BMC Public Health 19(1):1–11. https://doi.org/10.1186/s12889-019-7117-2

Senauer B, Sahn D, Alderman H (1986) The effect of the value of time on food consumption patterns in developing countries: evidence from Sri Lanka. Am J Agr Econ 68(4):920–927

Sirinivasan M (2020) Turmeric shortage sparks concern in Sri Lanka. The Hindu. Available at https://www.thehindu.com/news/international/turmeric-shortage-sparks-concern-in-sri-lanka/article32499264.ece

Udagedara MD, Sugirtharan M (2018) Status of micro irrigation systems adopted for vegetable cultivation in Polonnaruwa district, Sri Lanka, Adoption of micro irrigation system in Polonnaruwa. AGRIEAST J Agric Sci 11(2). Available at https://doi.org/10.4038/agrieast.v11i2.34

Udari UR, Perera MDD, Wickramasinghe WR (2019) Home garden fruit cultivation in Sri Lanka: determinants and the role of government intervention. Hector Kobbekaduwa Agrarian Research and Training Institute

Vidanagama J, Piyathilaka MD (2011) Enhancement of product quality and resource efficiency of Sri Lankan banana supply chain. In: 10th Asia pacific round table for sustainable consumption and production, pp 1–11

WHO (2015) Guideline: sugars intake for adults and children, World Health Organization

Wijeratne MA, Anandacoomaraswamy A, Amaratunge MKSLD, Ratnasiri J, Basnayake BRSB, Kalra N (2007) Assessment of impact of climate change on productivity of tea (*Camellia sinensis l.*) plantations in Sri Lanka. J Natl Sci Found Sri Lanka 35:119–126

Wijeratne TL, Costa WAJM De, Wijeratne MA (2014) Carbon sequestration potential of tea plantations in Sri Lanka as an option for mitigating climate change ; a step towards a greener economy carbon sequestration potential of tea plantations in Sri Lanka as an option for mitigating climate change ; a step. In: Fifth symposium on plantation crop research

Wijesekere G (2015) Changing patterns of food consumption in Sri Lanka: 1985–2009. The Australian National University, Australia South Asia Research Centre, Canberra

Wijewardene BA (2020) Where have all the coconuts gone? Surely, not out of the country? Daily News. Available at http://www.ft.lk/columns/Where-have-all-the-coconuts-gone-Surely-not-out-of-the-country/4-708299

Withanachchi SS, Köpke S, Withanachchi CR, Pathiranage R, Ploeger A (2014) Water resource management in dry zonal paddy cultivation in Mahaweli River Basin, Sri Lanka: an analysis of spatial and temporal climate change impacts and traditional knowledge. Climate 2(4):329–354

World Bank; CIAT (2015) Climate-smart agriculture in Sri Lanka. CSA country profiles for Africa, Asia, and Latin America and the Caribbean series. Washington D.C., The World Bank Group

Chapter 6
Managing Climate Change Impacts on Food Production and Security in Sri Lanka: The Importance of Climate-Smart Agriculture

Mohamed Mujithaba Mohamed Najim, V. Sujirtha, Muneeb M. Musthafa, Mirza Barjees Baig, and Gary S. Straquadine

Abstract Sri Lanka is increasingly concerned about the impacts of climate change on food production, food security, and livelihoods. This has been mostly discussed in terms of climate impacts on crop productivity (*food availability*), with little emphasis on other key aspects, namely food access and use. This chapter, based on existing literature, adopted a food system model to obtain a better perspective on food security issues in Sri Lanka. These issues include diminishing agricultural productivity, food loss and wastage along supply chains, low rural poor subsistence resilience, and the prevalence of high under-nutrition and infant malnutrition. This review indicates that ensuring food security requires actions beyond climate-resilient food production systems to take an integrated approach that can promote the climate stability of the entire food system, while addressing nutritional issues emerging from climate change impacts. There is, therefore, an urgent need for settlers to work towards a climate-smart agricultural framework that will tackle all aspects of food security. Besides the output of a few crop species, our study displays a lack of research

---

M. M. M. Najim (✉)
South Eastern University of Sri Lanka and Professor in Water Resources Management, Department of Zoology and Environmental Management, Faculty of Science, University of Kelaniya, Kelaniya, Sri Lanka
e-mail: mnajim@kln.ac.lk

V. Sujirtha · M. M. Musthafa
Department of Biosystems Technology, Faculty of Technology, South Eastern University of Sri Lanka, University Park, Oluvil, Sri Lanka
e-mail: sujirtha@seu.ac.lk

M. M. Musthafa
e-mail: muneeb@seu.ac.lk

M. B. Baig
Water and Desert Research, Prince Sultan Institute for Environmental, King Saud University, Kingdom of Saudi Arabia, Riyadh, Saudi Arabia
e-mail: drbaig2@yahoo.ca; mbbaig@ksu.edu.sa

G. S. Straquadine
Utah State University – Eastern Campus, Logan, Utah, USA
e-mail: gary.straquadine@usu.edu

© The Author(s), under exclusive license to Springer Nature Switzerland AG 2022
M. Behnassi et al. (eds.), *Food Security and Climate-Smart Food Systems*,
https://doi.org/10.1007/978-3-030-92738-7_6

into the consequences of climate change on Sri Lanka's food system. More such studies are required to examine how climate change can affect other components of food system, including the productivity of a diverse range of food crops, livestock, and fisheries, and to focus attention on the avenues of an environmentally-induced nutritional insecurity.

Keywords Climate · Food security · Livelihood · Nutrition · Productivity

1 Introduction

In Sri Lanka, there is a growing concern about the effect of climate change, variability and severe extreme events on food production, food security, and livelihoods. According to the Intergovernmental Panel on Climate Change's (IPCC) fifth assessment report (AR5), the temperature in the South Asian region is expected to rise by more than 2 °C by the mid-twenty-first century. The number of warm days will rise rapidly and rainfall will likely become more unpredictable with an increase in heavy rainfalls (Hijioka et al. 2014). Sri Lanka has already witnessed more intense severe extreme events, such as five major floods and four droughts during the last four years with significant negative effects on agricultural production and rural livelihoods (Perera 2015).

Sri Lanka is an island in the South Asian region with a total land area of 65,610 km^2. The country consists of 9 provinces and 25 districts. Sri Lanka is an agriculture-based developing tropical country, and agriculture remains key to the island's modern rural economy. The agriculture industry plays a huge role in the socio-economic background of the economy, adding 7.87% to the national GDP in 2018 (The Global Economy 2018). Furthermore, the agricultural sector is a significant source of employment; about 27% of the active population is involved in agricultural production (World Bank 2019). The extent of agricultural land in Sri Lanka has increased progressively in the past decade and, with the end of a thirty-year civil war in the country previously, unreachable regions have been transformed into productive cropland (World Bank 2015).

Sri Lanka's climate is a tropical monsoonal one and it has great rainfall variation seasonally. There are three principle climatic zones across the island: the wet zone (WZ), dry zone (DZ), and the intermediate zone (IZ). Most of the agro-ecological regions under these climatic zones are vulnerable to rainfall seasonality and variability (World Bank 2015). Major sub-sectors of the country's agricultural production systems are food crops, plantation crops and dairy cattle. Food crops such as rice, maize, pulses, and vegetables are cultivated in varying scales and intensities using intensive, extensive and input-dependent practices (World Bank 2015). Although paddy production was the major staple food crop in Sri Lanka, plantation agriculture crops also have a high contribution to agricultural exports (Bandara et al. 2014). However, agricultural production of the country must face climate change together with many other sector-specific challenges.

Climate change poses a serious threat to food security, particularly in developing nations, and it has broad consequences for agriculture and food production. Furthermore, it is difficult to interpret predictions of climate change to better understand its possible impacts on food security. About 820 million people worldwide are still suffering from chronic malnourishment in 2018 (FAO et al. 2019) and the 2030 Sustainable Development Agenda aims at mitigating hunger and achieving food stability (UNDP 2015). However, achieving this in the face of a changing climate will require increased technology adaptations.

The aim of this chapter is to provide a perspective about the impacts of climate change in terms of sustainability on food and nutritional security along with livelihoods. Further it addresses the potential solutions to reduce the dramatic effect of climate change in agriculture with a view to maintaining food production and security.

2 Impacts of Climate Change on Sustainable Development in Sri Lanka

2.1 Effect of Climate Change on Food Production and Food Security

Food security for an increasing world population is one of the main future challenges. While such a challenge is acute for developing countries, developed ones are increasingly concerned as well. Food insecurity is related to many uncertainties, but climate change will be the greatest threat to future food production. Although climate change will be slow, more important is the anticipated rise in extremes which will affect food production (Droogers 2003).

Climate change will continue to influence the agricultural sector in general, but will be particularly crucial for Sri Lanka's smallholder farmers in terms of resource availability (Chandrasiri 2013) and fluctuations in yield (Wijeratne et al. 2007; Punyawardena 2011). Based on climate predictions, Sri Lanka's wet cultivated lands are becoming wetter, while the dry arable lands are becoming drier, directly affecting agriculture practices in these areas (Marambe et al. 2013). It is predicted that annual rainfall will increase by a minimum of 57 mm, with up to 121 mm in the WZ, while droughts will be prolonged, predominantly in the IZ. Further, temperature surges will range between + 1 and + 1.2 °C, with greater impact in the DZ and IZ (World Bank 2015).

The agricultural sector in Sri Lanka has felt the consequences of severe weather events and climate change, including: a gradual, but steady increase in air temperature (0.01–0.03 °C/year); high-intensity rainfall resulting in soil and coastal erosion; salinity intrusion into soils and aquifers; tornado-like winds; and increasingly intense droughts and flash floods (World Bank 2015). Climate and agriculture are both related in technologies for crop production. Every pattern of crops varies within their climatic zone. Consequently, climate is a major factor in crop growth, distribution and species richness, and will impact food production (Ginigaddara n.d).

Farmers in Sri Lanka are gradually facing the deleterious outcomes of climate change. Farmers in the Kalpitiya region recently abandoned more than 600 acres of cultivated land devoted to vegetables and subsidiary crops due to heavy rain and flooding (Daily News 2020). Cordell et al. (2017) reported variations in the climate coupled with a low level of water availability, including late monsoons and more intense drought and floods directed to increased occurrence of pests and diseases, compact crop yields, and stern financial consequences. Furthermore, it results in an agronomic uncertainty and a risk of crop failures. Climate change makes crop selection difficult in terms of the types of crops that can be grown and the viability of cropping systems (Iizumi et al. 2013). Increased post-harvest decay of fruits and vegetables is allied with unpredictable weather during storage, and food price fluctuations are associated with climate-induced crop catastrophes in exporting countries leading to increased food insecurity among urban populations in low-income countries (Esham 2016). Climate change forecasts specify paddy yields could fall by up to 42%, which is estimated to affect about 70% of farmers, and one third of paddy farmers will turn into poverty levels (Zubair et al. 2015; Esham 2016).

Sri Lanka fulfils about 85% of its annual food rations within the country, with a low dependence on imports (Department of Census and Statistics 2014). However, the local food system is still tied to global markets. The country has almost attained self-sufficiency in rice (Department of Census and Statistics 2011). The local market is highly vulnerable to unfavourable weather conditions requiring the short-term importation of rice and other important products to make up for the shortage and to stabilize rising prices. Sri Lanka, for example, imported a record amount of rice in 2014, when supply fell by 27% due to adverse weather conditions (Esham et al. 2017).

The staple grain, rice, is particularly vulnerable to increased temperatures. Exposure of the rice plant's growing spikelets (flowers) to temperatures above 31 °C decreases pollen productivity and paddy yield (Dharmarathna et al. 2014). Although the amount of annual rainfall has not changed significantly, delays in the onset of rainfall and changes in the distribution of rainfall have caused disruptions to the regular crop calendar, frequent crop failure, and loss of yield (Esham and Garforth 2013). The prevalence of severe climate events has increased in recent decades, resulting in negative effects on food production, especially rice production, resulting in the need to import significant quantities of rice to meet national requirements (Esham et al. 2017; Gunda et al. 2016). It is projected that the main rice-growing region, the DZ where nearly 72% of rice is grown, will receive less rainfall in the years to come, leading to reduced rice production (De Silva et al. 2007).

2.2 Effects of Climate Change on Nutritional Security and Livelihood

Various factors, including diverse food and farming systems, establish nutritional security and livelihood (FAO 2009) in Sri Lanka. Food systems in the country are influenced to varying degrees by natural calamities with greater spatial variability linked to climate change. Traditional farming systems in Sri Lanka have evolved over hundreds of years, with farmers in varying agro-ecological regions maintaining production systems to better balance local environmental conditions. This has contributed to the island's rich agro-biodiversity in terms of food production such as rice, cereals, pulses, potatoes, tuber and root crops, spices, and fruits.

Regarding nutritional security, self-sufficiency in rice production has been the dominant agricultural policy tactic since Sri Lanka gained independence in 1948. It has supported livelihood generation and rural poverty abolishment. In the year 2010, Sri Lanka accomplished the reported target of rice self-sufficiency mainly due to research and development activities. The rice research results in Sri Lanka throughout the last couple of decades further contradict this claim in that, on average, rice production increased by 0.37% for every 1% increase in investment in rice research with an internal rate of return of 174% in a tariff-protected structure and a benefit/cost ratio of over 2,300 (Niranjan 2004).

Poverty, climate change, decreasing farmland, and booming population pressure are the key flaws that make the situation in Sri Lanka more difficult in terms of achieving national food and nutrition protection (Marambe 2012; Weerakoon 2013). The World Food Program (WFP) (2011) estimated that 12% of Sri Lanka's total population was seriously food insecure, 82% of which are in the Northern and Eastern Provinces. Extreme climate events, such as the severe drought that occurred over a 5–6 month time frame in 2012, will pose their own hurdles to food security. Sri Lanka ranked 60th out of 107 countries by the 2013 Global Food Security Index (The Economist 2013). This index helps people understand and compare the key food cost, availability, access and quality issues across countries.

Droughts, floods, cyclones, increased farmland losses, and sea-level rise in recent decades have posed major threats to agricultural production and food stability. Additional pressure from increasing population, poor terms of trade, weak infrastructure, and limited access to modern technology and markets restrict the options available for people to cope with the negative impacts of climate change. Sri Lanka's main food crops are rice and other field crops, fruits, vegetables, and animal products (milk, meat, eggs, and fish). Diminishing arable agricultural land, along with the growing population, renders these problems more difficult to tackle. Most crops in Sri Lanka are likely to be adversely affected by climate change, e.g., coarse grains, legumes, vegetables, and potatoes (Titumil and Basak 2010).

The varied climatic conditions in Sri Lanka's farming systems have resulted in a wide variety of crop species and plant breeds suitable for different atmosphere, rainfall, and altitude conditions. Species diversity is especially high among rice, other cereals, cucurbits, and vegetables, such as tomatoes and eggplant, suggesting the potential for improvement of crops in the face of natural disasters, especially the climate change induced ones. For decades, poor farmers have judiciously used genetic diversity to cultivate varieties that are suited to their own environmental stress conditions. Strategic adaptation strategies can be improved through: deliberate crop comparison programs in various agroecological regions; enhancing seed germ plasm selection programs undertaken by the Department of Agriculture; and providing access to and drawing on new genetic materials through intergovernmental initiatives.

The intensely managed animal production sector is also vulnerable to the environment. The heavily controlled animal production sector is solely dependent on rainfed farming systems and subject to climate fluctuations. The sector is composed of mainly smallholders spread through diverse climatic empires. Regional climate change effects are more evident in areas where smallholder farming and subsistence farming are practiced because they are particularly vulnerable to climatic changes. Moreover, the high genetic diversity that exists, in particular among indigenous animal populations, and their adaptability to the country's complex and localized climate regimes make it a positive incentive to construct resilience in the smallholder sector by managing the impacts of climate diversity within the farming system. As can be seen by the importance of smallholder systems to Sri Lanka's animal production sector, awareness about the country's climate change and animal interaction are not sufficiently enough to address the sector's potential growth challenges.

Agriculture and food systems are experiencing many obstacles, making it harder to attain the primary target of fulfilling global food demand. An alarming collection of unparalleled challenges and threats including intensified competition from non-agricultural sectors for land, water, and other natural resources affect the food security worldwide. Food security is multifaceted, and the rise in population and changes in consumption habits associated with increasing incomes drive increased production of food as well as other agricultural goods. Productivity developments, improvements in production processes, and improved food and environmental protection are three targets for countries to attain, and Sri Lanka will be no exception.

3 Agricultural Adaptation to Climate Change in Sri Lanka

Adaptation approaches to climate change have been extensively studied and reported in Sri Lanka (Esham and Garforth 2013). Possible strategies include application of indigenous and traditional knowledge, advanced and high throughput technologies, and an integrated approach. These have been practiced worldwide and some of them are yet to be applied in Sri Lanka. This section analyses many possible practices that could be used to overcome issues in food security, nutritional status and livelihood improvement towards sustainable development of the country in the context of climate change.

3.1 Climate-Smart Agricultural Practices in Food Production and Security

Climate-smart agriculture (CSA) is a growing paradigm that is being embraced internationally to address climate change in farming. Enhancing agricultural production and profits, adapting and building resilience to climate change, and reducing greenhouse gas emissions are widely applied sustainably. Some scholars describe CSA as a strategy for transforming and reorienting agricultural systems under the new realities of climate change in support of food security (Lipper et al. 2014). Agricultural production is synonymous with several other words, but CSA is unique in its emphasis on a variety of climate behaviour (Ginigaddara n.d.).

Indeed, CSA technologies and practices present opportunities to address the challenges of climate change while promoting economic growth and agricultural development. For the purpose of this profile, technologies and practices are considered CSA oriented if they maintain or increase agricultural productivity and contribute to lower greenhouse gas emissions (FAO 2013).

3.1.1 Adoption of Indigenous Practices to Mitigate Climate Change Effects

Indigenous knowledge has accumulated through generations, built on experience. Sri Lanka has a history of over 2,500 years, and this history is linked to the traditional paddy and chena production from ancient times. Indigenous or traditional knowledge includes cultural traditions, rituals, beliefs and methods of local people in different fields that give them a unique identity (Magni 2016).

Strategies used by home gardeners vary considerably across the country, depending mainly on the household head's education, farming experience, and perception of the impacts of climate change. Home gardeners choose these strategies to improve household food security and income, and to mitigate the adverse effects of climate change (World Bank 2015). Specific activities for domestic gardeners include the introduction of new technologies such as new varieties and irrigation equipment (55%); soil and water conservation measures (41%); agronomic practices (39%); and new planting cycles (37%). Animal sector is increasingly being included in CSA to improve food and nutrition security, and may contribute to increasing biodiversity (Weerahewa 2012). Furthermore, focusing on home gardens relieves the pressure exerted on natural forests and enhances conservation by providing a wide range of habitat niches, thus contributing to reduced deforestation and a wide range of landscape mosaics (Pushpakumara et al. 2012).

Though paddy cultivation is mostly practiced in lowlands, undulating lands, sloped lands and even highland slopes are also used for paddy cultivation. Sloped lands can be terraced for rice cultivation so that even highland slopes can be used. This is a well-organized way to use water from top to bottom paddy fields through a gravitational water management system (Dharmasena, 2010). Traditional paddy cultivation with low rainfall in dry areas was focused primarily on irrigation systems and developed with experience and knowledge of people related to temperature, rainfall patterns and soil behaviours (Panampitiya 2018). In traditional cultures, farmers would have a direct impact on the fertility and quality of seeds if they were well aware of soil fertility. First, the farmers clean the fields in order to remove the debris. This debris gradually decomposes and provides the soil with nutrients (Irangani and Shiratake 2013).

Shifting culture for Chena cultivation (Abeywardana et al. 2019) represents another significant aspect of traditional Sri Lankan agriculture. Grains such as Sesame, Mung Beans, Cowpea, Maize, Finger Millet, Millet, Gingelly and Mustard and other kinds of vegetables (Pumpkin and Green Chilies) are mainly grown in this type of field, as they are best suited for dry crops. Chena cultivation includes intensive two to three seasons of forest clearance and cultivation, leaving native vegetation to regenerate and regenerate soil fertility. It also never threatens nature in terms of soil, vegetation or bio-diversity fertility (Panampitiya 2018).

Having considered this knowledge, it is crystal clear that traditional agricultural methods and techniques in Sri Lanka result from proper interrelations with nature and natural phenomena. Most traditional agricultural systems were built with a deeper understanding of human-nature interactions. These agricultural practices are more useful in maintaining ecological stability and survival of ancient people's socio-economic activities. This suggests that conventional agriculture is based on observation of natural phenomena such as rainfall patterns, temperature, humidity, and soil behaviour.

3.1.2 Adoption of Advanced Technologies to Mitigate Climate Change Effects

Climate change may have a variety of negative consequences for agriculture, such as reducing precipitation and increasing infiltration of saltwater into aquifers and freshwater bodies. Flooding and storm damage can also lead to reduced crop yields. Increases in mean temperature and the timing and distribution of rainfall can also negatively affect agricultural output (Battisti and Naylor 2009). These phenomena alter the population of insect pests and the spread of diseases.

The effect of climate change on agriculture in Asia is likely to differ considerably by area (Cruz et al. 2007). The occurrence of diseases affecting rubber plants could rise and the production of rubber, along with the output of rice and coconut, could go down. A 0.5 °C increase in temperature is expected to reduce rice production by 5.91% (UNFCCC 2000).

Adaptive agricultural technologies can address climate change in many ways. A range of methods, including management strategies, should be considered for successful adaptation to climate change in the agriculture sector. The list of technologies described here is not exhaustive and is intended to demonstrate the range of technologies which can minimize vulnerabilities.

(A) *Crop breeding*

Crop breeding programs use both conventional and modern biotechnology techniques to recognize strains with climate-relevant traits. They require the expansion of existing traits or the transfer of traits to other plants. Marker-assisted selection (MAS) focuses on increasing plant breeding precision by controlled plant crossing based on phenotypic characteristics or other markers associated with desirable characteristics, such as enhanced stressor tolerance. For example, recent work has used technology to measure thermo-stability of the cell membrane (CMT) in plants (Choudhury et al. 2012). Researchers have been able to recognize CMT as a marker linked to heat tolerance by using the percentage of relative cell injury, which can then be used to distinguish crop varieties that can thrive at higher temperature. Zhengbin et al. (2011), in their study in Northern China, show "the significant impact of plant breeding on crop yield and water-use efficiency (WUE), in particular by growing the number of kernels per spike in wheat and maize". Crop breeding technologies are becoming increasingly popular in Asia and around the world. Potential investors in breeding sector still have a wide market opportunity open to them.

(B) *Fungal symbionts*

Fungal symbionts are fungi that exist in a symbiotic relationship with plants for mutual benefits. Even though this technology is in its infancy, it has the potential to be very successful in the field of agricultural adaptation. Recent studies have explored the concept of exposing vulnerable species to fungal endophytes found to improve stress tolerance in other species to see whether the endophyte can transfer similar benefits to vulnerable species. Rodriguez et al. (2004) reported that mutualistic fungi may bring multiple benefits on plants such as drought resistance (Read 1999), disease and temperature tolerance (Redman et al. 2002), and nutrient accumulation (Read 1999). This technology could help extend the existing range of useful food crops, in addition to increasing crop resilience.

(C) *Laser and levelling technology*

A significant part of the water loss in agriculture is attributed to excessive field runoff. A key solution to minimizing runoff is ensuring that agricultural fields are as level as possible. Recent innovations, including the use of laser technology, have increased field levelling precision. This precise, flat land helps manage runoff and allows effective use of water. Several experiments have shown that water quality and crop yield in laser-level fields have improved (Lybbert and Sumner, 2012). A 2006 study cited a 20% increase in wheat yield, with water savings of 25% achieved by laser land levelling (Akhtar 2006). Laser land levelling is most successful when used in conjunction with other agricultural water management methods.

(D) *Pressurized irrigation technologies*

Irrigation has been used for centuries to conserve water in agriculture (Kornfeld 2009), but developments in irrigation technology will become much more relevant as climate change puts heightened stress on water resources. In general, pressurized irrigation using sprinkler, drip, mini-sprinkler, or high-efficiency drip systems shows great promise to provide efficient water management and minimize convection losses. Unlike flooding techniques, drip systems allow farmers to supply water directly to the roots of the plants, drop by drop, thus reducing waste (Buyukcangaz et al. 2007). Palada et al. (2010) compared low-cost drip irrigation (LCDI) with hand watering and reported that LCDI significantly increased yield and water-use efficiency (WUE) for various crops: cucumber (13% higher yield and 41% higher WUE); eggplant (38% and 113%); and bitter gourd (121 and 35%). Micro-irrigation and underground irrigation systems are excellent options over large irrigation schemes.

(E) *Floating agriculture*

Floating agriculture involves planting crops on soilless rafts. Historically these rafts were made of organic composted material, including water hyacinth, algae, water wort, straw, and herbs. This technique can be extremely effective, particularly for crops damaged from flooding. With little to no chemical fertilization, this simple production method has the potential to increase productivity per unit of land. In the case of lettuce, the trials conducted in Thailand achieved productivity close to that of high-input soil farming, and even higher productivity for cabbage (Sterrett 2011). This farming technology has been used to grow leafy vegetables (e.g., lettuce), tomatoes, turmeric, okra, cucumbers, chilies, melons, roses, pumpkins and many varieties of gourds, beetroot, papaya and cauliflower (Tran 2013). Floating agriculture is being widely promoted by NGOs as an adaptation practice primarily to provide direct seasonal benefits (such as household nutrition) for the neediest populations (Irfanullah 2013).

The scope of implementation of these analysed agricultural technologies is very innovative and many of the innovations rely on scientific, industrial, or federal researchers to create or provide training on these new methods. There are also direct benefits from widespread use by individual farmers.

3.2 CSA Practices in Nutritional Security and Livelihood

Estimates suggest that 2.4 billion people will be living in developing countries in 2050, centred in South Asia and sub-Saharan Africa. Agriculture is a key economic sector and a major source of jobs in these regions, but more than 20% of the population is currently food insecure on average (Wheeler and von Braun 2013). Around 75% of the poorest people live in rural locations and farming is their main source of income (Rural Poverty Report 2011). Yet, agriculture production is already hampered by climate change. The IPCC states that climate change impacts food production in many parts of the world, with more frequent negative effects especially in developing countries (IPCC Summary for Policymakers 2014).

In Sri Lanka over the past three decades, food security has continued to be part of the national policy agenda. The goal of achieving sustainable food stability, however, has been difficult. Although climate change is one of many critical factors that can influence food security, our analysis highlights multiple climate-induced food security challenges in Sri Lanka. Such concerns include declining performance in agriculture, food loss along supply chains, low rural poor livelihood resilience, and prevailing high rates of undernourishment and child malnutrition.

Achieving food security needs urgent action over and above current efforts to develop climate resilient food production systems via climate-resilient agricultural system (Rai et al. 2018). Improving food security will take a systemic approach that ensures that the entire food supply is responsive to climate change, while addressing nutritional problems resulting from its impacts. To resolve all aspects of food security, therefore, the establishment of a CSA program is essential. Such a strategy will comply with the priority food security steps listed in the Sri Lankan National Adaptation Plan for Climate Change Impacts (Ministry of Mahaweli Development and Environment 2015).

In this regard, CSA's approach has plenty to offer for the emerging climate realities to change and reorient food systems. CSA is characterized as agriculture that sustainably increases productivity, increases resilience, reduces/removes greenhouse gas (GHG) emissions, and enhances national food security and development goals (FAO 2010). The concept can be extended to different facets of food systems, from agricultural activities to food intake and human nutrition. CSA is not completely new to Sri Lanka, as some practices considered climate-smart techniques were already implemented by some farmers in food production (World Bank, CIAT 2015). Steps should be taken to ramp up these activities by promoting their application via a broader range of tasks in the farming industry above the production power. These initiatives include developing an integrated and funded knowledge-management network and enhancing access by food system stakeholders to financing, technology, and investment in supply chain infrastructure.

In addition, CSA activities should have a long-term outlook and step beyond gradual adaptation to transition, because existing strategies may not be adequate to withstand the expected increase in air temperature and extreme weather. For example, temperature rise in the country is expected to surpass tolerable limits for cereal crops such as rice and maize (Malaviarachchi et al. 2014). This may force farmers to grow crops in protected environments and require new food crops such as quinoa to be introduced.

With climate change, food shortages, like those witnessed in 2008, are expected to arise, which will cause higher food prices (Cribb 2010). Food price increase may be driven by higher agricultural input prices. Throughout the 2008 food crisis, fertilizer prices rose to record levels, aggravating the impact of food shortages; the price of phosphate fertilizers rose by 800% (Cordell et al. 2011). These circumstances will push smallholder farmers to the edge of poverty and intensify food shortages unless measures are taken to emphasize efficient resource use, which is a prerequisite of the CSA approach. This can be accomplished by improving the national self-sufficiency of inputs such as fertilizer, improved productivity in input usage, use of bio- and organic fertilizer, and promotion of the recovery and reuse of nutrients (such as phosphorus) from food production and consumer systems.

The prevalence and scope of global hunger are still alarming despite significant strides in global food production. Over 820 million people are suffering from persistent hunger and that figure has been growing steadily over the past three years (FAO 2019). This situation emphasizes the importance of implementing effective policies that exploit the benefits of globalization, while mitigating the risks for achieving the Sustainable Development Goals (SDGs) related to terminating hunger, strengthening food security and nutrition, and supporting sustainable farming. Persistent starvation and child malnutrition pose threats for the ability of households to access food. The effectiveness of welfare state initiatives aimed at supporting these vulnerable groups should be improved, especially in rural areas. The National Poverty Alleviation Programme, which offers a monthly amount for small-income households and a savings incentive system, has struggled to produce anticipated results as it is plagued by mistakes, inefficiencies, and political intervention (Gunatilaka et al. 2009). Social safety net services can be strengthened by better targeting and improving initiatives to help the needy rise out of poverty through involvement in income-earning services and the creation of skills through the design of an intervention package to address cumulative shortcomings (World Bank 2007).

Another concern that calls for policy-makers to intervene is child malnutrition with potentially long-term consequences for adult wellbeing and labour productivity. A collection of cost-effective measures, such as breastfeeding, complementary feeding for infants over six months of age, enhanced hygiene habits, food security enriched with essential nutrients, and therapeutic feeding for undernourished, spectacular-food babies, can be effective.

There is a need to develop research on climate change impacts on farming industry activities outside the main crops of rice, maize, and a few plantation crops. Further effort is required to understand the effect of climate change on all agricultural activities, including fruits and vegetables, livestock and fisheries. Further investigation is needed in Sri Lanka into the causal pathways of nutritional insecurity under climate change.

3.3 CSA Adaptation and Conceptual Framework for Sri Lankans

As mentioned above, CSA is a strategy that transforms agricultural systems in the current context in order to improve farming livelihoods and ensure food security. It has the potential to mitigate the worst climate change impacts on agricultural livelihoods and help make people less vulnerable to food shortages and poverty. CSA's key goals are a sustainable increase in agricultural production, adaptation to climate change, and reduction or complete elimination of GHG emissions (Lipper et al. 2014). Throughout its context, the CSA model seeks to combine these three goals in order to boost agricultural production and food safety in a changing environment. Agricultural activities that fit into the CSA system or profile were classified as CSA technologies in addition to developments and technologies that are being harnessed for CSA implementation (Wekesa et al. 2018). This part explores agricultural practices defined as CSA and the level of adoption of such practices, especially among small-scale farmers in Sri Lanka.

Adopting agricultural innovations or technologies among farmers, especially small-scale farmers, is not instantaneous; the small-scale farming system has vital players and actors involved. The characteristics of small-scale farmers, combined with CSA technology solutions and knowledge agents accessible to farmers, are key factors to consider when introducing and implementing CSA. These factors not only impact adoption, but they may also affect the degree and intensity of adoption. The characteristics of farmers may have a large impact on the level of acceptance and growth, but the technical options available for CSA may also be an important element in the level of CSA acceptance. Figure 1 summarizes how this research conceptualizes the adoption of CSA among small-scale farmers.

Having recognized the positive effect of small-scale farming on economic development, many countries, including Sri Lanka, are making significant efforts to improve this vital area. But small-scale farming's contribution to Sri Lanka's national economy is still small in comparison with the region's other economic sectors. CSA practices need to be adapted to the particular characteristics of local farming systems and local socio-economic conditions. These practices need to form a well-articulated knowledge management system coupled with enhanced access to finance, services, and markets for smallholder farmers.

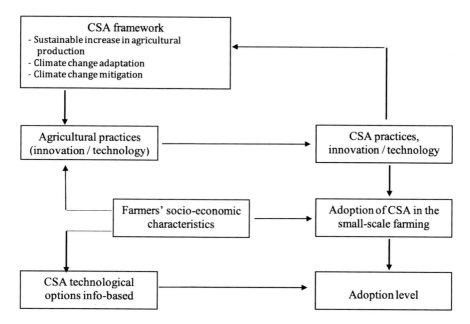

Fig. 1 Conceptual framework—Adoption of CSA practices among small-scale farmers

4 Conclusions and Future Directions

Extreme weather events and rapidly evolving climatic trends are becoming increasingly frequent, posing serious threats to agricultural production and food security in Sri Lanka. The country has taken many significant measures to increase productivity, strengthen the resilience of primary production systems, and conserve water resources. Sri Lanka has concentrated on enhancing adaptability with the introduction of improved varieties of plants through molecular breeding and better livestock breeds. It also is committed to maintaining genetic diversity through the protection of indigenous species. Climate adaptation approaches should be coupled with more comprehensive and detailed mitigation activities that encourage the protection of soil and water. Many of the best practices found for staple production systems in Sri Lanka apply to small- and large-scale producers alike.

Policy-making will be important for designing and implementing a climate adaptation strategy. Scaling up CSA would include well-articulated information management systems and better access to financial instruments, services and markets for smallholders. Building bridges among stakeholders, promoting information and expertise sharing, creating opportunities through social, legal, institutional or economic and market processes, encouraging long-term strategic infrastructure investments and engaging in public–private partnerships are important to ensure the achievement of CSA goals and long-term benefits for sustainable crop production.

Acknowledgements The authors are extremely thankful to Dr. Michael R. Reed, Emeritus Professor, and former Director—International Programs for Agriculture at the University of Kentucky, USA and Dr. R. Kirby Barrick, Emeritus Professor and former Dean, Agricultural Education and the Communication, University of Florida, USA, for reviewing the initial drafts, making helpful comments and offering valuable suggestions. Their sincere cooperation is highly appreciated.

References

Abeywardana N, Schütt B, Wagalawatta T, Bebermeier (2019) Indigenous agricultural systems in the dry zone of Sri Lanka: management transformation assessment and sustainability. Sustainability 11(3):910. https://doi.org/10.3390/su11030910

Akhtar MR (2006) Impact of resource conservation technologies for sustainability of irrigated agriculture in Punjab-Pakistan. J Agric Res 44(3):239–254

Bandara EGCD, Jayasinghe-Mudalige UK, Udugama JMM, Attanayake AMCM, Edirisinghe JC (2014) Has the food and agriculture sector played its intended role in socio-economic development of Sri Lanka? an empirical investigation. J Agric Sci–Sri Lanka 9(2)

Battisti DS, Naylor RL (2009) Historical warnings of future food insecurity with unprecedented seasonal heat. Science 323(5911):240–244

Buyukcangaz H, Demirtas C, Yazgan S, Korukcu A (2007) Efficient water use in agriculture in turkey: the need for pressurized irrigation systems. Water 32(S1):776–785

Chandrasiri WACK (2013) Farmers' perception and adaptation to climate change: a case study in vulnerable areas of Kurunagala District. ASDA 15:13–23

Choudhury DR, Tarafdar S, Das M, Kundagrami S (2012) Screening Lentil (Lens culinaris Medik.) Germplasms for Heat Tolerance. Trends Biosci 5(2):143–146

Cordell DJ, Dominish E, Esham M, Jacobs B (2017) Towards phosphorus and climate smart agriculture in Sri Lanka

Cordell D, Rosemarin A, Schroeder JJ, Smit AL (2011) Towards global phosphorus security: a systems framework for phosphorus recovery and reuse options. Chemosphere 84(6):747–758

Cribb J (2010) The coming famine: the global food crisis and what we can do to avoid it. University of California Press, Berkeley, CA

Cruz RV, Harasawa H, Lal M, Wu S, Anokhin Y, Punsalmaa B, Honda Y, Jafari M, Li C, Huu Ninh N (2007) Asia. In: Parry ML, Canziani OF, Palutikof JP, van der Linden PJ, Hanson CE (eds) Climate change 2007: impacts, adaptation and vulnerability. contribution of working group II to the fourth assessment report of the intergovernmental panel on climate change. Cambridge University Press, Cambridge, UK

Daily News (2020) How does climate change impact food production? Available Online: http://www.dailynews.lk/2020/02/19/finance/211831/how-does-climate-change-impact-food-production. Accessed 8 April 2020

De Silva C, Weatherhead E, Knox JW, Rodriguez-Diaz J (2007) Predicting the impacts of climate change—a case study of paddy irrigation water requirements in Sri Lanka. Agric Water Manag 93(1):19–29

Department of Census and Statistics (DCS) (2011) Self-sufficiency rate of rice 2005–2010. Colombo, Department of Census and Statistics, Sri Lanka

Department of Census and Statistics (DCS) (2014) Food balance sheet, 2012/2013. Colombo, Department of Census and Statistics, Sri Lanka.

Dharmarathna WRSS, Herath S, Weerakoon SB (2014) Changing the planting date as a climate change adaptation strategy for rice production in Kurunegala district. Sri Lanka. Sustain Sci 9(1):103–111

Dharmasena PB (2010) Indigenous agricultural knowledge in present context. Econ Rev.

Droogers P (2003) Climate change impact and adaptation on water, food, and environment in walawe basin. IWMI Working Paper. International Water Management Institute, Sri Lanka, in press

Esham M (2016) Climate change & food security: A Sri Lankan Perspective. University of Technology Sydney, Institute for Sustainable Futures

Esham M, Garforth C (2013) Agricultural adaptation to climate change: insights from a farming community in Sri Lanka. Mitig Adapt Strat Glob Change 18(5):535–549

Esham M, Jacobs B, Rosairo HSR, Siddighi BB (2017) Climate change and food security: A Sri Lankan perspective. Environ Dev Sustain 1–20

FAO (2009) Declaration of the world summit on food security, WSFS 2009/2. Available at http://www.fao.org/wsfs/wsfs-list-documents/en/

FAO (2010) '"Climate-smart"' agriculture: Policies, practices and financing for food security, adaptation and mitigation. Food and Agriculture Organization, Rome

FAO (2013) Climate-smart agriculture sourcebook. FAO, Rome

FAO (2019) State of food security and nutrition in the world. http://www.fao.org/3/ca5162en/ca5162en.pdf

FAO, IFAD, UNICEF, WFP and WHO (2019) The state of food security and nutrition in the world 2019. Safeguarding against economic slowdowns and downturns. Rome, FAO

Ginigaddara GAS (n.d.) Climate Smart Agriculture (CSA). Available online: http://www.rjt.ac.lk/agri/agri_onlinematerials/agriculturalsystems/Climate%20Smart%20Agriculture%20GAS%20Ginigaddara.pdf . Accessed 8 April 2020

Gunatilaka R, Wan G, Chatterjee S (2009) Poverty and human development in Sri Lanka, vol 340. Asian Development Bank, Manila

Gunda T, Hornberger GM, Gilligan JM (2016) Spatiotemporal patterns of agricultural drought in Sri Lanka: 1881–2010. Int J Climatol 36:563–575

Hijioka Y, Lin E, Pereira J, Corlett R, Cui X, Insarov G et al (2014) Chapter 24: Asia. In: Working group II contribution to the IPCC fifth assessment report climate change, pp 1858–1925

Iizumi T, Sakuma H, Yokozawa M, Luo JJ, Challinor AJ, Brown ME, Sakurai G, Yamagata T (2013) Prediction of seasonal climate-induced variations in global food production. Nat Clim Chang 3:904–908

IPCC Summary for Policymakers Climate Change (2014) Impacts, adaptation, and vulnerability. In: Field CB et al (ed) Part A: global and sectoral aspects. Cambridge University Press

Irangani MKL, Shiratake Y (2013) Indigenous techniques used in rice cultivation in Sri Lanka: an analysis from and agricultural history perspective. Indian J Tradit Knowl 12(4)

Irfanullah HM (2013) Floating gardening: a local lad becoming a climate celebrity? Clean Slate 88:26–27

Kornfeld IE (2009) Mesopotamia: a history of water and law. The evolution of the law and politics of water. Springer, Netherlands

Lipper L, Thornton P, Campbell BM, Baedeker T, Braimoh A, Bwalya M, Caron P, Cattaneo A, Garrity D, Henry K et al (2014) Climate-smart agriculture for food security. Nat Clim Chang 4:1068–1072

Lybbert TJ, Sumner DA (2012) Agricultural technologies for climate change in developing countries: policy options for innovation and technology diffusion. Food Policy 37(1):114–123

Magni G (2016) Indigenous knowledge and implications for the sustainable development agenda: background paper prepared for the 2016 Global Education Monitoring Report, UNESCO

Malaviarachchi M, De Costa W, Fonseka R, Kumara J, Abhayapala K, Suriyagoda L (2014) Response of maize (Zea mays L.) to a temperature gradient representing long-term climate change under different soil management systems. Trop Agric 25(3):327–344

Marambe B (2012) Current status and issues of food and nutrition security in Sri Lanka. In: National food and nutrition security conference Sri Lanka—towards a nutritious and healthy nation. Colombo, p 43

Marambe B, Pushpakumara G, Silva P, Weerahewa J, Punyawardena BVR. (2013) Climate change and household food security in homegardens of Sri Lanka. In: Gunasena HPM, Gunathilake HAJ, Everard JMDT, Ranasinghe CS, Nainanayake AD (eds) Proceedings of the international conference on climate change impacts and adaptation for food and environmental security. Colombo, pp 87–99

Ministry of Mahaweli Development and Environment (2015) National adaptation plan for climate change impacts in Sri Lanka 2016–2025. Climate Change Secretariat, Colombo

Niranjan F (2004) PhD thesis, Postgraduate Institute of Agriculture, Peradeniya

Palada M, Bhattaral S, Roberts M, Baxte N, Bhattarai M, Kimsan R, Kan S, Wu D (2010) Increasing on farm water productivity through farmer-participatory evaluation of affordable microirrigation vegetable-based technology in Cambodia. Zeitschrift Für Bewässerungswirtschaft 45(2):133–143

Panampitiya WMGN (2018) A review of indigenous knowledge related to the traditional agriculture in Sri Lanka. Int J Multidiscip Res Rev 4(10):113–118

Perera A (2015) El Nino creates topsy turvy weather in Sri Lanka. Inter Press News Agency, Rome

Punyawardena BVR (2011) Country report. Seoul, Republic of Korea, Sri Lanka. Workshop on Climate Change and its Impacts on Agriculture

Pushpakumara DKNG, Marambe B, Silva GLLP, Weerahewa J, Punyawardena BVR (2012) Review of research on Homegardens in Sri Lanka: the status, importance and future perspective. Trop Agric 160:55–125

Rai RK, Bhatta LD, Acharya U, Bhatta AP (2018) Assessing climate-resilient agriculture for smallholders. Environ Dev 27:26–33. https://doi.org/10.1016/j.envdev.2018.06.002

Read DJ (1999) Mycorrhiza: the state of the art. In Varma A, Hock B (eds) Mycorrhiza. Springer, Berlin

Redman RS, Roossinck MJ, Maher S, Andrews QC, Schneider WL, Rodriguez RJ (2002) Field performance of cucurbit and tomato plants colonized with a non-pathogenic mutant of colletotrichum magna (teleomorph: Glomerella magna; Jenkins and Winstead). Symbiosis 32:55–70

Rodriguez RJ, Redman RS, Henson JM (2004) The role of fungal symbioses in the adaptation of plants to high stress environments. Mitig Adapt Strat Glob Change 9:261–272

Rural Poverty Report (2011) International fund for agricultural development

Sterrett C (2011) Review of climate change adaptation practices in South Asia. Melbourne. Climate Concern, Australia. http://www.oxfam.org/en/grow/policy/review-climate-change-adaptation-practices-south-asia

The Economist (2013) Global Food Security Index 2013. Available online: https://www.eiu.com/public/topical_report.aspx?campaignid=FoodSecurity2013. Accessed 8 April 2020

The Global Economy (2018) Sri Lanka: GDP share of agriculture. Available Online: https://www.theglobaleconomy.com/Sri-Lanka/Share_of_agriculture/. Accessed 8 April 2020

Titumil RAM, Basak JK (2010) Agriculture and food security in South Asia. A historical analysis and a long run perspective. The Innovators, Dhaka

Tran M (2013) Aquatic agriculture offers a new solution to the problem of water scarcity. The Guardian. http://www.theguardian.com/global-development/poverty

UNDP (2015) Sustainable Development Goal 2. Available at http://www.undp.org/content/undp/en/home/sdgoverview/post-2015-development-agenda/goal-2.html.

United Nations Framework Convention on Climate Change (UNFCCC) (2000) Sri Lanka: Initial National Communication under the UNFCCC. Final draft

Wekesa BM, Ayuya OI, Lagat JK (2018) Effect of climate-smart agricultural practices on household food security in smallholder production systems: micro-level evidence from Kenya. Agric Food Secur 7:1–14

Weerahewa J, Pushpakumara G, Silva P, Daulagala C, Punyawardena R, Premalal S, Miah G, Roy J, Jana S, Marambe B (2012) Are home garden ecosystems resilient to climate change? an analysis of the adaptation strategies of home gardeners in Sri Lanka. APN Sci Bull 2:22–27

Weerakoon WMW (2013) Impact of climate change on food security in Sri Lanka. In: HPM Wheeler T, von Braun J (2013) Climate change impacts on global food security. Science 341:508–513

Wheeler T, von Braun J (2013) Climate change impacts on global food security. Science 341(6145):508–513. https://doi.org/10.1126/science.1239402

Wijeratne MA, Anandacoomaraswamy A, Amarathunga MKSLD, Ratnasiri J, Basnayake BRSB, Kalra N (2007) Assessment of impact of climate change on productivity of tea (Camellia sinensis L.) plantations in Sri Lanka. Natl Sci Found, Sri Lanka 35(2):119–126

World Bank, CIAT (2015) Climate-smart agriculture in Sri Lanka. CSA country profiles for Africa, Asia, and Latin America and the Caribbean series. The World Bank Group, Washington, DC

World Bank (2007) Sri Lanka—strengthening social protection. World Bank, Washington, DC

World Bank (2015) Climate-smart agriculture in Sri Lanka. CSA country profiles for Africa, Asia, and Latin America and the Caribbean series. The World Bank Group, Washington, DC

World Bank (2019) Climate smart agriculture to improve resilience and productivity in Sri Lanka. Available online: https://www.worldbank.org/en/news/press-release/2019/03/07/climate-smart-agriculture-to-improve-resilience-and-productivity-in-sri-lanka . Accessed 8 April 2020

Zhengbin Z, Ping X, Hongbo S, Mengjun L, Zhenyan F, Liye C (2011) Advances and prospects: biotechnologically improving crop water use efficiency. Crit Rev Biotechnol 31(3):281–293

Zubair L, Nissanka SP, Weerakoon WMW, Herath DI, Karunaratne AS, Prabodha ASM, Vishwanathan J (2015) Climate change impacts on rice farming systems in North-western Sri Lanka. In: Handbook of climate change and agroecosystems: the agricultural model intercomparison and improvement project integrated crop and economic assessments, Part 2, pp 315–352

Chapter 7
Impacts of Covid-19 Pandemic on Agriculture Industry in Sri Lanka: Overcoming Challenges and Grasping Opportunities Through the Application of Smart Technologies

Prasanna Ariyarathne S. M. W., W. M. Wishwajith W. Kandegama, and Mohamed Behnassi

Abstract COVID-19 is currently a global pandemic with worldwide magnitude and transversal implications. In Sri Lanka, during several island-wide curfews imposed since March 2020 and up to now, the agriculture industry's stakeholders and consumers have been coming across several challenges such as food scarcity, dramatic price fluctuations of commodities, and difficulties in searching of market for both product buying and selling. Also, communities have faced challenges related to the timely purchase of planting materials and other agro-inputs (fertilizers and pesticides), loss of income, inadequacy of reliable advices and directions, and market uncertainties, which have demoralized them with no clear way forward. This blockage created a sudden imbalance of the entire value chain of agriculture industry affecting almost all stakeholders. Electronic representation of physical entities—such as local fairs and mega trading hubs driven by computer intelligence services—are believed in this study to be viable solutions to overcome most of the above-mentioned challenges. Therefore, this study proposes a smart-agriculture support system naming it as Electronic Partner for Agro Services (EPAS), which would seamlessly connect farmers, consumers, and other relevant stakeholders of the agriculture value chain in the virtual space electronically. The system intends to regulate price for goods and services while organizing a balanced supply and demand in more informative and intelligent manner, thus provisioning electronic financial accounting facilities

---

P. Ariyarathne S. M. W.
Faculty of Commerce and Management Studies, University of Kelaniya, Colombo, Sri Lanka

W. M. W. W. Kandegama (✉)
Faculty of Agriculture and Plantation Management, Wayamba University of Sri Lanka, Makandura, Gonawila, Sri Lanka
e-mail: wishwajith@wyb.ac.lk

M. Behnassi
International Politics of Environment and Human Security, College of Law, Economics and Social Sciences, Ibn Zohr University of Agadir, Agadir, Morocco
e-mail: m.behnassi@uiz.ac.ma

Center for Environment, Human Security and Governance (CERES), Agadir, Morocco

© The Author(s), under exclusive license to Springer Nature Switzerland AG 2022
M. Behnassi et al. (eds.), *Food Security and Climate-Smart Food Systems*,
https://doi.org/10.1007/978-3-030-92738-7_7

for subscribers. Timely dissemination of knowledge, advice, financial services and linking agro-input suppliers are also embedded into the proposed model. The Design Science Research (DSR) Methodology was adopted in this study in developing the EPAS conceptual model.

Keywords Agriculture · Value chain · Digitalization · Fair trading · Information Communication technology Conceptual model (artefact) · Design science research

1 Introduction

COVID-19 is currently a global pandemic with worldwide magnitude and transversal implications. It has become the mostly heard and uttered word in the globe since the dawn of 2020 threatening the entire world and becoming one of the most fearful global pandemics that challenges communities to the extent of changing usual living norms. The pandemic has spread worldwide in harmful way regardless of political borders, social classes, size of economy, and nature of industry. The real impact of the pandemic started distressing Sri Lanka in mid-March 2020 after the detection of some infected overseas returnees which created a threat of community transmission and cross infections, resulting in a wider virus spread across the country. The closure of schools, imposing of curfews, lockdowns, and inter-provincial and inter-district mobility restrictions are some of the government preventive measures which caused the isolation of communities. The macro, medium, and small-scale enterprises have been all affected by the pandemic and the degree of the impact varies from one to another depending on the nature of industry and, particularly, on the levels of physical interactions required in doing business (Gunawardana 2020).

Agriculture industry is also among many hard-hit sectors such as tourism, logistics, and apparel which are severely affected and continuously facing several challenges. Newly enforced restrictions due to the pandemic are causing further disconnection in the agriculture value chain in various stages (The Netherlands and you, February 7 2020). At the same time, the pandemic is an event which is increasingly sensitizing the society about the lost opportunities and the new avenues with regard the use of modern Information Communication Technology (ICT) to improve several segments of the agriculture value chain in this tropical island.

This study analyses the current ICT initiatives taken place in the area of Sri Lankan agriculture industry and present the challenges that agriculture communities are facing with sudden shifts of the usual lifestyles. Then, the study continues to identify the challenges caused by the pandemic and proposes alternative solutions to overcome the negative impacts and grab new opportunities. Finally, the study explains the potential use of modern technology to implement a support system in the form of a conceptual model on digital platform to collaborate, transact, and carry out other value chain activities virtually on the electronic space. The model is named Electronic Partner for Agro Support services (EPAS).

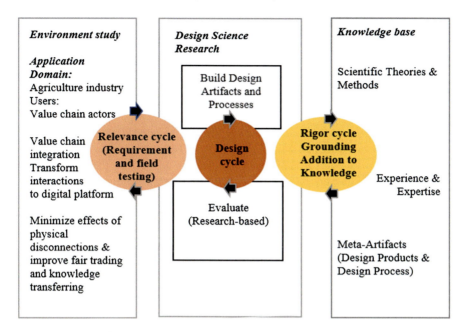

Fig. 1 Three cycle view of design science research. *Source* Hevner (2007: 2)

The study followed the design science research (DSR) methodology in developing the EPAS conceptual model. The three-way cycle view of DSR explained by Hevner (2007) was adapted in this study considering the knowledge as closely applicable in this area of agriculture value chain management in a digital platform. The research design cycle shown in Fig. 1 was considered in developing the EPAS conceptual model.

The scope of this research is limited to the study area and development of EPAS conceptual model. The design model evaluation, which is the second part of design cycle, and the rest of activities in rigor cycle are proposed for future research works. The following parts of the chapter analyse the main issues related to the COVID-19 pandemic context and the EPAS solution which is proposed to be developed on ICT platform.

2 COVID-19 Issues and Opportunities

As briefly mentioned above, the COVID-19 pandemic has generated multifaceted challenges and risks in agribusinesses affecting all segments of agriculture value chain. In this chapter, such key challenges and their implications for agribusinesses are analysed in detail. Also, the chapter explores the new opportunities created by the pandemic situation which could be capitalized to improve the value chain activities for more resilience operation, thus improving the performance of all segments with more efficiency, effectiveness, and quality.

2.1 COVID-19 Related Challenges in Agriculture Industry

During several lockdowns and island wide curfews imposed since March 2020 and up to now, the key challenges and issues mentioned below have been experienced by the public with no adequately proven remediation in overcoming them. Despite that several electronic solutions have been developed and became online, as explained in the next section, the use of such platforms for immediate and sustaining solutions is generally still unpersuasive according to the preliminary tests conducted and observations made during this study. An estimated 2.1 million agricultural households are at a risk of losing their livelihoods despite various measures taken by the government to safeguard agricultural supply chains (The Netherlands and you 2020).

The agribusiness in Sri Lanka, like in many parts of the developing world, faces several challenges preventing it to reach the potential growth which is proportionate to the wealth of agriculture support resources in the country. The constrains posed by the pandemic has worsen the situation by creating interruptions across almost all segments of the value chain, further weakening its equilibrium. The following are the top ten key challenges which worsened the agribusiness across the country, including export trades (Arumugam et al. 2020):

- Scarcity of subsistence foods and dramatic price fluctuations.
- Interruptions in purchasing seeds and planting materials.
- Disturbances in handing over the harvests from farms to local dealers or collectors.
- A deficiency of long-distance transportation due to inter-district lockdowns.
- Shutdowns in the retail farming markets and national agriculture wholesale trading hubs.
- Disconnection from the financial resources and related service.
- Scarcity of fertilizers, pesticides, other essential agriculture inputs, and agro-machineries.
- Limited access to consultations, advisory, and other field support services.
- Biosecurity and animal health.
- Growing demotivation due to uncertain future.

2.2 Potential Opportunities

Sri Lanka is an agricultural country which produces varieties of food crops (rice, fruits, vegetables, field crops, and spices), plantation crops (tea, rubber, coconut, sugar, and oil palm), floriculture and ornamental crops, livestock, fisheries and many more. The cascaded tank-village system (*Ellangu* system), the hill side cascade paddy fields (*Helmalu*), and the irrigation systems developed in the historical past were remarkable agro-biodiversity systems in Sri Lanka with regard to water management efficiency and adaptation to climate change; they also are buffers against natural disasters such as floods, droughts, and pest attacks. The mountains and the southwestern part of the country, known as the 'wet zone', receive ample rainfall (an annual average of 2500 mm). Most of the southeastern, eastern, and northern parts of the country comprise the 'dry zone', which receives between 1200 and 1900 mm of annual rainfall. Much of the rain in these areas occurs from October to January. Another intermediate zone demarcates the area, which receives a mean annual rainfall between 1,750 and 2,500 mm with a short and less prominent dry season.

Sri Lankan agricultural history is dating back to more than 2500 years. Over centuries, paddy cultivation in this island was not merely an economic activity but also a factor which shaped the culture, religion, and values of the nation. Traditional agriculture evolved through the accumulation of farmers' knowledge and experiences, managed to self-feed the entire population for many centuries in the past. Indirectly, the COVID-19 pandemic has been an opportunity to remember the great history of agriculture in the country and to reiterate the importance of paying more attention to this sector within the perspective of ensuring self-sufficiency and food security for a developing nation which has developed a heavy dependency on agricultural and food imports. The average import expenses on foods and beverages during last 6 years (2014–2019) is equally distributed around US$ 2,156.3 million, which is about 10% of total exports and 2.5% of GDP throughout this period. The agriculture industry is an utmost important segment in Sri Lankan economy, with an obvious potential to grow and achieve self-sufficiency in food crop cultivation, thus saving massive foreign exchange on imports and going further reaching global markets (Central bank of Sri Lanka 2020a).

2.3 Impact on Economy

The major components of the Sri Lankan economy are agriculture, industry, services, and tax less subsidies. However, the agriculture contribution to the national GDP was low as 7.3% in 2019, and it was further contracted to 5.6% during the first quarter of 2020. The Department of Sensors and Statistics, Ministry of Finance of Sri Lanka (2020), also predicts further contraction during the beginning of 2021 since the challenges remain the same or becoming more severe on daily basis.

2.3.1 Impact on Households and Livelihoods

The drifting attention and loss of focus on farming and home gardening have weakened the village-level production streams which contributed to localized independencies maintained in the past, particularly within rural communities. Growing trends of dependency on agro-product imports and interruption of import supply chains adversely affected the consumers, resulting in scarcity and unexpected price fluctuations during the cut-offs. However, it was observed that there is a sudden increase in the interest in small-scale farming and gardening as response mechanisms, which in turn result in creating considerable self-employment opportunities if those rising motivations are wisely converted into self-implements and sustain. Due to the pandemic related restrictions, families' members often stay at home throughout the day and most of the households showed an interest in small-scale farming and agriculture-based product development. However, according to various news, many families faced difficulties in managing both resources within their premises—such as water, land space, vertical space, crop management, pest and diseases control—and time. Therefore, the majority of gardening initiatives were unsuccessful over a short period of time.

2.3.2 Impact on the Macro-Economy

Despite that the country possesses a wealth of agriculture resources, the present contribution of the agriculture industry to the national GDP is below the potential given the resources and potentials of this tropical country. Indeed, this contribution is only around 7% in 2019 according to latest Central Bank statistics, and given the pandemic related challenges such a contribution is predicted to be further lower if the situation is not regulated cautiously. The total labour force in Sri Lanka is 8.39 million in 2019 to which the agriculture sector contributes around 25%; however, given the pandemic situation this contribution may be lower, thus worsening the unemployment problem in the coming years, especially among rural people that represent 70% of the national population (Central Bank of Sri Lanka 2020a).

According to the press release issued by the Department of Sensors and Statistics on national account estimates on the first quarter of 2020, the Sri Lankan economy has recorded an economic contraction of 1.6% whereas 3.7% of economic growth was recorded in the same quarter in 2019. The impact of the pandemic is predicted to be worse in quarter two and onwards during 2021. The official statistics regarding the rest of 2020 are not yet ready due to data collection challenges also posed by the pandemic situation. The contractions of agriculture sector in the first quarter of 2020 compared to the same quarter of 2019 were mainly driven by marine fishing (8.1%) and activities related to oleaginous fruits production, including coconut (12.8%), tea (27.5%), rubber (8.7%), forestry and loggings (3.3%). However, activities related to the production of cereals (12.9%), rice (4.1%), vegetables (1%), fruits (7.4%), spices (1.5%), and animal products (0.4%) recoded an expansion in the first quarter of 2020 compared to the same quarter of 2019.

2.3.3 Impact on Export-Oriented Agriculture

Sri Lanka is a reputed exporter of agriculture products. It has been particularly a renowned supplier of Ceylon tea and spices to most parts of the world for decades. There are significant demands for fresh and processed fruits and vegetables from

7 Impacts of Covid-19 Pandemic on Agriculture Industry in Sri Lanka …

Table 1 Agricultural export revenue in Sri Lanka, 2015–2019

Year	2015	2016	2017	2018	2019
Revenue US $Mn	2,421	2,335	2,704	2,222	2,522.31

Source Sri Lanka Export Development Board (2020)

Maldives, Middle East, Europe, and the United States. However, the COVID-19 pandemic has made a significant impact on agricultural exports, particularly through the disruptions of supply chain and the changing consumer's preferences of destination countries. Agricultural exports of Sri Lanka were growing slowly and the contribution to the total export revenue made in 2019 is only around 6% which is US$ Mn.43,596. Table 1 depicted the growth over the last five years, and the situation in 2020 may further deteriorate with the challenges mentioned above (Sri Lanka Export Development Board 2020).

3 Technology Readiness and ICT Initiatives in Agriculture Industry

The ever-evolving digital technologies are effectively enabling industries to improve the functionality of existing business processes across value chains for optimal gains and more added values. In some industries, ICTs have been deployed in reinventing the business models whereas others continue the same businesses through transforming processes into digitalized platforms. The modern digitalization concepts started to make smart shifts in the agriculture industry as well by providing improved access platforms and enabling the integration of traditional value chain activities. Parameters such as the growth of ICT infrastructure, the rate of internet penetration, and the use of social media are evidences to confirm the rapid technological adaptations of the country. As a result of such developments, there are several ICT-driven initiatives in agriculture industry available online which have been introduced by both the government and private sector in recent years. The following section explores and briefly assesses the existing technology developments and the modern initiatives evolved recently and growing on digital platforms to achieve various objectives.

3.1 Assessment of the ICT Infrastructure in Sri Lanka

Over past decades, the Sri Lankan telecommunication coverage and the number of user subscriptions have been significantly increasing surpassing the population of the country, particularly with rapid growth in cellular mobile networks. Figure 2 shows the growth of the cellular mobile coverage in past years. Table 2 estimates

Cellular Mobile Telephone Subscriptions

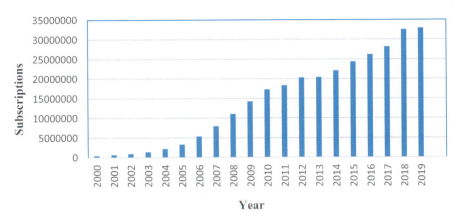

Fig. 2 Cellular mobile telephone subscriptions (2000–2019). *Source* Sri Lanka telecommunication regulatory commission

Table 2 Distribution of cellular mobile subscriptions in provinces as of year 2019

Province	Estimated population (2019)	Mobile cellular mobile subscriptions	% to estimated population
Central province	2,766,000	3,788,450	137
Eastern province	1,729,000	2,385,549	138
North central province	1,377,000	2,137,777	155
Northern province	1,143,000	1,476,480	129
North western province	2,551,000	3,632,076	142
Sabaragamuwa province	2,058,000	2,709,400	132
Southern province	2,654,000	3,980,780	150
Uva province	1,376,000	1,908,117	139
Western province	6,149,000	10,865,461	17

Sources Sri Lanka telecommunication regulatory commission and department of sensors and statistics of Sri Lanka

the provincial distribution of the mobile telephone subscriptions in Sri Lanka during 2019 against the population of each province based on sensors carried out in 2012.

The percentage of cellular mobile subscription in Sri Lanka is 68% of the total telephone subscriptions of the country (32.8 million). The geographic distribution of cellular mobile subscriptions in the country shows some variations from one province to another; however, in each province the numbers have surpassed respective populations. According to the central bank annual report in 2019 of Sri Lanka, cellular mobile usage in the country was 161% overall whereas fixed access line use was 10.5% (Central Bank of Sri Lanka 2019). By reference to this report, the internet penetration in the country was 61.5%, which is 13.4 million of the estimated of 21 million. The numbers of social media subscribers have reached 6.4 million in January 2020; therefore, the social media penetration has increased at a rate of 30% in this period (Global Digital Insights 2020).

3.2 ICT Technology Maturity Potentials in Agriculture Industry

The above data provide valid evidence on the richness of ICT infrastructure in the country and the distribution of use of technology across provinces. Rapidly growing technologies, digitization, and automation trends are good opportunities for the modernization of agriculture value chain, handling, and management on digitized platforms.

In parallel to this evolution, the following are some potential modern concepts and techniques which can be used to improve the agribusiness activities. Some of the evolving technologies started penetrating into agribusiness worldwide are Internet of Things (IoTs), Natural Language Processing (NLP), Machine Learning (ML), Artificial Intelligence (AI), Sensors, Robotics, Big data analytics, and Block-chain. The right combination of traditional agriculture practices with smart technologies could make a revolutionary shift in the agriculture industry towards 'Smart Agribusiness'. Such technology adaptation into agriculture value chain could potentially leverage the readiness of the business continuity of the industry with higher degree of resilience (Matthieu et al. 2018; Manoj and Nimesha 2019).

3.2.1 Potential Smart Technologies to Be Used in Agriculture Value Chain Integration

Advanced technologies like Interactive Voice Responses (IVRs), Natural Language Processing (NLP), and Machine Learning can be used to create a user-friendly communication platform enabling bilateral interaction between human voice and a digital system in native languages. This platform can facilitate many communication needs such as market information exchange, questions and answers (Q&As) platforms, knowledge sharing, and other various interactions which will be discussed later (Teixeira 2020). The use of internet, cloud, World Wide Web (WWW), and modern mobile applications (Apps) are becoming popular among farming communities as well as other segments of agriculture industry. These technologies can be utilized to help disseminate information on agriculture such as sales, markets, banking, insurance, and other support services such as agriculture inputs, including various advisory services in the form of general broadcasts or on demand requests (i.e., instant notification on the supply need of tomatoes for vegetable farmers and other actors in the value chain).

Satellite and Global Positioning System (GPS) techniques can be used for high end remote-sensing and areal images to detect the weather patterns and to receive remote views of the farm fields for variety of quick decision making. Satellites remote-sensing techniques produce crucial data required for farming such as soil condition, snow cover in cultivation areas, soil–water level in a drought and crop development, pest infestation, relative humidity (RH), and light intensity required for farming. Analysis of rainfall patterns by satellites and providing real-time data help farmers plan irrigation cycles for their crops and manage resources efficiently. Further, these techniques can help scientists and authorities evaluate the crop health based on the weather conditions and estimate the crops' anticipated yield. The images given by satellites can be analyzed for crop cultivation along geographical distribution; hence the potential of disease spreading, and pest infestation can be forecast, and preventive methods implemented.

Modern devices such as drones, robots, and sensor networks can be used for improving precision agriculture with higher accuracy and reliability of data for intelligent information processing. Sensing micro-climatic condition in both soil and atmosphere in cropping areas helps determine the requirements needed to maintain optimum conditions (moisture and temperature) and plant health for higher productivity. They also contribute to the protection of the environment with some possible applications (Central Bank of Sri Lanka 2020b).

There is a massive amount of data already saved and is being isolated in individual computer or network clouds. Big data analytics techniques can effectively be employed to use those data banks insightfully to formulate accurate business intelligence (BI) information for decision making. Big data sources provide essential data for farmers on climatic condition, agronomic information, early warning of pest, diseases infestation, and product demand. Therefore, farmers can take smart decisions in terms of choosing the right crops to plant, thus ensuring a better profit. A cohesive data pool could be developed by connecting all scattered data across different sources and merge with international sources to communicate with global value chains.

The connected devices mentioned above and many more in the future are useful information sources and interfaces for smarter interactions in the electronic space. Techniques mentioned above can be used to coordinate and command those devices intelligently in order to get the desired inputs and outputs, thus facilitating the industry value chains in many ways (Hyea and Vikas 2017). The use of digital platforms for payment transactions and processing are still limited in the country, and these modern techniques are potential opportunities to improve the situation.

3.3 Present ICT-Based Initiatives in the Agriculture Industry

Over recent years, the Sri Lankan government and some private sector organizations have taken several initiatives to promote the use of ICTs in the agriculture industry, particularly focusing on facilitating direct marketing and information dissemination. The following section closely analyses such initiatives in order to understand the kind of technologies being used as solutions and the extent to which the issues are being addressed by such applications.

3.3.1 E-Agriculture Strategy in Sri Lanka

In 2016, the Ministry of Agriculture through the Department of Agriculture had taken an initiative to develop the e-agriculture strategy with the primary objective of achieving self-sufficiency in food crops, which may grow locally and save foreign exchange on imports of those food items. This strategy has been developed under the guidance and recommendations jointly published by the Food and Agriculture Organization (FAO) and the International Telecommunication Union (ITU) in 2016. This document provides compressive and insightful framework on adaptation of modern ICT in agriculture industry (FAO and ITU 2016). The strategy is also in line with the agriculture policy framework and the national food production programme designed and published for the period 2016–2018 by the Ministry of Agriculture with the following key objectives (NAICC 2016):

- Achieve self-sufficiency in food crops, which may grow locally and save foreign exchange on imports of those food items;
- Increase availability of safe food by promoting ecofriendly practices and minimizing agro chemicals and pesticides in food crop production;
- Ensure food security through appropriate management of buffer stocks;
- Introducing and implementing agro-ecological region-based food crop cultivation programs; and
- Increase the productivity of crop production through appropriate technologies.

3.3.2 Agriculture Supportive Websites, Programs, and Systems

Over the past years, several novel concepts and initiatives emerged in the agriculture sector, which have added value to agribusinesses in various ways, enhanced farmers' knowledge, and help disseminate important market information. The following list summarizes some live projects in the sector and their objectives, respectively:

- www.doa.gov.lk: website of the Ministry of Agriculture which provides information for farming community, other stakeholders, and general public on policies, guidance, and administrative services.
- *Wikigoviya web*: Cyber agriculture Wikipedia which provides learning and education materials.

- *AgMIS*: Short Message Service (SMS) for market information requests.
- *Crop forecasting (boga purokathanaya)*: Farmer advisory (SMS) on crop selection for better pricing.
- *SL paddy fertilizer app*: Mobile application (App) which provides fertilizer recommendation.
- *Govipola app and web site*: Mobile application (App) for market linkage and improvement of price awareness.
- *Rice knowledge bank website (IRRI)*: Country information system for rice cultivation hosted by the International Rice Research Institute (IRRI).
- *Agriculture videos on the internet*: Documentaries, awareness programs on timely and relevant technologies, including success stories.
- *Govi Mithuru project*: A project to strengthen the technology transfer in agriculture to farming community using mobile technology.
- *Market price information systems*: Telephone dial-in service to provide daily market price information over mobile phones by two mobile networks.
- www.b4fn.org/countries/sri-lanka: A web application offering information on healthy food items and preparations together with the respective nutrition values (Information System for Biodiversity of Food and Nutrition System).
- *doaseed.lk*: Seed and planting material management information system for real-time seed and planting material information to monitor their productions, certifications, distributions, sales, and available stocks.
- *Progress monitoring system for national food production program*: A software solution to monitor the progress of the National Food Production Program (NFPP).
- *QR code system for GAP certification program*: A program to promote Good Agriculture Practices (GAP), to certify agricultural products, and meet international standards for export market (Sri Lanka E agriculture Strategy, 2016, pp. 38–39).

4 Electronic Partner for Agro Services (EPAS)

As discussed above, there are several web-based and mobile-based agriculture information systems which have been developed to assist farmers to carry out some of their agriculture value chain activities. Those systems can be classified as knowledge and information sharing systems, including market and price information. In addition to those systems, Ginige et al. (2016) propose a digital knowledge ecosystem for sustainable agriculture production in Sri Lanka. This model conceptualizes a holistic information flow model connecting all major stakeholders in the agriculture sector in the form of close-loop mobile-based information ecosystem. However, the functionality of the system is limited to information management and dissemination of information among actors of the agriculture value chain. The concept is depicted in Fig. 3.

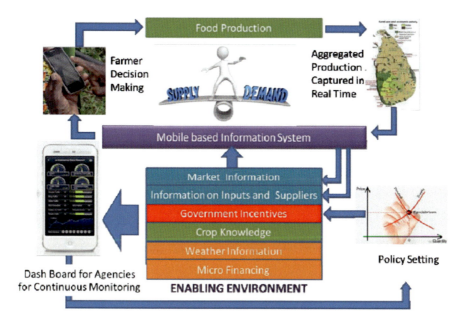

Fig. 3 Conceptual Agriculture product management system. *Source* Ginige et al. (2016)

4.1 Integration of Value Chain Actors in E-Space

Onboarding of agriculture value chain users into an electronic space and enabling the users to overcome the above-mentioned issues are the prime objective of this proposal. There are visible gaps in cohesive integration of value chain users to provide sustainable solutions for such issues, and this has been an ongoing concern in agriculture sector throughout many decades. The impact of unavailability of such system became more obvious during the COVID-19 pandemic. The interruption of physical interaction across the value chain activities appeared as a more critical issue impeding the functioning of all segments of the value chain. Therefore, the proposed model in EPAS is a rather focus on coordinating and facilitating the activities across the value chain through insightful and actionable information processing which are gathered from various sources. The system will transform data into actionable information through cognitive intelligence supported by modern technologies.

The proposed model as a solution which will be designed on existing ICT platforms utilizing evolving digitalization concept is a self-intelligent virtual agriculture service centre which integrates both voice and data devices to form interactive communication platform. The novel system intends to provide bilateral human to machine interaction seamlessly using native languages and English. The system will be designed with a full fledge resilience in its operation providing continued access to farming community and other stakeholders of the agriculture value chain to go on undertaking their activities even in emergency and disastrous situations.

The primary concept of the design is to implement electronic representation of local fairs and economic centres/national agro-products trading hubs (mega trading hubs) around the island and integrate them with other elements of the agriculture value chain.

The study believes that onboarding farmers, consumers and other stakeholders of agriculture sector into a digital system will help improve the performance of their activities and foster coordination and standardization across the segments of the value chain. This integration will also help the centralized system ensure intelligent decision making and coordination of actors and activities of agriculture value chain which could provide practical and holistic solutions for the above-mentioned issues. The agriculture value chain activities are diverse according to crop types, volumes, seasonality, and geography. The study analysed the existing trading ecosystem of agriculture products in the country, which has been evolved over years, and accordingly proposes the development of EPAS electronic platform. The main segments of such a trading network are depicted in Fig. 4.

The route of agriproducts from farm gate to consumers takes place in different channels as shown in Fig. 4. The activities can be categorized into three main areas: near farm filed, long distance transportation, and sales outlets (near end-users). There are several traditional stakeholders in each segment of this network, and, in addition, there is emerging stakeholders taking part in the business particularly during the pandemic, mentioned in the diagram as delivery agents. The latter represent regular motorbikes, three-wheelers, and other stablished logistics and courier service companies. Also, there is a growing trend of supermarkets directly reaching out to farmers integrating their supply chains backwards. The EPAS intends to link all these stakeholders to facilitate the business through the following key objectives, which help foster resilience and agility in emergency situations:

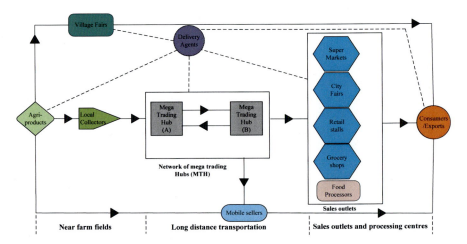

Fig. 4 Segments of Present agriculture trading network in Sri Lanka. *Source* Developed by the authors

- Achieve more supply and demand balance through intelligent coordination;
- Leverage fair trading through price standardization of products and associated services;
- Optimize product outputs through improving farmer knowledge and providing actionable information, including early warning and early action alerts;
- Minimize the gaps in payment processes and improve credibility of transactions across the value chain;
- Interconnect agriculture support services to value chain for better information, facility, and transaction flow; and
- Produce agro intelligence information for decision makers and researchers.

A sketch of technical architecture of the EPAS model is depicted in Fig. 5, which shows the basic technology and devices to be used to enable bilateral interaction with voice and data networks. The proper design and technical implementation of the system require in-depth research on such systems which is beyond the scope of this preliminary study.

The technology being proposed to use in the design are Internet, interactive voice response (IVR), natural language processing (NLP), machine learning (ML), artificial intelligence (AI), and mobile apps. It is valid to assume with existing data (Fig. 2 and Table 2), that primary communication means of farming community and other users is voice communication, mostly through regular cellular mobile phones

Fig. 5 Sketch of the proposed EPAS technical architecture. *Source* Developed by the authors

presumably using native languages either in Sinhala or Tamil. Implementation of fully digitized interactive voice communication system with native language support has become a reality with latest technologies such as natural language processing, machine learning, and AIs. Those newest technologies are proposed in this solution to facilitate user-friendly interaction with the system for confident communication with clarity and understanding for end users who may not have adequate literacy of handling smart functions of modern cellular mobile devises or using English language.

The following sections describe the modules of EPAS in detail, including how the system will handle interactions, process data, and provide services to subscribers.

4.2 Modules of Electronic Partner for Agro Services

The proposed solution is basically centred on the concepts of transforming value chain activity management and coordination to a central electronic entity (EPAS). This primally represents e-fair and e-trading hubs along with other existing physical entities in the trading network and other support services. This electronic representation is expected to improve performers of agribusiness activities and related processes in terms of efficiency, effectiveness, standards, and credibility of transactions while strengthening business continuity providing seamless interaction even in emergencies such as pandemic situations.

The conceptual artefact of designing EPAS was segregated into six main modules, which are defined based on the findings of this primary study. These modules require further research-based evaluation to finalize the contents and matching it with the context and practical business needs. The Fig. 6 describes the components of EPAS in details.

4.2.1 Attract and Empower Users to Use the System

As in many digitalized systems of this nature, attracting subscribers and sustaining their subscription for the provided services are perceived as a key challenge in the proposed model. Such a challenge needs to be strategically addressed through an appropriate implementation of modules based on extensive field research to help identify the influencing factors. Empowering users and motivating them to use the system as well as building trust for and credibility of this system have been identified as important elements of the EPAS model. In Sri Lanka, online retailing is estimated to US$40 including services, and this represents less than 1% of the total annual retail sales of the country (Dailymirror online 2020). This information shows in general the slow adaptation of online platforms by Sri Lankan communities which could be argued as worse in the agriculture sector given the various hardships stakeholders are still facing on the ground. The literature presented above on the use of existing online platforms in agriculture also supports the existing of such a gap. A potential module,

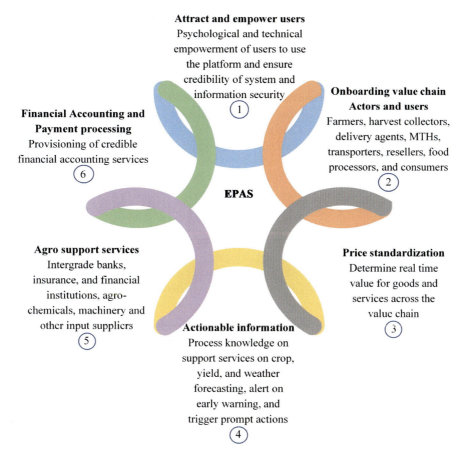

Fig. 6 Proposed Modules of EPAS development. *Source* Developed by the authors

which will be designed to address this key issue of user motivation and empowerment on the use of EPAS system, will require in-depth research on behaviours, perceptions, and attitudes of users who will be interacting with the system in various segments of the value chain.

4.2.2 Onboarding Value Chain Users to the System

Despite the existence of some platforms which have been developed and operationalized with the primary objective of facilitating direct trading of harvest from the farmer gate to the end user, the use of such systems seems limited. The consequences were revealed by the public media during the hard lockdowns and curfews in several occasions. Loads of pumpkins, watermelon, onions, and many other perishable fruits and vegetables were stranded on farm fields and roads with no market destinations.

Generally, all marketplaces, including regular village fairs and local retail stalls, were also unreachable due to shutdowns. Therefore, this disconnection of basically each segments of agriculture trading network (Fig. 6) was observed as an issue with no appropriate solution so far.

A functional virtual platform, primarily representing village fairs and mega trading hubs (MTH) and other segment of the trading network, would enable subscribers to meet effectively in situations where physical interactions are impossible or constrained. Given the valid assumption that most users in this trading network will regularly interact using voice communication, which is the most convenient and viable means for them, EPAS will allow users to interact with the system using regular voice calls and the system will then transform the conversations into digital format for further processing and follow-up functions. The end users can enter EPAS with a subscription and can place orders for products through voice, text, mobile app, or using Internet web pages. The system will analyse the request and coordinate with the nearest agent to deliver the ordered goods to the doorstep of respective end user. The local harvest collectors can get notifications on supply availability by contacting EPAS. Also, EPAS can send automated supply and demand notifications for selected users determined based on different criteria like price, proximity, quality, and volume with live information. The algorithms on such decision making have to be formulated through in-deep research findings of the relevance and usefulness of such requirements.

As argued in literature, the fluctuations of supply and demand of agribusiness is an unmitigated issue in Sri Lankan agriculture sector, and the situation became bad to worst during the COVID-19 pandemic. Such fluctuations have been caused by various reasons such as impulse unplanned purchases, change in consumers' preferences due to health reasons, and overly price awareness due to future uncertainty. The provincial and mega trading hubs, long distance transporters, and wholesale buyers faced enormous complexities and challenges in stock planning, sales, and transportations due to the above-mentioned abrupt fluctuations in supply and demand.

A functional virtual trading hub would have been a viable solution to mitigate such irregularities in supply and demand through an insightful coordination between users with the right information in the right time. The users of the trading network EPAS services can retrieve market information on supply and demand with real-time prices and volume needs. This facility will help facilitate harvesting planning and stock management, thus minimizing the yield waste while optimizing the income. The transporters can plan the deliveries optimally by selecting the right varieties lumping together, thus minimizing transport losses and delays, which will result in a better cost efficiency and the improvement of the safety of transported goods. The markets and retailers can reach out to EPAS for their stock replenishments based on the real-time demands minimizing delivery lead times. In many ways, the EPAS can seamlessly interact with subscribers to coordinate and guide them, thus ensuring the best outcome in all segments of the agriculture value chain.

4.2.3 Standardization of Pricing of Goods and Services of the Value Chain

In general, unregulated pricing of the agriculture products often lead to unfair deals, and often farmers and consumer are mostly affected by the ad hoc deals. The deep investigation on the root causes of this disparity is beyond the scope of this primary study. During the pandemic, the pricing fluctuation of agricultural products deteriorated further due to the various reasons mentioned above in addition to lockdowns and curfews. As a solution to control the dramatic fluctuations of supply and demand, the proposed system will run intelligent processing through taking various data into account to determine the prices of goods and associated services of all segments. Specific criteria and mechanisms of pricing for agriculture products and services will be developed using secondary data and expertise information from relevant sources.

In brief, the system will assess market dynamics taking various factors into consideration from various other integrated sources on supply and demand patterns to government laws and regulations on pricing controls and taxations. Factors—such as contextual and seasonal transaction patterns of farmers and buyers, geographical and seasonal consuming patterns, and distribution trends, nature and shelf life of products, transportation and handling expenses, and other service premiums—will be considered to determine the prices of products. Standardized service premiums will be determined by the system for support services such as local collection, long distance transportation, and doorstep deliveries based on the weight, distance, and the level of precaution and safety required in transports.

4.2.4 Actionable Information and Alerts

The pandemic-driven uncertainties have demoralized the farmer community and other agribusiness stakeholders. This situation could have been minimized if proper guidance and advice have been provided on time. Initiatives made by farmers in rush—such as the cultivation of various crops in farms and home gardens during the peak of the pandemic—had been often considered ineffective in producing the anticipated outcome. Instead, such initiatives result in waste of resources with no proportional return on investments. Providing credible advice and trustworthy warning and alerts in a timely manner to the right clientele would indeed have helped manage the emergency situation better. Appropriate planning of cultivation and upcoming harvesting to determine the suitable crops to cultivate in right quantities in a right season, geographical location, climate, and ideal time to start the harvesting are useful information and advices for the farmer community.

Early warning, prompt actions, and remote advisory services seem as highly important services during the pandemic, especially with the sudden shutdown of services and the interruption of physical access to fields. The system's accumulated institutional memory and intelligence will formulate and broadcast credible warnings and alerts on actions to the relevant user in the right time as the situation demands.

Also, the platform can be used for quick emergency warnings and alerts on unpredicted calamities such as pandemics, floods, and tsunamis. The list of the warning and action services are to be developed through a study on the relevance and usefulness of contents to end users.

Advice on seed and planting material selections, matching the right crop to right season, nature friendly use of agro-chemicals, knowledge on how to optimize the harvest while minimizing the waste, understanding the protection of environment, mitigation actions, and optimizing water and other resources management are some of the facilities the system can provide to the farming community and others for a meaningful value addition. The list of activities in this system is not exhaustive and can be expanded further along with the growth of EPAS. Similarly, other advices on transportation planning, safe handling of items, stock management and healthy postharvest storage, and optimal rout to the market are services the system can disseminate to other users in the agriculture value chain. Advices on banking and other financial and insurance services will be relevant and beneficial to all users across the value chain. Also, the system will have the needed intelligence to provide advice with credible statistics to banks, financial service organizations, insurance companies, education and research institutes, media, and other agriculture related vendors who supply agro-chemical and machineries and various other agro inputs.

The authors recommend the development of a flowchart/s showing information flow among various stakeholders in a EPAS system.

4.2.5 Onboarding Agro Support Services

Banking support of agricultural development, project financing contribution from financial institutes, and insurance schemes are very vital elements in the agriculture industry to up lift the sector in developing countries. However, the inadequacy of such services was not seen as an immediate driving force impeding the industry dynamics during the pandemic. However, the appropriate establishment of such services would have provided much resilience to the agriculture sector by empowering the farming community as well as other stakeholders to overcome the challenges more successfully with elevated courage and confidence.

In longer-term perspectives, these support services will play a key role in achieving the national agriculture policy goals, especially a self-sustained crop production in Sri Lanka. The integration of such services in the system will facilitate for all parties credible and expedited actions on field inspections, fact verifications, information validations, processing of supporting documents, and payments on time. Obviously, the module is needed to be developed adhering to the financial and banking regulations of the country according to legal frameworks. The modern-day farming is heavily dependent on fertilizers, chemical pesticides, other inputs, and various agro machineries. Market accessibility to these essential agro inputs were blocked during lockdowns, therefore the direct purchasing option through call ordering facility via essential service channels would have been a relief for farmers since inputs are time critical and weather sensitive in the cultivation process.

The physical disconnection of relationships among communities and institutes created unpredictable delays on activities such as field progress monitoring and verifications and situation analysis of incidents and damages. These delays caused a financial impact for all stakeholders of the value chain. The proposed system provides a platform to receive site images, videos, and audio narratives with live field information to central locations of those institutes assuring the credibility of materials which are verified and endorsed through locally appointed agents. In addition, the system can propose recommendations on behalf of users through bank loans and micro-financing assistance based on the historical credibility and trust developed and maintained within the EPAS. Banks, insurance companies, and other financial institutions can retrieve business intelligent information from the system when necessary for decision making without violating the privacy of subscribers.

4.2.6 Payment Processing and Accounting

Financial transactions on electronic platforms remain less used in many sectors in Sri Lanka. Such transactions are probably the less used in agriculture industry compared to other industries. Defaulted or elongated credit periods among parties were seen common during the pandemic in rural areas due to various reasons described above. Also, the use of physical currency has been a fundamental threat in any industry since currency notes and coins can easily spread the COVID-19 virus.

EPAS proposes virtual accounts attached to subscriptions. For more reliability and confidence, we propose here the implementation of this facility by preferably integrating with a reputed state bank. The system will handle financial accounting of transactions across subscriber base itself in collaboration with the affiliated bank. The consumers, markets, and other retailers could purchase credit vouchers from affiliated bank and maintain EPAS account for purchasing goods and services. The farmers, sellers, and other service providers can convert the saving surpluses of their EPAS accounts into cash through the affiliated bank on their preference and vice versa. Financial services such as payment credit transfers and loans can be facilitated through these EPAS's accounts.

5 Conclusions

The COVID-19 pandemic has obviously decelerated the development of the Sri Lankan agriculture sector. The physical disconnection in value chain is the main issue identified in this analysis, and the authors believe that the situation can be improved through digitalizing value chain activity management and coordination. Even though there are various systems developed and operational in digital platforms, those systems were not adequately efficient in overcoming the challenges induced by this crisis. Assuring fair trading of agriculture products has been an ongoing challenge over years and the pandemic has further deteriorated the situation.

The potential of reaching self-sustain subsistence crop production in Sri Lanka is obviously a feasible target given the country's rich wealth of agricultural resources. The island is situated near equator with blessings of monsoon rains twice a year. However, the young generation's interest in agriculture industry keeps diluting during last decades and such a strategic shift towards new technology adaptation would help regain the youth's attraction to the industry. This will be possible while preserving the historical agricultural values and knowledge the country inherits form its long agriculture history.

The proposed solution in EPAS primarily focuses on providing fair trading platform for farmer communities, consumers, and other players in the agriculture value chain while assuring transparency, trust, and credibility of their transactions. Also, the EPAS system assures the business continuity with higher resilience through insightful coordination and communication across the users of the entire value chain. This helps make an agile and smart balance in supply and demand. In addition to the above main objective, the solution is a strategic adaptation of modern technologies and concepts to bring value added services for users such as dissemination of technology know-how and provision of agro intelligence information to agribusiness communities. Credible and instant financial accounting, linking banking, insurance, and financial support services, connecting agro input suppliers and vendors, and data sourcing for educational institutes and researches are also targeted value additions incorporated in the solution.

Also, this solution would lay a solid digitalization foundation to eliminate major acute issues in any segment of the agriculture value chain. Such a shift towards smart agriculture through this solution would be a strong supporting pillar to achieve targets of envisioning self-sustain agriculture industry in Sri Lanka and going beyond by entering into bigger export market. The system has the potential of seamless integration of overseas supply chain elements widening the export market.

The generous willingness and support of government authorities, institutions, regulatory bodies, universities and private sector are instrumental to make EPAS a functioning reality. Also, the study highlights the timely requirements of cohesive integration of existing agricultural support services and projects developed in e-space into EPAS to bring best integrated and harmonized value chain activity management system assuring fair trading and above-mentioned value addition to agribusiness.

The pandemic has brought some indirect opportunities, especially through an awareness on lost opportunities in developing nations. The interest in self-sustaining agriculture industry has become a top priority in many national agendas, and the EPAS would be a trustful partner in this journey providing the modern and comprehensive digital platform while preserving the traditional agricultural values and knowledge.

This study is limited to the environment analysis and designing of primary building blocks of the EPAS conceptual model through a research design cycle which was established according to DSR methodology as explained above. As a perspective, carrying out both research-based evaluation and research-based pilot feasibility test of the proposed model are future research areas identified by the authors in the research design cycle.

References

Arumugam U, Kanagavalli G, Manida M (2020) Covid-19: impact of agriculture in India. Accessed from file:///C:/Users/pariyarathna/Downloads/SSRN-id3600813%20(2).pdf

Central Bank of Sri Lanka (2020a) Annual report 2019.

Central Bank of Sri Lanka (2020b) Adoption of Modern Technologies in Agriculture. Accessed from https://www.cbsl.gov.lk/sites/default/files/cbslweb_documents/publications/annual_report/2019/en/13_Box_04.pdf

Central Bank of Sri Lanka (2019) Economic and Social Infrastructure. Accessed from https://www.cbsl.gov.lk/sites/default/files/cbslweb_documents/publications/annual_report/2019/en/7_Chapter_03.pdf

Dailymirror online (2020) Redefining e-commerce in Sri Lanka: Prospects post COVID-19. Accessed from http://www.dailymirror.lk/features/Redefining-e-commerce-in-Sri-Lanka-Prospects-post-COVID-19/185-187711

Department of Sensors and Statistics, Ministry of Finance of Sri lanka (2020) National account estimate of 1st quarter of Sri Lanka. Accessed from http://www.statistics.gov.lk/PressReleases/press_note_english

FAO and International Telecommunication Union (ITU) (2016) E-Agriculture Strategy Guide Piloted in Asia-Pacific countries. Accessed from http://www.fao.org/3/a-i5564e.pdf

Ginige A, Walisadeera AI, Ginige T, Silva DL, Giovanni PD, Mathai M (2016) Digital knowledge ecosystem for achieving sustainable agriculture production: a case study from Sri Lanka. In: 2016 IEEE international conference on data science and advanced analytics, pp 602–611

Global Digital Insights (2020) Digital 2020: Sri Lanka. Accessed from https://datareportal.com/reports/digital-2020-sri-lanka

Gunawardana DP (2020) The Impact of COVID19 on the MSME Sector in Sri Lanka. Accessed from https://sustainabledevelopment.un.org/content/documents/26277Report_The_Impact_of_COVID19_to_MSME_sector_in_Sri_Lanka.pdf

Hevner AR (2007) A Three Cycle View of Design Science Research. Accessed from https://aisel.aisnet.org/cgi/viewcontent.cgi?article=1017&context=sjis

Hyea WL, Vikas C (2017) Agriculture 2.0: How the Internet of Things can revolutionize the farming sectors. Accessed from https://blogs.worldbank.org/digital-development/agriculture-20-how-internet-things-can-revolutionize-farming-sector

Manoj, T. and Nimesha, D. (2019). Farm Smart! Developing Sri Lanka's Agriculture Sector in the 4IR. Accessed from https://www.ips.lk/talkingeconomics/2019/10/21/farm-smart-developing-sri-lankas-agriculture-sector-in-the-4ir/

Matthieu DC, Anshu V, Alvaro B (2018) Agriculture 4.0: The Future of Farming Technology: World Government Summit. Accessed from https://www.worldgovernmentsummit.org/api/publications/document?id=95df8ac4-e97c-6578-b2f8-ff0000a7ddb6

National Agriculture Information and Communication Centre (NAICC) Sri Lanka (2016) Sri Lanka E-agriculture Strategy. Accessed from https://www.doa.gov.lk/ICC/images/publication/Sri_Lanka_e_agri_strategy_-June2016.pdf

Sri Lanka Export Development Board (2020) Export performance Indicators 2019. Accessed from https://www.srilankabusiness.com/ebooks/export-performance-indicators-of-sri-lanka-2010-2019.pdf

Teixeira T (2020) Navigating COVID-19 with proactive risk management. Accessed from https://www.cutter.com/sites/default/files/DA_DT/2020/dadtu2003.pdf

Telecommunications regulatory commission of Sri Lanka (2020) Statistics 2020 June. Accessed from http://www.trc.gov.lk/2014-05-13-03-56-46/statistics.html

The Netherlands and you (2020) Impact of COVID19 on food supply chains in Sri Lanka. Accessed from https://www.netherlandsandyou.nl/latest-news/news/2020/06/02/impact-of-covid19-on-food-supply-chains-in-sri-lanka

Chapter 8
Integrated Transition Toward Sustainability: The Case of the Water-Energy-Food Nexus in Morocco

Afaf Zarkik and Ahmed Ouhnini

Abstract Known for being a climate change hotspot, Morocco is at the forefront of a climate disaster. Consequences are already being felt, whether in the form of increasing temperature or a downward trend in precipitations, which directly threaten the water security and, by extension, the social-ecological systems of the country. The systems by which food, energy, and water are produced, distributed, and consumed heavily depend on one another. Their implicit feedbacks and links are not linear, but are rather highly complex, and are at the intersection of numerous disciplines. It is a complicated and delicate equation of multiple inputs that guarantees certain outputs. With the current conditions of water being no longer those that prevailed during the past century, when water resources were much more abundant and much less demanded, it has now become essential to manage it carefully. This chapter first outlines the physical limitations of available resources in Morocco, and determines physical areas of intersection between the food-energy-water sectors. It also explores the existing sectoral policies and adaptation strategies employed in the three sectors in order to evaluate them and assess whether they are evolving separately or co-evolving as part of a nexus system approach. Finally, it investigates seawater desalination as a drought-proof and economic alternative that Morocco can utilize to face climate change-induced water scarcity.

Keywords Morocco · Climate change · Water-energy-food nexus · Water scarcity · Seawater desalination

---

A. Zarkik (✉) · A. Ouhnini
Researcher in Economics, Policy Center for the New South, Rabat, Morocco
e-mail: a.zarkik@policycenter.ma

A. Ouhnini
e-mail: a.ouhnini@policycenter.ma

1 Introduction

Imagine you wake at dawn, switch on the light, turn on the tap to quench your thirst and wash your face, then click on a button on your espresso machine to make a delicious cup of coffee. Each involves a separate action drawing on separate resources. A utility company supplies your electricity via power lines, a water company supplies water via a hydraulic system, and you probably bought your organic fair-trade coffee online. Then imagine a city-wide power failure. Bizarrely, not only will you be unable to turn on the light, but also your tap water may also run dry, you will have to buy ordinary coffee from the store with cash, and when you turn on your espresso machine, nothing. Food, water, and energy systems have become ever more interconnected.

On a more macro level, the systems by which food, energy, and water are produced, refined, distributed, and consumed are intimately tangled and intertwined. To illustrate: energy is needed to pump water; irrigation is required in agriculture; and biowaste can be exploited to make biofuels that could be used to transport food to the market. However, the implicit feedbacks and links between these sectors are not linear, but highly complex, and are at the intersection of bio-geochemical cycles (carbon, nitrogen, phosphorus, among others), the anthroposphere, technology, and physical and social sciences (i.e. economy and politics) (Higgins and Najm 2020). It is a complex and delicate equation of multiple inputs that guarantees certain outputs. As resources become scarce, or if the demand expands uncontrollably, it becomes vital to manage these resources carefully and delicately in order to strike a balance.

In Morocco, it is increasingly challenging to meet the needs of a growing population, especially in conditions, made worse by climate change, that threaten water and food security. The Moroccan population is expected to grow from 36.9 million in 2021 to nearly 46 million by 2050, a 25% increase (UNICEF 2019). In this context, recognizing the dynamic relationship between food, energy, and water systems is essential for the sustainable well-being of future generations. Such a systems thinking will ultimately bring supplies and demands into a sustainable balance.

This chapter outlines the physical limitations of available resources in Morocco in the context of climate risks, to determine areas of intersection between the three sectors (food, energy, and water). It then explores sectoral policies and adaptation strategies employed in the three areas in order to determine whether they are evolving separately or co-evolving as part of a systems approach. It also investigates seawater desalination as a drought-proof and economic alternative that Morocco can utilize to face climate change-induced water scarcity.

2 Climate Change Impacts on Water Resources in Morocco

Located in the Mediterranean area, Morocco is considered a 'hotspot' zone where climate change effects are expected to be intense. Global warming will expose the country to frequent droughts, threatening water and food security. The regions in the country where water demand is highest could experience the largest drop in rainfall.[1] Over the past decades, a pattern of increasing temperatures and a downward trend in precipitation have been observed throughout the country. Intense heat has been particularly noticeable over the past 30 years, with an average rise of + 0.42 °C/decade since 1990—i.e. 0.25 °C/decade above the global trend. There has been an upsurge of the number of annual hot days.[2] Morocco will also experience an increase of 2.3–2.9 °C (around 3 °C) in annual average temperatures by 2050 (Zeino-Mahmalat et Bennis 2012).

Rising temperatures can interfere with the water cycle and alter the delicate balance of evaporation and precipitation (and all of the steps in between). Warmer temperatures increase the rate of evaporation of water into the atmosphere, in effect increasing the atmosphere's capacity to 'hold' water. Increased evaporation may dry out some areas and fall as excess precipitation on others. Warmer air can hold more water vapor, which can lead to more intense rainstorms, causing major problems including overflows and extreme flooding, especially in coastal areas. At the same time, other areas can experience droughts and drier lands and soils, creating a vicious cycle of hardened lands with low absorption rates, resulting in more floods. In the meantime, the water needs of natural systems and agriculture are increasing, and water demand is rising.

Morocco's rainfall patterns are erratic. The county has received an average of 400 mm per year since 1980. This volume is expected to decrease because of climate change-induced drought, associated with the rise of temperature that the country is expecting in the near future. In fact, the FAO predicts a 13% to 30% decrease in rainfall by 2050. A significant decline in cumulative precipitation has already been observed, in the order of −20% of the annual average between 1960 and 2018. When looking at the evolution of the yearly average precipitation over 40 years, the particularly dry nature of the Moroccan climate appears quite clearly.

Morocco is also confronted with dwindling groundwater reserves. Population growth and human activities are increasingly exerting pressure on hydraulic resources. In these constricted conditions, the quantity and quality of available water has seen a sharp decline in the past decades and fresh water is increasingly a cause of rivalries and conflicts over its different uses. Overall, water demand already exceeds renewable water supply.[3]

---

[1] Up to 50% drop in rainfall in the north east region.

[2] For instance, in the regions of Oujda and Marrakech, there has been a staggering six additional hot days per decade since the 1960s. In those regions, the number of annual hot days now exceeds 50, compared to 20 days in the 1960s.

[3] Water supply is estimated at 12.90 billion cubic meters (1 BCM = 1000 l) while water demand is 19.90 BCM (average values from 2010 to 2017). Under existing policy, demand is projected to continue growing from 15.90 BCM in 2017 to 24.4 BCM in 2040. Future water supply–demand imbalances will be 15.68 BCM/year (Hssaisoune et al. 2020).

3 Water Scarcity Challenges for Food Security and Growth in Morocco

3.1 The Water Challenge for Agriculture in Morocco in a Context of a Changing Climate

Climate change will in general have several negative impacts on the agricultural sector. It will reduce yields of major crops and increase the variability of agricultural production. In Morocco, the agricultural sector heavily relies on rain, making it particularly vulnerable to climate change. This exposure stems from the disproportionate volume that cereals occupy within the agricultural sector and its critical role for food security and livestock survival. Wheat and other rain-fed crops are among the crops most affected by climate change. They occupy nearly 60% of useful agricultural land, of which 90% is located in dry zones. This is counterintuitive since grains are almost entirely rain-fed. Consequently, dry years have a significant impact on overall agricultural output. Moreover, cereals' share of agricultural added value (AAV) is disproportionate compared to the area it occupies (16% of AAV compared to 60% of land used) (DEPF 2019).

However, even irrigated lands can be vulnerable to climate change. Irregular weather patterns in 2019, which brought hotter than expected conditions in some citrus-producing areas in the south of the country, caused production of some fruit to fall by 32% in the 2019/2020 season (Oxford Business Group 2020). Moreover, irrigated lands are also an essential component of the agricultural sector.[4] Currently 20% of agricultural land is irrigated,[5] representing an area of more than 1.5 million ha. This is the result of dam and drip-irrigation policies that have mobilized large-scale irrigation infrastructures (Ennabih 2020). In the last two decades, between 51 and 75% of the investment budget for agriculture has been devoted to irrigated areas. The agricultural sector nevertheless continues to grapple with major challenges related to the depletion of water resources: while the surface of the irrigated perimeters fitted out by the state expands by an average 2.3% a year, supplies of water fall by 2%, on average, a year.

The World Bank and the FAO carried out a first study in 2009 on the impact of climate change on the agricultural sector in Morocco, in collaboration with national institutions (Gommes et al. 2009). It showed that in the next decades, water stress will persist and increased aridity will particularly reduce agricultural yields, especially from 2030 onwards. Two scenarios of climate change future impacts were developed,[6] relating to six groups of crops. Rain-fed crops will be particularly affected

---

[4] Irrigation accounts for 99% of sugar production, 82% of vegetable output, 100% of citrus production, and 75% of milk production.

[5] A significant improvement compared to 2010, when this number stood at 4.55%, according to the FAO.

[6] Scenario A1 is a pessimistic scenario which describes a world in which the global population increases rapidly, with strong economic growth based on polluting technologies in a world that has

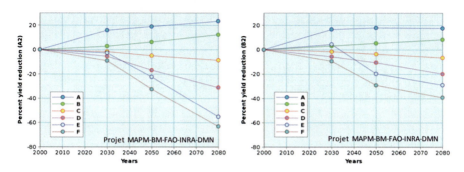

Fig. 1 Percentage of yield reduction, according to scenarios A2 and B2, by 2100.[7] *Source* Gommes et al. (2009)

by climate change effects (Fig. 1). Agricultural yields are expected to remain more or less stable up to 2030, and then will drop rather quickly beyond this date, more markedly in the case of scenario A2 than in that of scenario B2 (see footnote 6). Not all crops will be equally vulnerable to the phenomena, and if irrigation water continues to be available in sufficient quantities, irrigated crop yields will continue to increase in spite of climate change.

In line with the above projections, the yields of the main crops are likely to face significant variations, and while 81% of arable land is still not irrigated, climate change effects are expected to reduce sharply the harvests of rain-fed crops that are predominant countrywide, where irrigation is not present to alleviate adverse climatic conditions.

3.2 Towards a Persistent High Dependency on Critical Food Imports

Cereals are the most widely cultivated crops in the world and an important component of food security policies. Morocco consumes more imported than domestically produced cereals, with a cereal dependency ratio that has exceeded 40% during the last decade, two-and-a-half times higher than the world average of 16%. This high

---

become more protectionist, with increasing disparities between North and South. There is continued use of fossil fuels and uneven regional economic growth. Scenario B2 is an optimistic scenario that describes a world in which the focus is on local solutions, from the point of view of economic, social, and environmental viability. The world population increases in a continuous way, but at a slower rate than in A2. There are intermediate levels of economic development, and technological advances are slower and more varied.

[7] Adaptation due to current technology trends is not taken into account here. The crops are gathered into "impact groups" shown as A to F which can be characterized as follows: A: irrigated maize and irrigated seasonal vegetables; B: irrigated fruits and vegetables; C: fodder crops and vegetables; D: rain-fed cereals and legumes; E: rain-fed wheat and barley; F: Other rain-fed crops.

dependency ratio is explained by the vulnerability of rain-fed cereal crops in the country. The ratio of imports to exports in the country is around 54% and about 20% of export revenues are spent on food imports, which is about four times higher than the world average (Ghanem 2015). High and volatile world prices pose a particular challenge to net food-importing countries such as Morocco.

In normal times, high dependence on imported food is not necessarily critical if a country has sufficient export revenues to cover its food import bill. However, high import dependence poses particular challenges for Morocco as the demand for food imports (particularly cereals) is highly inelastic, meaning the country is unable to reduce imports in response to price increases and, therefore, has to bear the full impact of high prices (El Youssfia et al. 2020).

Morocco has been able to meet its food needs through a combination of domestic production and imports. Moreover, policies to address climate change effects on agricultural performance are already being put in place to prepare for future constraints on agricultural productivity. Moroccan agriculture has adapted through diversification, promoting the expansion of high-value irrigated crops to the detriment of cereal production, even though cereal production remains predominant and vulnerable to climate variability. There has been an increasing trend towards the development of horticulture and livestock production under the previous agricultural policy, *Morocco Green Plan* (2008–2020), through a package of incentives for farmers. Those incentives play a significant role in raising the competitiveness of the Moroccan agriculture as a key growth driver sector. They are also expected to be maintained in the new Moroccan agricultural policy called Generation Green 2030 that aims to upscale small farming systems and foster youth agricultural entrepreneurship.

3.3 Structural Drought is Reducing Morocco's Economic Growth

Given that AAV to GDP has been around 12.5% on average for the past twenty years (World Bank database), through its macroeconomic transmission, drought can significantly impact Moroccan GDP.[8] The impact of the variation in cereal production during years of severe drought induced a sharp decline in GDP in those particular years. In 1995 (a year of drought), an 81% drop in cereal production led to a significant 41% drop in AAV. This contributed to an overall GDP decline of 5% that year. In 2016, however (comparable to 1995 in terms of precipitation), a 71% drop in cereal production only generated a slight decrease of 13.7% of AAV, and therefore influenced GDP to a lesser degree (World Bank database).

On the other hand, although agriculture accounts, on average, for 12.5% of GDP, its socio-economic impact is far greater. Agriculture provides about 38% of national

---

[8] According to the world bank, the average value of agriculture as a share of GDP from 1965 to 2019 was 15.28%, with a minimum of 10.68% in 1981 and a maximum of 23.45% in 1965. For comparison, the world average in 2019 based on 161 countries was 10.19% (World Bank database).

employment and nearly 74% of rural employment, with smallholders representing almost 70% of the agricultural workforce. At the same time, unpaid employment of family workers represents more than 50% of the agricultural workforce (2019 figures). Furthermore, a significant share of harvested crops is still used for self-consumption (CESE 2019).

Given the nature and structure of farming in Morocco, a drought-induced agricultural shock will automatically turn into an income shock for farmers, in the form of reduced crops for self-consumption or reduced income normally derived from the sale of the remaining share. A drop in farmers' incomes would lead to a drop in all other activities (agricultural and non-agricultural), especially locally, given the weight of agriculture in the economy. On a larger scale, an agricultural shock can effectively provide an added shock to GDP through income formation and household consumption (DEPF 2019) (Figs. 2 and 3).

In sum, the consequences of water scarcity in Morocco are very dire: as surface and groundwaters dry up, so do livelihoods. Water insecurity represents a threat to the socio-economic fabric of the country and can lead to spikes in food prices and social unrest. That tension can erupt in ways that governments cannot contain, such as in the 'onion demonstrations' in India (2013), 'pasta protests' in Italy (2007), and 'tortilla riots' in Mexico (2007) (Jones et al. 2017).

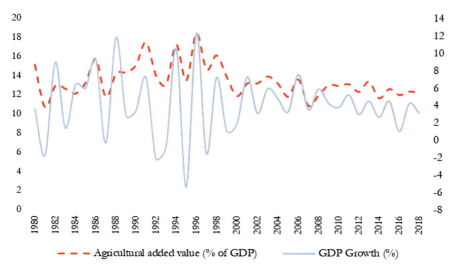

Fig. 2 GDP growth rate and AAV (% GDP). *Source* World Bank Database

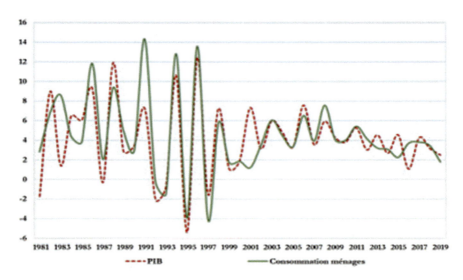

Fig. 3 GDP growth rate and household consumption. *Source* DEPF (2019)

4 How Does Agriculture Contribute to Water Scarcity?

Agriculture can also have a negative impact on water supply and water quality. Perhaps the most important impact linked to irrigation in Morocco is the overexploitation of aquifers (around 1 BCM/year on average), which causes a worrying drop in water tables. As mentioned earlier, groundwater supplies 40% of the overall water used for irrigation. This proportion exceeds 70% in the southern basins. Groundwater accounts reveal that no less than 4.22 BCM/year of groundwater is extracted, against a renewal potential of about 3.40 BCM. This indicates an annual groundwater deficit of about 862 MCM for the entire country, but the situation varies between different basins and from north to south. This imbalance is responsible for a noticeable decline in groundwater levels (20–65 m) over the past 30 years for the totality of the basins, and a worrying drop in water tables that may exceed 1−2 m/year.

This depletion is caused by two main factors: overexploitation and degradation are the main contributor with numerous consequences, including drying up of water resources, disappearance of natural lakes, the irreversible degradation of oasis ecosystems, and desertification; and to a lesser degree, climate change-induced drought, which causes low natural recharge.

Moreover, total water withdrawals in Morocco are divided between three main usages: irrigation (87%), industry (6%), and potable water (8%), making agriculture the bulk consumer of water resources. Agricultural practices in the country are unvaryingly described as wasteful, most of the blame being put on traditional gravity irrigation, which wastes large amounts of water and does not allow accurate evaluation of the quantities applied. Overall, the efficiency of water use in irrigation, including the conveyance of water and application on the fields, does not reach 50%, which means that less than half of the water delivered to farms currently reaches the crops.

5 Morocco's Agricultural Policies at the Forefront of Climate Change Adaptation

Agricultural policies in the country, over the last decades, have mainly focused on the establishment of hydraulic infrastructure for the storage of water and on crop irrigation. Since the 1960s, a continuous hydraulic development effort has made it possible to have a portfolio of large reservoirs behind dams. These hydraulic infrastructures play an essential role in the economic development of the country. They contribute mostly to the development of access to drinking water, protection against floods, stabilization of agricultural production through the irrigation of more than one million hectares, and, to a lesser extent, electricity production.

The 2009 National Plan to Fight Against Climate Change provides a detailed list of measures, including the assessment of the impacts of climate change on different crops, which aims to strengthen farmers' adaptation capacities through identification and development of appropriate technical, institutional, and policy options, in particular the application and adaptation of direct seeding techniques in semi-arid zones[9]. The country has also implemented programs in different regions to enable dissemination in rural areas of selected varieties of crops, supplemental irrigation, and direct seeding techniques, which help improve the efficiency of the use of rainwater, increase the presence of organic matter in soils, and prevent their erosion. Other initiatives have already been put in place in terms of adapting to climate change in agriculture, including raising awareness about the impact of climate change on agriculture, adapting agricultural techniques to aridity (soil and water conservation), and securing agricultural income through climate risk insurance.

Moreover, to face the challenges of water scarcity, the country set out measures in Morocco Green Plan to mobilize financial and technical resources with the aim to modernize irrigation practices by encouraging the adoption of drip irrigation. As of early 2020, as much as 585,000 ha had been equipped with drip irrigation systems across the country, facilitating effective water savings of over 1.6 BCM per year (as of 2018) (Oxford Business Group 2020). These achievements would benefit, however, from being strengthened and generalized, since a large area is still irrigated essentially through gravity irrigation (nearly 63%) (DEPF 2019).

Aware of the vulnerability of small-scale agriculture to climate change, the Ministry of Agriculture integrated the climate change component into Pillar II projects of the Morocco Green Plan. The Project for the Integration of Climate Change in the implementation of Morocco Green Plan (PICCPMV) is one of these projects which helps strengthen the capacities of small farmers in terms of adaptation

---

[9] ACCAGRIMAG project funded by ehe French Facility for Global Environment (FFEM), https://www.ffem.fr/en/international-partner-working-global-environment-and-sustainable-development.

to climate change through the dissemination of appropriate technologies that have been developed by national agricultural research institutions. Additionally, the fifth building block of the Morocco Green Plan supports the use of renewable energies in the agricultural sector (solar and wind energy, biofuels).

However, despite big effots deployed in this field, the implementation of some of the policies described above have to be reinforced. For instance, despite significant investment, the agriculture sector remains very dependent on rainfall. At the same time, despite scarcity of water resources, Morocco is still encouraging the expansion of irrigated areas. Irrigation network expansion might not be an adequate solution since water is already used beyond renewal levels in many basins, and agriculture, which uses 87% of water resources, faces competition from industrial and urban demands. In marginal areas, drought and medium to long-term declines in agricultural yields will affect the livelihoods of rural communities, making poverty reduction more difficult. Lower and more variable yields will translate into greater dependence on imported food, making Morocco more vulnerable to increasingly volatile international food prices.

Box 1 National Program for Irrigation-Water Saving and the Solar Pumping Project (an example of the WEF nexus)

The National Irrigation-Water Saving Program, a part of the 2008–2020 Morocco Green Plan, was initiated to allow irrigated agriculture to cope with water scarcity and to make the best use of the country's limited water resources. One of the hallmarks of this program was to promote drip irrigation as an efficient alternative to surface irrigation. The program's broader goals aimed at modernizing irrigation networks around large dams, equipping agricultural properties with localized irrigation with financial aid from the State, supporting farmers in their irrigation water system enhancement, introducing high value-added cropping systems, and capacity building (farmers, professional organizations, technical services, etc.)

The energy component of the WEF Nexus can be seen at the level of water pumping for irrigation. The first solar pumping project was launched in 2013, later supplanted by the Global Environment Facility (GEF) solar pumping. The project was intended to promote photovoltaic-energy based pumping systems for irrigation in the agricultural sector, as a partnership between the United Nations Development Program (UNEP) and the Moroccan Agency for Energy Efficiency (AMEE). This program was created to deal with several anomalies, including the use of subsidized butane in water pumping systems for agriculture or clean water which put additional projects on the Compensation Fund. The program is also part of Morocco's ambitious plan to source 42% of its power needs from renewable energies by 2020, and 52% by 2030

6 Water and Energy Feedbacks in Morocco

Agriculture is not the only sector influenced by water scarcity. In fact, water is essential throughout the cycle of providing energy, from mining to direct power generation from coal, natural gas, oil, biomass, and concentrating solar power (CSP) plants. Given its lack of energy natural resources, Morocco depends on imports for most of its primary energy needs. Thus, we will not dwell much on the mining part, but rather focus on water and energy feedbacks in power generation.

Electricity represents around 20% of primary energy in Morocco. Faced with risks related to the increasing cost of fossil-fuel imports and the subsequent high burden on public finances, Morocco adopted a new energy strategy in 2009 that was based on two main axes: strengthening power generation capacity through renewable energy deployment and energy efficiency. Therefore, Morocco planned to increase the share of renewables in its power generation mix to 42% by 2020 and 52% by 2030. Moreover, the agency for energy efficiency, AMEE, identified an energy savings potential of 12% by 2020 and 15% by 2030.

By the end of 2020, the electricity mix was still dominated by thermal energy which comes in part from steam thermal (44%), such as fuel and coal, and other thermal power plants (21%) such as gas, combined cycle and diesel. Renewable energy accounted for 35% of installed capacity, hydropower stations and pumped storage power stations (STEP) accounted together for 1770 MW—i.e. 17% of total share—while solar energy capacity reached 711 MW in 2019—i.e. 6.7% of total capacity—and wind reached 1220 MW—i.e. 11.4% of total capacity (ONEE 2019). This means that Morocco missed its 42% target.

Water requirements of power plants depend on their efficiency. The literature is not unanimous on water requirement for each type of power plant. However, to have an order of magnitude, in inefficient fuel power plants, cooling water ranges between 3,000 and 7,000 L/MWh. This compares to 2,000 L/MWh for new coal-fired power plants and 1,000 L/MWh for more efficient natural gas combined cycle power plants (Rodriguez et al. 2013).

For its solar energy technology choice, Morocco has mostly pursued CSP plants instead of photovoltaics, as a dispatchable form of solar energy with consequential storage capacities. The country now counts in total four CSP power plants either commissioned or still under construction, and one PV plant under construction with a total capacity of 2000 MW according to the Moroccan Agency for Solar Energy (MASEN).[10]

Some renewable energy technologies such as wind and photovoltaics require very little water. However, as a thermal energy generating power station, CSP has more in common with thermal power stations such as coal, gas, or geothermal. In fact, in terms of water consumption, CSP plants spark some controversy since they are usually located in areas where water is scarce because of their radiation needs. Moreover, because of the lower efficiency of solar thermal plants compared to conventional

---

[10] https://www.masen.ma/sites/default/files/documents_rapport/Masen_NOORM_SESIA_NTS.pdf.

fossil fuel power plants, larger amounts of water are required for cooling purposes. Some concentrating solar technologies need to withdraw as much as 3,500 L/MWh generated (World Bank 2013) (Rodriguez et al. 2013).

According to the Moroccan Agency for Sustainable Energy (MASEN), the water needs of its Noor Midelt hybrid CSP and PV plant currently in development will require more than 2 billion liters of water in its lifetime (2 million cubic meters): 300 million liters during the whole construction phase and 70 million litter of water a year during the plant's 25-year operating phase, used mainly to top up the steam cycle and wash the CSP (MASEN 2019).

While the large-scale implementation of renewables such as CSP can have a positive impact on greenhouse gas emissions and can improve energy security, it could negatively impact water resources.

Moreover, thermal electric power plants and CSP plants alike can have an adverse effect on water quality as they alter water temperature and cause thermal pollution and changes in oxygen levels, which is why these water streams should be treated before being returned to their water source (Rodriguez et al. 2013).

Additionally, Morocco should commit further to energy efficiency, which could be an important lever in water saving. Its objectives were difficult to achieve given the limited human and financial resources of AMEE, in addition to the legislative and regulatory texts that are slow to emerge. For example, the Decree on the mandatory energy audit was not published until 2019, 10 years after the adoption of Law 47–09 on energy efficiency.

Beyond the impact of energy technologies on water resources, water-related climate impacts will affect the energy sector. Water scarcity is already exerting a destructive effect on energy production and reliability in Morocco; further constraints may call into question the physical, economic, and environmental viability of future projects. For instance, hydropower is especially vulnerable to water scarcity. Morocco has been investing in hydropower generation since the 1960s, and the segment accounted for 1770 MW in 2019, or about 17% of total installed capacity. In addition to being a key pillar of clean energy transition, hydropower is also an important source of system flexibility that can enable higher shares of more variable renewables. Unfortunately, according to the International Energy Agency (IEA), Morocco will certainly undergo a significant drop in load factors of hydropower plants between 2020 and 2090. Another challenge facing Morocco in the future is the increased year-to-year variability in hydropower capacity factors caused by climate change (IEA 2020). Oddly, while IEA estimates that this energy source will only represent 12% of the mix as early as 2030, Morocco's energy strategy plan aimed at stretching its part of the power generation mix to 20% (an objective that should have been reached in 2020)(Fig. 4).

Hydropower also comes with its fair share of environmental risks. Although hydropower plants do not consume most of the water, but merely divert it to generate electricity, they change the distribution and movement of groundwater in the area in which they are located by converting free-flowing rivers into reservoirs, thus altering the timing and flow of the water. This stored water affects water quality and aquatic life, as rivers and lakes can fill with sediment and baseline nutrient

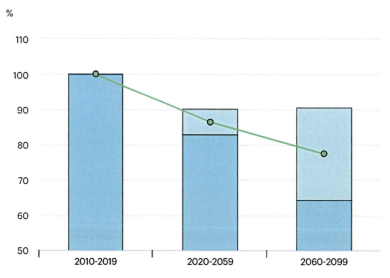

Fig. 4 Changes in hydropower capacity factor in Morocco, 2020–2099, relative to the baseline 2010–2019. *Source* IEA (2020)

levels can be altered. Moreover, water rushing through turbines can increase the presence of dissolved oxygen in the water, affecting the ecosystem. Hydropower stations may also alter water quality because they slow the river's flow by dams, thus potentially increasing the temperature of the water body. According to the United States Geological Survey (USGS), reservoirs also contribute to water loss through evaporation, unlike flowing rivers.

Finally, diminished freshwater resources can lead to a greater reliance on energy-intensive sources of water supply, such as desalination (Walton 2020).

7 Seawater Desalination as a Promising Solution for Water and Food Security in Morocco

Our overall examination of the WEF sectors in Morocco revealed two interesting findings. First, although agriculture, water, and energy are all in a transition phase, significant efforts are still required to consider the dynamic interactions between them in order to reach a co-evolution that guarantees greater benefits, ranging from economic and resource efficiency to greater innovation. This will further legitimize and reinforce their reciprocal development. Second, Morocco law implementation has to be enhanced. Although climate change-induced drought and water scarcity are accounted for in agricultural transition plans via the National Plan to Fight Against Climate Change and the Morocco Green Plan's National Irrigation-Water Saving Program, these policies need to be reinforced. Indeed, despite significant investment, agriculture remains heavily dependent on rainfall. In fact, some of those projects have had the opposite desired effect, for even with water being used beyond its renewal levels, the expansion of irrigated crops is still encouraged. Finally, the law on energy efficiency could have had a tremendous water saving impact, however results are still ambiguoys.

Therefore, as climate change significantly increases water deficits, the need for additional water resources for industrial and agricultural purposes has become a necessity in Morocco. In the near future, construction of dams may no longer be a viable solution to increase water *per capita* to meet domestic water demand for freshwater and irrigation, especially under current and future climatic conditions. The construction of a large number of hydraulic devices for water storage and irrigation does not automatically prevent the country from running out of water, as drought has become structural.

However, with a coastline of 3500 km, abundant wind and intensive solar radiation, seawater desalination combined with renewable energy is now perceived as the main drought-proof alternative so that Morocco can face climate change effects and meet water and food security challenges. The supply of desalination plants with renewable energy represents a pioneering solution to reduce production costs, making desalination more economically feasible in a country with great renewable energy capacity, such as Morocco.

The first desalination plants were installed in 1975 to address the shortage of potable water in southern regions.[11] Subsequent schemes have all contributed to water supplies in southern Moroccan areas with shortages of fresh water and insufficient brackish water availability. Small-scale desalination plants were established later in 1995 and Morocco had completed 15 (mostly relatively small) installations by 2016, reaching a desalination capacity of 132 Mm^3/year (Global Water Intelligence 2016; World Bank 2017). Morocco is preparing to build major seawater desalination infrastructure, in close collaboration with public institutions including the Office National de l'Electricité et de l'Eau Potable (ONEE)—the Moroccan National Drinking Water Authority—and certain ministries such as the Ministry of Agriculture and Fisheries, in relation to desalination projects intended for irrigation.

As part of its sustainable development strategy ahead of COP 22 (2016, in Marrakech), Morocco chose seawater desalination as one of the solutions to deal with the shortage of water in certain regions. Changes in legislation in 2016[12] were intended to support such large-scale desalination of seawater projects to tackle Morocco's strategic challenges in water and food security. The National Water Plan (NCP) foresees the construction of seawater desalination plants to produce nearly 515 million cubic meters per year in 2030, and Morocco is currently establishing Africa's biggest seawater desalination station in Agadir using reverse osmosis (RO) technology, to be coupled with a wind power plant, with a capacity of 275,000 m^3 per day, expandable to 450,000 m^3 to serve both daily drinking water and irrigation needs. Another seawater desalination plant intended to provide irrigation for 5,000

---

[11] The first desalination plant was built in Tarfaya using electro dialyses (ED) technology.

[12] Water Act No. 36–15 of 25 August 2016.

farmers (business farmers with contracts with the Agricultural Development Agency) was launched in 2015 in Dakhla region, where groundwater depletion is very critical, reducing the potential for growing high-value early season crops usually intended for exports.

Though seawater desalination through renewable energy seems to be the go-to solution to deal with water scarcity in Morocco, it could be questionable in terms of its contribution to food security in the country, as this technology is intended to promote mainly agricultural exports through high-value crops that require high water consumption. Indeed, the focus is still on the development of an export-oriented agriculture, while the cereal sector on which Morocco depends for its food security, remains unstable and highly dependent on rainfall. Moreover, desalination is potentially harmful to the environment. The United Nations Principles for Responsible Investing (UNPRI) expressed concerns about the implications of seawater desalination on sea life, including from concentrated brine after desalination that are usually discharged back into the same body of water that serves as the intake source. However, the environmental impacts, of brine especially, can be fully controlled by designing long-term and sustainable solutions for management of discharges. Finally, water from desalination plants is often criticized for being more expensive than freshwater, however when freshwater becomes scarce, its cost will eventually go up, making desalination increasingly economically viable. In the meantime, desalination will send the right price signal to promote water conservation.

Given the issues raised above, while seawater desalination is an excellent solution when faced with the imminent threat of water scarcity, other policies such as water conservation, reuse and storage must be reinforced simultaneously when it comes to increasing water supplies.

To quote His Majesty King Mohamed VI: *"The time has come for us to radically change our perceptions and our attitudes towards water, through managing the demand for this resource and the rationalization of its consumption. Moreover, it is necessary to follow up on efforts that are engaged to mobilize all the water resources possible"*.[13]

References

Abdellatif A, Taoufik O, Aziz L, Karima T, Antoine G, Marie-Noelle W, Said EK, Khalid ER, Idriss HM, Riad B (2020). Le Maroc à l'épreuve du changement climatique: situation, impacts et politiques de réponse dans les secteurs de l'eau et de l'agriculture. Poliy Brief. Direction des Etudes et des Prévisions Financières (DEPF), Ministry of Finance of Morocco. https://www.finances.gov.ma/Publication/depf/2020/PolicyBrief18.pdf

Conseil Économique, Social et Environmental (CESE) (2019) Annual report 2019

Direction des Etudes et des Prévisions Financières (DEPF) (2019) Le secteur agricole marocain : Tendances structurelles, enjeux et perspectives de développement. Direction des Etudes et des Prévisions Financières, Ministry of Economy and Finance of Morocco

---

[13] The speech of His Majesty the King Mohammed VI at the opening of the 9th session of the High Council for Water and Climate (Morocco, 2013).

El Youssfia L, Doorsamy W, Aghzar A et al (2020) Review of water energy food nexus in Africa: Morocco and South Africa as case studies. E3S Web of Conferences 183, 02002. https://www.e3s-conferences.org/articles/e3sconf/pdf/2020/43/e3sconf_i2cnp2020_02002.pdf

Ghanem H (2015) Agriculture and rural development for inclusive growth and food security in Morocco. Working paper 82. Global Economy and Development at Brookings

Gommes R, El Hairech T, Rosillon D, Balaghi R, Kanamaru H (2009) Morocco study on the impact of climate change on the agricultural sector

Higgins CW, Najm MA (2020) An organizing principle for the water-energy-food nexus. Sustainability 12(19):8135. https://doi.org/10.3390/su12198135

International Energy Agency (IEA) (2020) Climate Impacts on African Hydropower. https://iea.blob.core.windows.net/assets/4878b887-dbc3-470a-bf74-df0304d537e1/ClimateimpactsonAfricanhydropower_CORR.pdf

Mohammed H, Lhoussaine B, Abdelfattah S, Ilham B, Abdelghani C (2020) Moroccan groundwater resources and evolution with global climate change. Geosciences 10(2):81. https://doi.org/10.3390/geosciences10020081

Moroccan Agency for Solar Energy (MASEN) (2019) Specific environmental and social impact assessment for solar power plant project NOOR MIDELT I. https://www.masen.ma/sites/default/files/documents_rapport/Masen_NOORM_SESIA_NTS.pdf

Office National de l'Électricité et de l'Eau Potable (ONEE), Branche Electricité (2019) Annual Report

Oxford Business Group (2020) Green Morocco Plan increases Morocco's agricultural output. https://oxfordbusinessgroup.com/overview/bearing-fruit-sector-development-plan-makes-concrete-gains-challenges-reaching-full-potential

Rodriguez Diego J, Anna D, Pat D, Antonia S (2013) Thirsty energy. Water papers, World Bank, Washington, DC. https://openknowledge.worldbank.org/handle/10986/16536

United Nations Children's Fund (UNICEF) MENA GENERATION 2030 report (2019) Investing in children and youth today to secure a prosperous region tomorrow. Morocco Country Fact sheet https://www.unicef.org/mena/media/4141/file/MENA-Gen2030.pdf

Walton M (2020) If the energy sector is to tackle climate change, it must also think about water. International Energy Agency (IEA). https://www.iea.org/commentaries/if-the-energy-sector-is-to-tackle-climate-change-it-must-also-think-about-water

World Bank (2013) http://www.fao.org/nr/climpag/pub/FAO_WorldBank_Study_CC_Morocco_2008.pdf

World Bank (2017) Managing urban water scarcity in Morocco. https://openknowledge.worldbank.org/bitstream/handle/10986/29190/122698-WP-P157650-SummaryReport-Urban-water-scarcity-in-Morocco-ENG-P157650-2017-12-25-0412.pdf?sequence=1&isAllowed=y

Zeino-Mahmalat Ellinor et Bennis Abdelhadi (2012) Environnement et Changement Climatique au Maroc Diagnostic et Perspectives. Konrad-Adenauer-Stiftung, https://www.kas.de/c/document_library/get_file?uuid=9d3195fc-dc16-9338-d9a5-39b16de7e3ff&groupId=252038

Chapter 9
Food Security in the Kingdom of Saudi Arabia Face to Emerging Dynamics: The Need to Rethink Extension Service

Mirza Barjees Baig, Khodran H. AlZahrani, Abdulmalek A. Al-shaikh, Ali Wafa A. Abu Risheh, Gary S. Straquadine, and Ajmal M. Qureshi

Abstract The Kingdom of Saudi Arabia is one of the world's most water-stressed countries. The Kingdom's harsh physical environment is characterized by high temperatures, low rainfall, limited arable land, lack of fertile soils, and diminishing water resources. In addition, over-extraction of underground water and population growth also continue to stress existing water resources. Specifically, climate change is expected to unfavorably impact the already depleted water resources and cropping systems. Unfortunately, the country is beset by prolonged droughts and devastating floods. As a result, the potential for increasing agricultural production to meet rising food demands is extremely limited, although it is being currently met through imports. While it is true that food import is the only alternative, uncertainty and unforeseen risks are a growing concern. Therefore, analyses of the prevailing situation indicate that the issues of climate change and food security have to be a top

---

M. B. Baig (✉) · A. Wafa A. Abu Risheh
Prince Sultan Institute for Environmental, Water and Desert Research, King Saud University, Riyadh, Kingdom of Saudi Arabia
e-mail: mbbaig@ksu.edu.sa

A. Wafa A. Abu Risheh
e-mail: aliwafa@ksu.edu.sa

K. H. AlZahrani
Department of Agricultural Extension and Rural Society, College of Food and Agriculture Sciences, King Saud University, Riyadh, Kingdom of Saudi Arabia
e-mail: Khodran@ksu.edu.sa

A. A. Al-shaikh
Director - Prince Sultan Institute for Environmental, Water and Desert Research, General Secretary of the Prince Sultan Bin Abdulaziz International Prize for Water, King Saud University, Riyadh, Kingdom of Saudi Arabia
e-mail: aasheikh@ksu.edu.sa

G. S. Straquadine
Utah State University (Eastern), Logan, UT, USA
e-mail: gary.straquadine@usu.edu

A. M. Qureshi
Senior Associate, Harvard University, Asia Centre, Cambridge, MA 02138, USA

Senior Advisor/Honorary Professor, Chinese Academy of Agricultural Sciences, Beijing, China

© The Author(s), under exclusive license to Springer Nature Switzerland AG 2022
M. Behnassi et al. (eds.), *Food Security and Climate-Smart Food Systems*,
https://doi.org/10.1007/978-3-030-92738-7_9

priority for the Kingdom and these topics need greater attention by all stakeholders. The role of appropriate extension and education systems/methodologies is crucial when addressing the impacts of climate change on agricultural and food systems – technology alone cannot solve this situation. There is urgent need to reform and upgrade the agricultural extension system so that it is better equipped to respond to the growing needs of farmers. With this perspective, this chapter provides ample evidence that strengthening the agriculture extension system through sound scientific curriculum will help realize food security and address climate change. Workable strategies are outlined to assist policy makers and planners to strengthen and make the extension system more productive and efficient in today's context.

Keywords Extension education system · Greenhouse crops · Climate change · Natural resources · Environmental issues · Ecosystems management

1 Introduction

The Kingdom of Saudi Arabia (KSA), as one of the world's most water stressed countries (Baig et al. 2019), cannot produce enough food to meet domestic demands due to its harsh and hot climate, low rainfall, water shortage, and limited arable land and fertile soils. Central to these limitations is diminishing non-renewable water resources (Baig et al. 2017). The review of existing reports and scientific literature points to the fact that the agricultural sector has evolved through many phases. Prior to the 1970s, food production was restricted by high temperatures and scarce water resources in the desert landscape. Agriculture-related activities occurred on a small scale in scattered geographic regions. In the 1980s, the Kingdom began to modernize its agricultural system with the goal to achieve food self-sufficiency by offering farmers interest-free loans, free land, and access to water for irrigation. This intensive farming policy, backed by attractive incentives, generated a substantial water deficit in the Kingdom (Al-Shayaa et al. 2011; Baig and Straquadine 2014; Baig et al. 2017). As a result, two-thirds of the ground water supplies were consumed by the nation's agriculture (Al-Otaibi 2015). Consequently, in the 2000s, in order to conserve water resources, a strategic move was made towards sustainable agricultural practices. The government encouraged farmers to stop planting crops with high water needs such as wheat, barley, and green fodder (GCC Food Industry 2017; Baig et al. 2017).

As the situation presents itself today, extension services are not adequately equipped to guide greenhouse farmers and are unable provide them with the needed information, particularly about marketing their commodities. Agricultural products need specific value chains with the longer shelf life of perishable commodities. Entire production systems have changed in the country due to the changing structure of agriculture and agribusiness. Hence, a complete overhaul and reform of the extension system are needed to address the emerging food security challenges. More specifically, the curricula need to be redesigned with a view to incorporate the climate change imperative so as to empower the extension workers as agents of change.

Fig. 1 Saudi Arabia on the map. *Source* FAOSTAT (2021)

Recommendations have been made for policy makers, planners, and academia to help reform the entire fabric of outmoded extension services.

The coordinates of the Kingdom are presented in the Fig. 1.

2 Overview of Challenges Faced by Agriculture in KSA

Agriculture in the Kingdom faces daunting challenges such as: climate change; over-extraction from aquifers and diminishing renewable water resources; low knowledge of environmental pollution; population growth; rising living standards; changing dietary habits; changing cropping patterns; and limited fertile arable land. The following sections analyze these challenges in order to propose workable solutions. However, factors – such as limited agricultural resources, over-reliance on food imports, generous subsidies, trade policies, water scarcity, food handling losses and the negative effects of climate change—threaten the Kingdom's initiatives aiming at achieving sustainable food security (Almazroui et al. 2017; Haque and Khan 2020; MEVA 2020).

3 Evolving Phases of Agriculture Sector in the Kingdom

3.1 Food Self-sufficiency—A Prime Challenge

Since the 1970s, the government launched many programs to encourage farmers to produce wheat (Karam 2008; Baig and Straquadine 2014; Baig et al. 2017) with the objective to meet domestic food demands. Several reasons, such as a rapidly increasing population, rising living standards, and the instability of food imports, forced the Kingdom to take immediate steps and move towards realizing self-sufficiency in agricultural production in the 1980s. Important measures, attractive incentives on fertilizers and water for irrigation and a 45% reduction in the cost for farm machinery further helped the Kingdom's cereal sector grow and prosper (Lovelle 2015).

In addition, farmers received substantial government support—directly through aid programs and indirectly through policy enhancement (Baig and Straquadine 2014). The government bought wheat at a high price and sold wheat at a much lower price. Farmers were paid approximately three times the world market price for wheat (Al-Shiekh 1998). The government was spending five times the market price to produce a tonne of grain (Nations Encyclopedia 2020). Saudi farmers were provided with a guarantee that the Saudi government would buy domestic wheat for SR 3,500 ($933.5) per tonne (Karam 2008).

The Kingdom achieved self-sufficiency in many food products, thus becoming the sixth largest exporter of wheat, but to the detriment of its water resources (Woertz 2011; Taha 2014; FAO 2016). The grain subsidy program led to poor water management and overuse of fertilizers, which ultimately led to the depletion and contamination of the groundwater supply. The intensive extraction of groundwater from non-renewable aquifers in recent decades has resulted in a significant reduction in groundwater level and the deterioration of groundwater quality.

By 1993, the country fully recognized the adverse effects of its wheat subsidy program on its meager water resources. Having a strong concern over the depletion of local water reserves, a decree was issued by the government in 2008 with the goal to phase out domestic wheat production each year and to completely halt wheat subsidies by 2016. In response to the decree, the hectares planted with wheat gradually decreased from 450,000 hectares in 2008 to around 83,000 hectares in 2014 (GIEWS FAO 2014). A similar decree was adopted to phase out fodder production by 2019, as some farmers had moved from wheat to even more water demanding forage crops (GIEWS FAO 2020). Table 1 shows the impacts.

One thousand tons of water (1000 cubic meters) are needed to produce one tonne of wheat (Allan 2001: 39). The water saved from the abandonment of the cultivation of 2.613 million tonnes of wheat and barley between 1993 and 1999 was used to produce 1.2 million tonnes of alfalfa. Six times more water is needed to produce alfalfa than to grow wheat (Altukhais 2002: 12; Elhadj 2004).

As the wheat subsidies program was completely terminated in 2016, the Kingdom offered subsidies to dairy farmers (another water-intensive industry) and sheep

Table 1 Cereal production in Saudi Arabia

Crops	2015–2019 (Average) 000 tons	2019	2020 (Estimated)	Percent Change
Wheat	163	200	500	150.0
Sorghum	150	148	152	2.70
Maize	85	89	88	−1.10
Others	29	33	40	21.20
	427	470	780	66.00

Note Percentage calculated from the unrounded data
Source FAO/GIEWS Country Cereal Balance Sheet (FAO-GIEWS 2020)

herders to use manufactured feed rather than growing barley and alfalfa, which require large amounts of water. In order to discourage forage production, the government re-authorized wheat production in 2018 (FAO-GIEWS 2020).

In May 2020, the Saudi Grains Organization (SAGO), the sole buyer of wheat on behalf of the government, was purchasing wheat at SAR 1,140 (equivalent to USD304) per tonne, much higher than the international price of USD 220 per tonne, all calculated without cargo and shipping charges. However, despite a high purchase price attached to wheat, many farmers still prefer cultivating high protein alfalfa as it brings higher price per hectare (FAO-GIEWS 2020).

3.1.1 Agriculture for Food Security—A Key Issue

Agricultural development in KSA over the past three decades has been very impressive. Vast desert areas have been turned into agricultural fields, a major achievement in a country that receives an average of four inches of rain per year, one of the lowest in the world (Embassy of the Kingdom of Saudi Arabia 2020).

The generous subsidies, sufficient technical guidance, and supportive extension services helped the country export wheat, dates, dairy products, eggs, fish, poultry, fruits, vegetables, and flowers to markets around the world. Dates, which used to be a staple of the Saudi diet, are currently cultivated primarily for feeding the hungry and starving. KSA makes them available to the less fortunates around the globe through FAO (Embassy of the Kingdom of Saudi Arabia 2020). The Saudi government allows dairy products—including milk—to be exported to neighboring countries. However, it would be better for Saudi Arabia to ban exports of agricultural commodities and produce only to meet its own domestic food requirements. The Kingdom needs to save its depleting water resources and grow only high value horticultural crops like fruits and vegetables.

The Ministry of Agriculture is primarily responsible for the formulation and launch of agricultural policy. Other government agencies include the Saudi Arabian Agricultural Bank (SAAB), which disburses grants and provides interest-free loans, and the Grain Silos and Flour Mills Corporation, which buys and stores wheat, builds flour mills, and produces animal feed. The government also provides programs for land reclamation and distribution and funds research projects (Embassy of the Kingdom of Saudi Arabia 2020).

The Kingdom achieved self-sufficiency in many agricultural commodities, food items, and other products with only 2% of arable land (PME 2005; Baig and Straquadine 2014). However, food security continues to be the hot challenge for the government.

According to the Food and Agriculture Organization (FAO) (2015), agricultural productivity (yields) in Saudi Arabia is still very low despite impressive agricultural development. The FAO claims that food security continues to be a major concern due to the scarcity of arable land and water resources, dry climate, inefficient food subsidies, reliance on imports, trade policies, and post-harvest losses. Saudi Arabia is a food secure country only so far as it feeds its people by importing foodstuffs.

To face such a challenge, the Kingdom has, among others, initiated innovative research programs, particularly in the fields of sustainable agriculture and organic farming. In addition, a number of studies and research projects have been carried out by government agencies and universities across the country MEWA (2020).

3.1.2 Aquaculture and Shrimp Production

Public and private sectors are making impressive investments in aquaculture. As their numbers grew, fish farms using both marine enclosures and onshore tanks became both a source of food and an export commodity. Shrimp are one of the most lucrative aquatic products that are being produced. The continued support and development of the aquaculture industry will enhance food security in the Kingdom (Lovelle 2015).

The Ministry of Environment, Water and Agriculture (MEWA) and FAO have together formulated and launched strategic national initiatives on sustainable food security in the framework of National Transformation Program 2020.

3.1.3 Efforts to Realize Sustainable Agriculture

At the beginning of this millennium, the government realized that intensive agriculture can only be sustained with heavy use of production inputs and water, resulting in a depletion of groundwater resources, especially from non-renewable aquifers (Al-Otaibi 2020; Baig et al.2017). In order to addresses these issues, MEWA (2020) launched its Sustainable Agriculture Rural Development Program (2018–2025) with the objective to diversify agricultural sector, enhance small farmers' incomes, create more job opportunities, contribute towards food security and promote sustainable development. This program focuses on eight sectors:

- Production of Arabic coffee and its marketing;
- Beekeeping and honey production;
- Production of Roses and their marketing;

- Fruit production and their marketing;
- Artisanal fishing and fish farming;
- Small-scale animal production; and
- Development and promotion of Rainfed crops.

3.1.4 Organic Farming

Organic farming involves the production of fruits, vegetables, cereals, and other food products without the use of herbicides, fungicides pesticides, or other chemicals. The Kingdom has been advocating and supporting the adoption of organic agriculture for many years as a part of sustainable agriculture given its potential benefits like sustainability and profitability. It is anticipated that organic production will help meet the country's nutritional needs, ensure public health, and improve the environment (Khan 2016). Moreover, organic farming uses 30% less water, which seems to be quite a valid option for a water-stressed country (Al-shahwan 2016; Khan 2016). Recently, MEWA (2020) signed 12-months agreements with farmers to promote organic agriculture and MEWA pays for the certification of these organic farmers.

Saudi Arabia: A Country with Diversified Agro-Climatic Regions

In general, the Kingdom has a desert climate. However, there are also large seasonal and regional differences in rainfall and temperature regimes, resulting in large differences in agricultural production across the country. It is partially due to its diversity of landscapes and climatic regions. The central regions are hot in summer and cool in winter. During the summer months, temperatures exceed 50 °C. In contrast, night temperatures in winter fall below freezing. The climate is dry all year round with an average annual rainfall between 50 and 100 mm. In contrast, the coastal areas are hot and humid in summer while winters are warm. In the mountainous region in the southwest of the country, the total precipitation is 250 mm/year, and conditions for agriculture are comparatively favorable.

With the exception of Asir Province on the west coast, Saudi Arabia experiences a desert climate that is characterized by extreme hot temperatures during the day, a sharp drop in temperature at night and very little annual rainfall. The average temperature in summer is 45 °C, and may also reach 54 °C. The temperatures go high and heat becomes intense shortly after sunrise and last until sunset, surprisingly followed by cool nights. In winter, the temperature rarely falls below 0 °C, but the almost complete absence of moisture and the high wind chill factor create a rather cold atmosphere. In spring and autumn, the temperatures are moderate with an average of 29 °C (Weatheronline 2020).

Asir region is really famous for growing a wide variety of fruits and its small-scale agricultural production on the terraces due to its mountainous landscape. Farmers in the village of Asir live in the mountains and earn their living by growing food commodities to support themselves. Asir is cold in the winter and hot in summer and the climate seems perfect for cultivation of citrus fruits, peaches and apricots.

In addition to the fruits, grains crops, and vegetables (potatoes and lettuce) are also grown on these terraces throughout the year (Al-Bakri 2018).

The Kingdom, predominantly an arid country, is affected in the south by the southwest monsoon from June to August. The Mediterranean depressions originating in the north from October to May, which bring some rains. While some areas experience precipitation each year, many other regions do not receive precipitation for years. The Kingdom generally experiences no rains between June and September, while in the southwest observes the rainy season from April to August (Hasanean and Almazroui 2015).

In addition to other factors, the climatic features of the region also impact the yields of the major crops. The mild coastal climate of the southwest allows the cultivation of tropical fruits. Wheat and vegetables are mainly grown in the interior of the Kingdom. For organic production, these varying local conditions and an easy access to market are equally important (Hartmann et al. 2012). The most important areas for agricultural production are Qassim and Al Kharj, which are located in the center and representing the agricultural heartland of the country. The main products of these regions include dates, vegetables (e.g., carrots, cucumbers, tomatoes and lettuce) and forage crops.

The main olive production areas occupy the northern region of Al-Jouf, which is also known for producing dates. The main centers of fruit production are in the southwest, along the Red Sea coast. The Jizan region is particularly famous for the production of tropical fruits like mangoes and papayas.

The Tabuk region in the northwest has fertile soil, mountainous terrain, mild climate and abundance of high-quality groundwater. It is famous for producing olives, alfalfa, and wheat (Al-Zaidi et al. 2014).

4 Water Challenge in KSA and Adopted Management Practices

4.1 Management of Scarce Water Resources

Saudi Arabia is one of the driest regions of the world, characterized by low rainfall and high temperatures and enhanced evaporation rates (National Environment Strategy 2017). Located on the mainland, where temperatures are high in summer and low in winter, the country has a desert environment. It is also characterized by a scarcity of annual precipitation and a lack of perennial rivers or permanent water bodies (Baig et al. 2019). Climatic conditions and the depletion of groundwater resources are constant challenges. Due to an acute deficit, water has always been treated as an extremely precious and important natural resource in the Kingdom. No doubt, water is a renewable resource, however its availability is enormously low compared to an increasing demand (Baig et al. 2019).

With an area of 2.15 million km^2, the Kingdom ranks among the top five countries facing water scarcity in the world (Malek 2019), suffering from a severe freshwater shortage (Chowdhry and Al-Zahrani 2013). Despite limited renewable water resources, it is still the third largest per capita water user in the world after Canada and the United States, consuming around 250 L per capita per day (Drewes et al 2012). The rapid increase of both population growth and water demand, especially in urban centers, poses a serious challenge to available water supplies. Saudi Arabia will likely have to double its drinking water supplies over the next two decades (US-SABC 2010). Surface water is generated by random flooding events; however, adequate groundwater is available from the shallow renewable and non-renewable groundwater aquifers that are being utilized for different purposes and supplemented by desalinated water (National Environment Strategy 2017).

The private sector has played an important role in the agricultural development of the Kingdom. The country made impressive achievements primarily due to generous supportive programs and farmers-friendly policies. Among the most utilized incentives are free seeds and fertilizers, water, fuel, inexpensive electricity, and duty-free imports of raw materials and machinery.

As previously mentioned, underground reservoirs and aquifers are the main source of water in the Kingdom. With the government's efforts to identify their location, mapping and estimation of these aquifers with available volumes of water, tens of thousands of deep wells have been dug in the most promising and suitable areas for agricultural and urban purposes.

With no permanent rivers or lakes and very little rainfall, the country's water resources are limited and scarce, but the demand for high-quality drinking water continues to grow. Growing agricultural and domestic needs are met by desalination. Globally, the Kingdom ranks number one in the world in the production of desalination water from the sea for drinking purposes. In addition, over 200 dams collect around 16 billion cubic feet of runoff water in their reservoirs each year. The water is collected in large dams, located at Wadi Fatima, Wadi Jizan, and Wadi Bisha for agricultural purposes and is made available to users through irrigation canals (Baig et al. 2019).

In order to alleviate its water shortage, the Kingdom is also using recycled water for irrigation of agricultural fields and city parks. The government's goal is to recycle up to 40% of the water used for domestic purposes in urban areas. Therefore, water recycling plants have been built in Jeddah, Riyadh, and other major urban industrial centres of the Kingdom (Market Research 2020).

Climate change and arid conditions (drought) impact the availability of fresh water and, in turn, threaten food security (MAW 1994; Alkolibi 2002). Therefore, understanding the effects of climate change on crop water requirements (CWR) is essential for better management of water resources. High temperatures and prolonged droughts increase the demand for water, thus depleting water supplies (DeNicola et al. 2015). Agricultural sector consumes almost 85% of total groundwater drawn from both renewable and non-renewable sources. The report warns if the current withdrawal rates continue, the available water resources reserves may deplete in the next 50 years (Frank Water 2021). Therefore, the Kingdom is investigating innovative ways to develop sufficient water resources to support its agricultural sector.

The staff and future (students) extension providers must be educated on the efficient use of limited water resources through modern technologies for growing crops. In the Kingdom, at many farms, advanced irrigation technologies have been installed that employ computer software and decision-support systems to reduce water usage by 50% compared to technologies used 20 years ago.

All stakeholders, including private companies and academic scholars, conclude that agriculture's consumption of water at very high volumes is a prime issue (Baig et al. 2020). Similarly, extension professionals, in order to address governmental concerns about the high consumption of water in agriculture, have made efforts by hosting agricultural water consumption awareness sessions through various campaigns. Therefore, the development of properly skilled agricultural extension employees is of utmost importance for agriculture (TADCO 2020).

Many professionals of private agricultural companies have also developed training programs in their specialized fields to enhance the skills and update their extension staff with the latest developments, trends, and technologies in the area of water use. These private companies provide their extension staff with in-house training and external training courses to upgrade their skills, thus enabling them to work with farmers effectively. Such in-service trainings are equally important for governmental extension agents and must be emphasized in annual performance reviews and goal setting (Fig. 2 and Table 2).

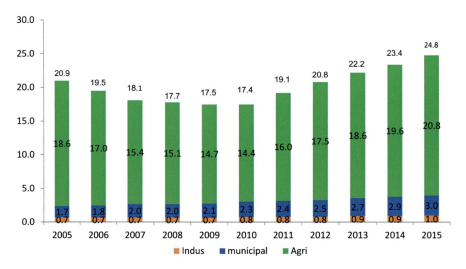

Fig. 2 Water usage by the agricultural, municipal and industrial sectors. *Source* Al-Subaiee (2018)

Table 2 Water Demand (million Cubic meters/year)

Use	Total demand 2017	Percent to total (%)	Total demand 2018	Percent to total (%)	Year-on-year variation (%)
Municipal purposes	3,150	14	3,392	13	+18
Industrial purposes	,000	4	1,400	5	+40
Agricultural purposes	9,200	82	21,200	82	+10
Total	23,350	100	25,992	100	+11

Source https://www.argaam.com/en/article/articledetail/id/1312492

4.2 Water Harvesting as a Climate Adaptation Practice with the Potential to Improve Water Availability

The KSA covers 75% of the Arabian Peninsula. The areas included in the Arabian Peninsula are arid and characterized by desert and semi-desert; dry, hot, and harsh climatic conditions for most months. However, for short periods, some arid regions can receive heavy rainfall, resulting in floods in the wadis and valleys. In these situations, it is necessary to take measures to minimize flood damage, reduce evaporation losses due to high temperatures, and tap the potential benefits of this floodwater. Şen et al. (2011) report that simple, cost-effective interventions can capture water runoff and store it for productive purposes. They are of the opinion that water harvesting is a technique that can be used to supplement the quality and quantity of existing water resources. Such harvesting is capable of providing water where other resources are either unavailable or are too costly to develop.

The harvesting and storage of rain and runoff water have become more important considering recent studies conducted by Al-AlShaikh and his co-workers (2021). They indicate that climate change will bring about substantial changes in precipitation patterns. Some areas are expected to experience reduced precipitation while others would experience heavy pours. These studies indicate that the central and eastern regions in KSA would receive 25% more rain during the next few decades due to the impacts of climate change. In order to tackle the potential situation, appropriate planning and strategies are needed for realizing rain and runoff water harvesting and its storage in the country (Al-AlShaikh et al. 2021).

Al-AlShaikh (2021) indicated that rainwater and runoff water harvesting techniques have been employed for centuries to obtain water for direct human consumption, livestock, and other agricultural uses. He maintains that methods of water harvesting can be broadly categorized as: (i) collection and storage of rainwater or runoff for direct use; and (ii) groundwater recharge, particularly recharging dams. He believes that more concerted and focused efforts are needed to convert the negative impacts of climate change into positive ones by employing appropriate water-harvesting technologies to ensure food security in the Kingdom.

5 Climate Change—An Emerging Issue

Climate change is a global phenomenon and all countries, including Saudi Arabia, are affected by its adverse effects. Many researchers (Abderrahman 2001; Alkolibi 2002; Chowdhury and Al-Zahrani 2012) have shown that climate change will have a negative impact on agriculture in Saudi Arabia, since a 1 °C increase in temperature may go beyond the thermal (heat or temperature tolerance) limits of some crops. This can lead to a 5 to 25% reduction in crop yields (Parry and Swaminathan 1993; Chowdhury and Al-Zahrani 2012). Similarly, the Hadley Center Global Model (HadCM3) confirms that arid regions are more likely to experience decreased agricultural crop yields (Parry et al. 1999).

Mestre-Sanchís et al. (2009) and Chowdhury and Al-Zahrani (2012) predict an increase in temperature between 1.8 and 4.1 °C from 2011 to 2050 in Saudi Arabia. Al-Zawad (2008) also predicts an increase of 2.5 to 5.1 °C for the period 2070 to 2100. However, Chowdhury and Al-Zahrani (2012) believe that an increase in temperature regimes by 1°C would increase evapotranspiration from 10.3% to 27.4%. They believe with an increase in temperatures and evapotranspiration rates, variable rainfall regimes and their interactions with other meteorological parameters would have negative influence on crop-water relationships (CWR).

Climate change will have pronounced impacts on water resources and cropping patterns while a minor increase in high temperatures would also significantly affect agricultural productivity. The Kingdom has experienced prolonged droughts and floods in the past correlated to climate change, with heavy economic losses. High variations in rainfall that increase the risk of flooding have negatively impacted many Arab countries. For example, severe floods in the KSA and Yemen from 2008 to 2009 caused an economic damage of approximately 1.3 billion USD.[1] Thus, climate change may significantly impact agricultural productivity and could have severe implications for food security, rural–urban migration, and social stability (FAO 2008; El Mostafa and Al Assiri 2010).

Farmers in KSA are at crossroads. They can either cease farming or move to modern technologies, such as greenhouse farming, to adapt to climate change and ensure food security in the country. It is imperative for the government to develop a more innovative and responsive agricultural extension service to face this challenge. There is a need to design extension programs to address existing challenges by placing stronger emphasis on food security and climate change adaptation. This can only be accomplished by changing the curricula of agricultural extension education, developing awareness programs, and launching in-service training courses to familiarize extension workers with scientific innovations adapted to appropriate pedagogical principles. Hasanean and Almazroui (2015) have noted that scientific information and knowledge have been shared in numerous publications. However, such information and publications relating to the climate and climate change in the Kingdom have not been compiled and remain scattered. There is a serious need to design, launch and integrate courses on climate-smart agriculture to meet the challenges facing

---

[1] EM-DAT, www.emdat.be

today's agriculture. Extension must play a significant role in the dissemination of such knowledge.

6 Changing Cropping Patterns Remains a Challenge

Rapid population growth and improved living standards have increased food and water demands in urban areas for domestic use, as well as in the agricultural and industrial sectors. Such increased urban demands result in more pressure on rural and manufacturing sectors to provide the goods necessary for urban living. This, in turn, puts increased pressure on natural water supply. If demands exceed water and food availability, the Kingdom could experience high levels of food and water insecurity.

In the past, Saudi farmers were facilitated by providing inputs and favorable polices to grow more food commodities. However, this achievement, as mentioned above, came at the heavy cost of depleting precious water resources. Consequently, the Kingdom has framed a policy to stop growing wheat, barley, and fodder crops with higher water requirements (Hartmann et al. 2012). As a result, the Kingdom has realized the imperative need to bring immediate changes in its cropping patterns to improve food security and ensure the sustainability of water resources. Farmers are now required to grow vegetables using modern, water-saving technologies, such as installations of greenhouses (FAO 2017).

Farmers must avoid growing Alfalfa, Sorghum, and other grain (barley) crops. Instead, they should produce high value horticultural crops requiring less water. Indeed, the Kingdom needs an immediate shift from water-intensive to water-efficient crops. In addition, large farms need to use decision-making computer software that could help reduce the water required to grow a crop up to 50%. Recognizing the importance of water saving technologies, the private sector is also playing an appreciable role in meeting consumers and government's concerns about high-water consumption in the agricultural sector (TADCO 2020).

7 Switching to New Crops Consuming Less Water

Keeping in view the water situation, the Kingdom has recently shifted its production systems towards sustainable agriculture, while leaving conventional farming. Farmers have been encouraged to grow high value crops, such as fruits and vegetables instead of producing crops requiring more water. It has been proven that such a switch will yield higher economic returns from scarce water resources (Hartmann et al.2012; Ouda 2014; FAO2016). Previously, the Kingdom has stressed the conversion of barren and dry lands into productive fertile lands to realize enhanced productions. This policy has changed and the current focus is to bring less area under the plow while consuming less water and producing more. The Kingdom is also promoting modern water-saving technologies, specifically the installation of high efficiency irrigation systems (FAO-GIEWS 2020). To conserve freshwater, the Kingdom also started to reduce agricultural lands and incorporate improved irrigation practices (MOEP 2010; FAO 2016). Production of the most water-intensive crops, such as wheat and alfalfa, have been reduced to the minimum between 2016 and 2019. The Saudi government has adopted a new strategy intending to end domestic wheat production (Drewes et al. 2012). Measures have also been taken to reduce green fodder production. Large corporate farms have also been directed not to produce green fodder in late 2018 (KSA Water Report 2020). Yet, areas previously covered with agricultural crops are increasingly threatened by desertification if they remain fallow for a long period. Therefore, the country needs to focus on innovative techniques to keep these areas under crop rotation or growing less water-using but high value crops (Baig et al. 2017).

8 Greenhouse Crop Production

The government is encouraging greenhouse production of fruits and vegetables along with high efficiency drip irrigation systems. This technology replaced open field farming where water evapotranspiration rates are much higher. Greenhouse technologies replace soil with nutrient-rich water for plants to grow. Hydroponic techniques help producing high-quality vegetables and fruits under controlled environmental conditions. This technology is environmentally friendly, more productive, and reduces pressure on land and water resources. Water-efficient hydroponic systems are also capable of producing sustainable fodder crops to feed the livestock of herders even in the desert like conditions (Econa.Art 2019).

Greenhouse cultivation has great potential to produce more food items and reduce hunger. Due to their inherent advantages, FAO, the United Nations Development Program (UNDP), and others have supported projects in 13 countries. The technology reduces the space required for crops by more than 75% and water consumption by up to 90%. Plant nutrients are added to the system and recycled so nothing goes to waste. No herbicides are used in the greenhouses and the pesticides are used in the form of sprays and natural products made from the plant sources (Brandlay and Marulanda 2001).

> **Box 1 Greenhouse Crop Production**
>
> - The material costs around USD6.42 per m^2 to prepare the seedbeds with the dimensions of 2 m^2. Simplified hydroponics systems, typically the production of edible feed, ranges from 0.11 to 0.23 kg per sq meter/day
> - Cash crops grown on an area of 40 square meters can generate an income of USD101.00 per month. Profits range from USD5.28 per square meter per year for cucumbers to USD40.26 per square meter per year for lettuce.

> The production also varies between 8.8 kg/m² per year for cucumbers and 64.9 kg/m² per year for lettuce
> - Farmers, on average realize about 5 tons of vegetables per acre in their fields. However, under greenhouse/hydroponics conditions, the yields can increase up to 200 tonnes per acre. Previous studies reveal that the hydroponic forage technique requires approximately 2 to 10% of the water to produce the equivalent amount of crop grown in the soil
>
> These facts and figures indicate that the greenhouse technology is economical, uses less water and produces more food in desert countries like Saudi Arabia
>
> *Source* Brandlay and Marulanda (2001)

KSA has initiated its greenhouse production in recent years to conserve water and offset harsh weather conditions. With a total of 73,542 greenhouses, they cover an area of 32,947,306 m². Occupying an area more than 12,607,632 m², tomatoes ranked first among all greenhouse vegetables in terms of number and area. Attaining the 2nd position, cucumbers were grown on an area of 8,727,132 m² (APSB 2019).

These crop production systems are new for farmers. Such systems are highly technical in nature and require a specific package of production practices for each crop. Therefore, farmers need continuous expert guidance, from planting to harvesting. Extension agents with adequate training can assist farmers with information on post-harvest technologies, including packaging, storage, and marketing of their products at the premium prices at the right timings.

9 Heavy Imports of Food Commodities—A Challenge

Limited land and rapidly diminishing nonrenewable water resources remain a prime barrier to grow sufficient food to meet domestic human and livestock needs, thus realizing food security (Baig and Straquadine 2014; Baig et al. 2017). It is for this reason that the country heavily relies on food and feed imports from foreign markets. Consequently, the Kingdom has increasingly become dependent on imports to meet domestic needs. The country's main bilateral trade partners are Ukraine, Russia, India, and Pakistan. Barley, wheat, rice, chicken and mutton are the country's main food imports, accounting for 80% of total food needs (CTCS 2013; FAO-GIEWS 2020). Wheat imports and exports are shown in Fig. 3.

However, such a heavy dependence on food imports creates more risks for national food security. In case the global food exports and supplies of food items are reduced due to climate change or other factors, prices will certainly go higher. In the situation, Saudi Arabia will be more vulnerable to food insecurity if immediate action plans are not developed. Recent severe reductions in oil prices and the COVID-19 pandemic

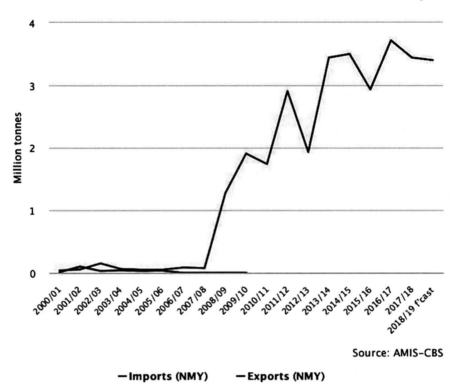

Fig. 3 Wheat imports and export. *Source* http://statistics.amis-outlook.org/data/index.html#

could reduce the country's revenues and weaken its ability to continue importing food. According to Lovelle (2015), food imports negatively affect the sustainability of Saudi Arabia's food security.

Therefore, it is vital for the Kingdom to continue securing a wide range of trading partners around the world. Of equal importance is the need for a variety of established shipping routes to avoid trade blockages and volatility in an increasingly insecure region (Lovelle 2015). In order to overcome trade issues and ensure sustainability in importing food items from partner countries and make trading process smooth, planners and policy makers need to collaborate with all relevant stakeholders, including farmers, experts on food and agriculture, and economists.

10 Natural Resources and Farming Systems

Forests and trees that surround farms have significant impacts on weather conditions. Farm trees deserve attention to improve the micro-climate at the farm level. Forests and trees on the farms help saving crops from the harmful effects of harsh temperatures and desiccating winds. Hence, there is a need to educate local communities on the benefits of forest management and the conservation of trees to avoid the expansion of farms at the expense of forests (Baig et al. 2020). Participation of rural communities in managing natural resources, including forests, is quite a beneficial option to follow. Both rural populations and urban dwellers need to perceive and appreciate the productive and protective roles of natural resources, including forest ecosystems to realize the countless benefits they do offer (Al-Subaiee 2016).

11 Environmental Challenges and Pollution Hazards

Food self-sufficiency continues to face serious challenges due to climate change, aridity, limited arable land and water (AFED 2014). Still agricultural departments encouraged farmers to adopt high input agricultural practices to improve food security and farmers' livelihoods by producing food commodities. However, intensified agriculture based on the adoption of new agricultural methods and the heavy use of chemical fertilizers and pesticides has resulted in negative impacts on the environment and natural resources, such as reduced soil fertility and water quality and soil erosion. Inappropriate and heavy applications of agricultural chemicals have also led to health hazards and environmental concerns, such as ground-water pollution (Baig et al. 2017).

Farm workers as well as employees at the governmental agricultural departments usually have limited knowledge when working with agro-chemicals. They ignore essential safety measures, including the use of protective equipment, and are unable to read the label precautions written on the product containers and bags (Al-Zaidi et al. 2011). To address this situation, farmers and extension agents must be trained on the application of pesticides and fertilizers to achieve desired results while keeping farmers safe, realizing healthy crops, enhancing yields, and keeping the environment pollution free.

Since input-intensive agriculture is practiced, previously non-existent diseases and pests can occur due to an imbalance in the natural ecosystem. As a result, traditional methods of agricultural production can conflict with the optimal and sustainable use of natural resources. As a result, Saudi Arabia could face deadly problems, especially as pesticides migrate through the soil, water and air, possibly reducing biodiversity and increasing pollution. Therefore, organic farming in Saudi Arabia must be viewed as a means of sustainable agriculture (MEWA 2019; Ramadan et al. 2020).

There is also a need to learn how and when to use the right chemical for each crop and pest. Threshold levels for each pest for different horticultural crops need to be diagnosed correctly to decide when it is appropriate to use pesticides to avoid waste and combat environmental pollution. To avoid the problems created by chemicals and address certain issues faced by farming, the Kingdom has also launched research programs on the use of organic products (made of plants): for example, research on the products of Neem tree to protect date palm trees has been initiated.

12 Growers, Post-Harvest Technologies and Marketing of the Produce

Serious attention has not been paid to post-harvest technologies to overcome losses. To raise farmers' income levels, all possible efforts are needed to reduce post-harvest losses and increase the shelf life of horticultural commodities and fruits. There is a need to equip greenhouse growers with the necessary skills to enable them to apply post-harvest technologies. Greenhouse farmers need the knowledge to increase the shelf life of the produce and should be provided with markets and marketing facilities to absorb the produced fruits and vegetables. Farmers need to understand market dynamics to maximize their profits. Public institutions may take proactive initiatives by arranging short courses with the objective to provide farmers with the relevant knowledge, thus enhancing their marketing and business management skills.

13 Farming on Overseas Lands

Saudi Arabia meets 80% of its food needs through imports. With strong domestic demand for food and feed, such as wheat, rice, barley, yellow corn, soybeans and green fodder, the government is encouraging agricultural investment abroad to produce crops that will later be imported into the Kingdom. This initiative of farming on overseas lands aims to close the gap between national food production and consumption while focusing on wheat, rice, barley, yellow corn, soybeans, and green forages (Lovelle 2015; FAO 2017; FAO-GIEWS 2020). To conserve national water resources and secure the food supply chain, Saudi government encourages its citizens to acquire fertile irrigated lands having adequate water resources in Asian and African countries.

According to Lovelle (2015), the country has had a policy of outsourcing its agricultural production by buying land in other countries more suitable for agricultural use and bringing the products back to Saudi Arabia. However, this initiative has been threatened by political turmoil, natural disasters, and inflation in commodity prices in these countries. Furthermore, the policy is highly controversial internationally, with some referring to the practice as 'land grabbing' comparable to the colonialism of two centuries ago (Global Agriculture 2020; Behnassi and Yaya 2011). At this point, it would be important for an advisory service such as extension to educate Saudi investors on various socio-cultural, economic, and technical aspects of farming before they go overseas to acquire lands for farming.

14 KSA as Signatory to WTO and Other Treaties

The KSA has been a member of the World Trade Organization (WTO) since December 11, 2005 (WTO 2020), and has signed several treaties to liberalize international trade. Agricultural graduates are to be educated and provided with the basics of these treaties and the needed information on the role and responsibilities of Saudi Arabia in international markets. By the early 1990s, due to its subsidized agricultural program, the country had become a grain exporter and ranked sixth in the world. During the country's accession, negotiations began with the WTO in 2005. It was conceded that KSA agreed to reduce its agricultural subsidies and make its trading system free to match with international markets' trading norms (Hartmann et al. 2012).

15 Applications of ICT in Farming and E-Extension for Technology Transfer

The use and applications of information and communication technology (ICT) in farming are vital to modern agricultural technologies. Farmers and agricultural extension staff need training to incorporate these technologies in their farming activities and to keep them updated with the latest scientific information. ICT-based knowledge would help extension workers perform their duties in a better way. Both extension workers and farmers must enhance their working abilities, upgrade their skills, and update their knowledge base to capture the benefits of ICT and realize elevated agricultural productions (Al-Shayaa et al. 2012).

Afzal et al. (2016) and MEVA (2020) consider e-extension as a modern means of communication, capable of enhancing the effectiveness and efficiency of advisory services to achieve agricultural sustainability. Electronic updating for the provision of renewal services depends on the use of Internet. E-extension uses the latest ICT-based tools to develop networks, online sharing, and collaboration between farmers. Afzal et al. (2016) conducted a study to assess employee attitudes towards using electronic extensions in KSA. They concluded that agricultural workers are a key factor in implementing effective agricultural growth plans. The study found that extension workers generally have a positive attitude towards using electronic extensions. Significant associations were identified between employees' general attitudes towards electronic improvement and their age, years of service, and IT experience. In light of these findings, the recommendations focus on: the encouragement of extension staff, particularly the elderly, to use the electronic renewal system through exclusive training programs and refresher courses; the use of combined workshops for extension staff with fewer or more years of service to bridge the generation gap and provide a better understanding of the electronic expansion system.

16 Strategies to Improve the National Extension Service

16.1 Practicing Agriculture Depending on the Suitability of the Area

Saudi Arabia's climate is predominantly semi-arid to hyper-arid, with very low precipitation (annual average 70.5 mm), coupled with extremely high temperatures. Extreme hot weather conditions cause water scarcity and reduced, poor and sparse vegetation cover (Al-Wabel et al. 2020). With a total rainfall of 250 mm/year, the mountainous region in the southwest of the country seems suitable for agricultural activities. Likewise, the region of Tabuk in the north-west, due to its mountainous location, temperate climate and high soil fertility and fresh underground water resources, is an area suitable for fruits, vegetables, olives, medicinal herbs, and high-quality wheat (Al-Zaidi et al. 2014).

With an ideal climate for agriculture, the province of Asir on the west coast is cold in winter and hot in summer. The mountainous landscapes of the region are ideal for small-scale agricultural production of fruits and vegetables (Pellikka and Alshaikh 2016). The area produces many crops throughout the year on these terraces, especially cereals, fruits (peaches and apricots), and vegetables like potatoes and lettuce (Al-Bakri 2018). To the southwest, along the Red Sea coast, are the main centers of fruit production. Jizan region enjoys a sufficiently favorable environment for tropical fruits, such as mangoes and papayas.

16.2 Enhancing Farmers' Participation

The Kingdom launched its comprehensive agricultural program in early 80 s to grow wheat at the fullest scale by implementing farmers' friendly policies and offering them generous and attractive incentives. Such incentives included provision of long-term interest-free loans, exemplary technical support services, free seeds and fertilizers, cheap water, fuel and electricity, and duty-free imports of raw materials and machinery (Embassy of the Kingdom of Saudi Arabia 2020). The government's programs have had a positive impact, making the Kingdom achieve food self-sufficiency and surplus to export. Al-Shayaa et al. (2012) believe that through the launching of updated extension and farmers' needs responsive programs, farmers' participation in the national food security initiatives can be greatly enhanced.

16.3 Developing Public–Private Partnerships for a More Productive Extension Service

The private sector played an important role in the agricultural development of the Kingdom. Private agricultural companies tried to assist farmers wherever they could. Many private companies engaged in agricultural business hosted events for farmers. For example, a private company awarded the 'Prince Fahad Ben Sultan' prize to the farmer with the highest production, as an incentive for other farmers to achieve better crop productivity and become part of the competition in the coming years. The goal of such an incentive was to improve farm productivity, increase farmers' incomes, and ensure their economic and social development. MEVA (2019) recognized the water stress situation and attempted to foster collaboration between public and private sectors in the development of technologies for cost-effective desalination and distribution mechanisms. The Ministry in charge indicated that public and private sectors are currently struggling to maintain food reserves to cope with food insecurity with little dependence on international markets (Muhammad 2014). Some private agricultural companies like TADCO, especially in the Tabuk region, provide agricultural education and orientation services to farmers to help them use agricultural inputs more effectively and efficiently. Farmers of the region are supported through advice and guidance on cost-effective crop production techniques, which in-turn increase farm income. Information is made available to school students in the Tabuk region, and some agricultural companies have continued to welcome school groups for visits during the academic year. Some private companies also offered training programs and services to the staff of Saudi Agriculture Bank to enable a better understanding of agricultural concepts. Such initiatives enabled bankers to better serve the farming community.

It is anticipated that the quality of the training programs can be enhanced if the Saudi National Extension Service also joins the private sector. This would make training more cost-effective and enable extension workers to reach a greater number of farmers. Launching environmental awareness programs is becoming equally important by promoting tree planting and helping farmers follow suit by making fruit trees available for transplant. MEVA and the private sector, including agricultural companies in the Kingdom, can strengthen and implement public awareness campaigns more efficiently, effectively, and cost-effectively by joining hands.

16.4 Fighting Pollution and Conserving Natural Resources

The widespread use of chemical fertilizers to increase yields has raised serious food security concerns. Saudi citizens are increasingly health conscious and organic food is gaining acceptance and popularity due to concerns about food security and safety (Al-Otaibi 2020).

The Ministry has made concerted efforts to protect the environment and conserve natural resources by developing guidelines and adopting the environmental protection policy in the Basic Law on Governance. Art. 32 of this law stipulates that: "The state must be responsible for the conservation, protection and improvement of the environment". Various laws to preserve everything that affects society, including water, air, land and space, plants, animals, and various forms of energy have been framed and disseminated.

Efforts have been made to promote the protection of wildlife, stop environmental degradation, stop the depletion of natural resources, save wildlife from extinction, protect biodiversity, and preserve wildlife habitats to ensure their populations remain in their natural habitat. The Ministry has succeeded in maintaining a network of protected areas to preserve biodiversity. This network aims to protect 103 territories and maritime zones which represent approximately 10% of the country. These initiatives would contribute to the conservation and renewal of natural resources for the benefit of humanity. There are 15 officially protected areas that cover 4% of the Kingdom's surface area and form the natural habitat and gene bank for biodiversity in aquatic and terrestrial ecosystems. The Ministry is working on the implementation of many national rules and strategies, as well as international agreements ratified by the Kingdom in the field of wildlife protection.

16.5 Maintaining a Balanced Rural–Urban Ratio

Maintaining a fair balance between rural and urban populations is extremely important from social, cultural, and economic perspectives. Maintenance of natural ecosystems existing in rural areas are also vital to the survival of human beings. Unfortunately, the rural fabric is fading quickly due to heavy migration, especially the youth, to urban centers. Farming is a professional occupation that breeds and generates many other enterprises and businesses. Rapid migration from rural areas not only causes a shortage of experienced farm workers, but also results in a forced closure of many rural businesses. Extension educators can help maintain a fair balance between rural and urban populations. Therefore, the National Agricultural Extension Service (NAES), with the involvement of academia, researchers, opinion leaders, representatives of farming communities, NGOs, agricultural organizations, farmers' organizations, experts from relevant ministries, policy makers and planners collectively must find ways and means of increasing the attractiveness of rural areas and presenting them as hubs of economic opportunities. The Kingdom needs a modern rural development strategy to save its professional and rich cultural heritage.

17 The New Roles of Extension Service

17.1 Renovation of Agricultural Extension Systems Through a Rethinking of the Curricula

The present curricula taught at agricultural colleges and universities seem unable to address emerging challenges in agriculture and cannot offer possible solutions and research-based strategies (Baig et al. 2017). Immediate changes to the existing curricula and degree programs offered at the universities are needed to address new challenges. For example, the subject of the biological control of insects and mites by using natural enemies in the production of greenhouse vegetables, or mites and Lepidoptera pests on date palm in Saudi Arabia deserves to be launched for the extension staff (FAO 2011). The curricula at different levels of the education system need to be updated, providing adequate information on topics such as the impacts of climate change, eco- and climate-smart agriculture, natural resource management, economics and marketing, emerging production technologies, and post-harvest management. Equally vital is the role of ICT in introducing modern techniques of agriculture extension, use of agrochemicals, composting and manures, organic farming, and water conservation technologies, etc. Trained extension staff must be kept abreast and equipped with adequate solutions to this set of challenges (Baig et al. 2017).

NAES needs to develop new mechanisms to empower farmers, thus meeting emerging challenges, especially climate change. New mechanisms need to include the development of appropriate technologies for efficient use of scarce water resources to grow various crops in different ecological zones. In addition, new information needs to address the design of training programs for farmers and extension agents (Baig et al. 2017). Alsaghan and Diab (2017) highlighted the need to bring reforms and radical changes in the working of NAES. The findings of their study have revealed that all aspects of the extension system in KSA deserve immediate reforms. They suggested that NAES should transfer some specific extension activities to non-governmental organizations (NGOs), farmer-based organizations (FBOs), and private firms at different levels. The authors believe that participatory extension, farming systems development, and training/education are the most suitable and effective approaches in working with Saudi farming communities. The participation of elected groups of farmers, as well as NGOs and FBOs, could be very appropriate for planning extension programs following a bottom-up approach. They also reported that extension personnel prefer provision of extension services through their governmental offices.

All stakeholders and specialists, including agronomists, irrigation engineers and water experts, IT experts, extension experts, sociologists, entomologists, plant pathologists, post-harvest technologists, marketing experts, and agricultural economists must devise new vibrant extension programs based on the prevailing comprehensive agricultural policy. These new extension programs form part of an integrated agricultural strategy to keep farming profitable and sustainable. In addition, policy makers and relevant authorities must extend their full support to experts, to enable farmers to realize better yields, higher farm incomes, and increased water-use efficiency through better irrigation technologies.

17.2 Focus Areas of the Future Agricultural Extension Education Curricula

Previously, the degree programs of Agricultural Extension and Education in KSA have been focusing on Extension methodologies and approaching farmers with the innovative ways to disseminate modern knowledge to grow more by practicing high input agriculture and applying heavy doses of chemical inputs. This kind of farming proved to waste of precious resources and a source of environmental degradation and health risks. Today farming has gone beyond producing food commodities and is seen as a way of improving and stabilizing the environment and enhancing public health. Farming in KSA has witnessed many phases, including: subsistence agriculture, food self-sufficiency, sustainable and climate-smart-agriculture, and greenhouse crop production. Today, 'Extension Education' and working with farmers have become a complex science that draws lessons, principles, and practices from all scientific disciplines. The curricula for degree programs need to be modernized, capable of meeting the various challenges faced by the extension staff and farmers. The following subjects are to be included to make the curricula more responsive, productive and useful:

- Adopt modern high efficiency irrigation systems (e.g., drip irrigation) to enhance the efficient use of water resources;
- Launch market-oriented extension services to elevate farm profits, thus helping farmers stay in the farming business;
- Enhance shelf-life of perishable agricultural commodities;
- Help food importers to ensure that KSA meets a part of its domestic food demands through imports. However, maintaining a fair balance between domestic production and imports will be ideal;
- Promote organic farming as a premium and sustainable practice;
- Scale-up extension workers' abilities to help greenhouse farmers grow vegetables and flowers;
- Educate farmers and extension agents on measures to prevent fertile lands from turning into barren lands;
- Tailor and develop varieties of crops that require less water inputs, using modern plant breeding tools like bio-technology;
- Provide trainings to farmers and other stakeholders in the value chain to manage the impacts of environmental and climatic changes;
- Sensitize farmers about the importance of moving from present production levels to self-sufficiency without damaging the environment and other natural resources;
- Enable farmers and investors to successfully launch overseas-farming ventures;

- Educate extension workers about social issues affecting both rural and urban communities;
- Increase trainings (professional development) for extension staff to keep them updated regarding recent developments in the agriculture sector;
- Equip extension workers with technologies and strategies to minimize food waste along the farm to fork paradigm;
- Upgrade the working proficiencies of extension agents by elevating their skills through the knowledge drawn from the modern, up-to-date and innovative research to enable them to offer workable solutions to the emerging challenges faced by farmers;
- Extension staff must learn about innovative farming techniques—such as the use of ICTs, basics of greenhouse farming – in addition to the ways of minimizing food losses by applying post-harvest technologies and marketing of agricultural commodities;
- Reform the existing curricula used in academic degree programs to make them more demand-driven and responsive to farmers' needs.

18 Conclusions and Recommendations

Farming in KSA has undergone significant changes through the course of time, from the Bedouin style of former years to the modern methods of agriculture, especially greenhouse crop production. Yet, farming businesses face increasing challenges. In the coming years, farmers may suffer from unfavorable situations while practicing agriculture, combined with changing policies and emerging challenges affecting agriculture, this may question their will to continue investing in agri-business (Al-Shayaa et al. 2012; Baig and Straquadine 2014). Greenhouse agriculture is likely to become an appropriate alternate farming system in the Kingdom.

Climate change, early warning systems, remote sensing, advancements in information transfer technologies (IT), and issues related to food security need to be part of the curricula in the preparation of extension agents, so they can be more effective in meeting the demands of modern agriculture. There is an urgent need to examine the curricula and adjust their contents to address solutions to such challenges. The subjects that frame the extension education curricula must be grounded in scientific studies and bias-free facts, keeping in sight available resources and the geopolitical situation of the region.

This study underlines that the issues of climate change and food security ought to remain on top of the government's agenda. To ensure food security sustainably, the government must reform its agricultural extension system and make it more responsive farmers' needs. Although KSA meets its food needs through imports, this constitutes a risky strategy in the long run. Therefore, it must maintain a fair balance between domestic production and imports from around the world.

For extension services to work efficiently and effectively while addressing the challenges of todays' farming, it is imperative that the skills of the extension staff are upgraded. Key areas include: the application of ICTs in farming and E-extension; the promotion of greenhouse farming; effective and efficient marketing of agricultural commodities; minimization of food losses by applying post-harvest technologies and enhancing shelf-life through appropriate packaging and proper storage; judicious use of agricultural chemicals without harming the environment and living beings; dealing with climate change and its negative impacts; and the use of high value crops with low water requirement.

The skills of the current extension staff should be upgraded through professional development programs, workshops, videos, self-directed, online learning, and exhibitions. Agricultural extension departments must launch refresher courses and arrange lectures on the 'changing environment and role of extension agents' on a regular basis every quarter. They need to be deputed to the university where they could have close interaction with researchers and the academia to upgrade their knowledge. Extension agents and economists need to develop marketing plans to get the prime prices for the agricultural commodities. Furthermore, it is time to redefine the roles of extension agents and delineate their responsibilities through collective wisdom of stakeholders, including researchers, farmers, extension staff, planners, policy makers, owners of superstores, loan officers of banks, economists, irrigation scientists, and greenhouse owners, under one umbrella. Extension agents need extensive trainings in order to perform their duties in this ever-changing agri-industry. These trainings should acquaint them with recent approaches and tools to face new challenges effectively and efficiently in the wake of local, regional, and global environmental shifts.

Acknowledgements The authors are extremely thankful to Dr. Michael R. Reed, Emeritus Professor, and former Director – International Programs for Agriculture at the University of Kentucky, USA for editing, reviewing, making helpful comments and offering valuable suggestions. His sincere cooperation is highly appreciated. The authors also want to express their deep appreciation for the Colleagues at the Policy Center for the New South, Rabat, Morocco for making useful suggestions and valuable edits.

References

Abdrrahman W (2001) Energy and water in arid developing countries, Saudi Arabia: a case sstudy. Water Resour Dev 17(2)

AFED (2008) 2008 Arab Forum for Environment and Development (AFED) Published with Technical Publications and Environment & Development magazine P.O. Box 113–5474, Beirut, Lebanon

AFED (2014) Arab environment: food security. Annual report of the arab forum for environment and development. In: Sadik A, El-Solh M, Saab N (eds) Beirut, Lebanon. Technical Publications

Afzal A, Al-Subaiee FS, Mirza AA (2016)The attitudes of agricultural extension workers towards the use of e-extension for ensuring sustainability in the Kingdom of Saudi Arabia. Sustainability 8:980. https://doi.org/10.3390/su8100980. www.mdpi.com/journal/sustainability.

Albalawi E (2020) Assessing and predicting the impact of land use and land cover change on groundwater using geospatial techniques: a case study of Tabuk, Saudi Arabia. A thesis is presented for of the Degree of Doctor of Philosophy at the School of Earth and Planetary Sciences. Curtin University

Allan T (2001) The middle east water question: hydro-politics and the global economy. I.B. Tauris & Co. Ltd., London and New York (xvii + 382 pages).

Almazroui M, Islam MN, Saeed S, Alkhalaf AK, Dambul R (2017) Assessment of uncertainties in projected temperature and precipitation over the Arabian Peninsula using three categories of CMIP5 multimodel ensembles". Earth Syst Environ 1(2):23

Alkolibi FM (2002) Possible effects of global warming on agriculture and water resources in saudi arabia: impacts and responses. Clim Change 54:225–245

Alsaghan BM, Diab AM (2017) Does the agricultural extension system in Kingdom of Saudi Arabia needs to be reformed? A paper presented at the 23rd European Seminar on Extension (and) Education (ESEE)—Transformative learning: new directions in agricultural extension and education held on 4th-7th July 2017 at the Mediterranean Agronomic Institute of Chania, and hosted by the Lab. of Agricultural Extension, Rural Systems & Rural Sociology; Dept. of Agricultural Economics & Rural Development, Agricultural University of Athens

Altukhis ABS (2002) The future of Water Resources in the Kingdom of Saudi Arabia. A paper presented by the Vice Minister, Ministry of Agriculture and Water affairs of Saudi Arabia at a conference organized by the Ministry of Planning in Riyadh

Al-Bakri D (2018) In pictures: Saudi Arabia's Asir region houses terraced farm fields between the mountains live the residents of Asir who depend on these farms for food and to build their economy. (Al Arabiya). Al Arabiya English Edition Tuesday 26 June 2018, Available at: https://english.alarabiya.net/en/life-style/travel-and-tourism/2018/06/26/PHOTOS-How-residents-of-Saudi-Arabia-s-Asir-region-rely-on-their-farms

Al-Bakri D (2020) In Pictures: Saudi Arabia's Asir region houses terraced farm fields Between the mountains live the residents of Asir who depend on these farms for food and to build their economy. (Al Arabiya). Al Arabiya English Edition Tuesday 26 June 2018, Available at: https://english.alarabiya.net/en/life-style/travel-and-tourism/2018/06/26/PHOTOS-How-residents-of-Saudi-Arabia-s-Asir-region-rely-on-their-farms

Al-Otaibi G (2015) By the numbers: facts about water crisis in the Arab World. Submitted by Ghanimah Al-Otaibi on the blog set up by the World Bank on Thu, 03/19/2015 and available at: https://blogs.worldbank.org/arabvoices/numbers-facts-about-water-crisis-arab-world

Al-Otaibi BA (2020) Farmers' perceptions of organic agriculture in southern Saudi Arabia. J Agric Extension 24(4). https://doi.org/10.4314/jae.v24i4.3

Al-Shahwan IM (2016) Vice Chairman of the Saudi Organic Farming Association (SOFA) made statements and published by the Daily Arab News. Available at: https://www.arabnews.com/saudi-arabia/news/876586

Al-Shayaa MS, Al Shenifi MS, Al Hadi HA (2011) Constraints to use Computers among Agricultural Extension Workers in Riyadh and Qaseem Regions of Saudi Arabia. J Anim Plant Sci 21(2):264–268

Al-Shayaa MS, Baig MB, Straquadine GS (2012) Agricultural extension in the Kingdom of Saudi Arabia: difficult present and demanding future. J Anim Plant Sci 22:239–246

Al-AlShaikh HMH (1998) Country case study-water policy reform in Saudi Arabia. Food and agriculture organization of the united nations, regional office for The Near East Cairo. In: Proceedings of the second expert consultation on national water policy reform in The Near East Cairo, Egypt

Al-AlShaikh AA et al (2021) Water harvesting and storage in Saudi Arabia with case studies (An article to be submitted)

Al-AlShaikh AA (2021) Rainwater and runoff harvesting and recharge in Saudi Arabia. A lecture delivered at the 9th Annual Conference (held 29–31 March, 20121) on water organized by the Prince Sultan Institute for Environmental, Water and Desert Research, King Saud University, Riyadh, Saudi Arabia

Al-Subaiee FS (2016) Local participation in woodland management in the Southern Riyadh Area: Implications for agricultural extension. Geograph Rev 105(4):408–428

Al-Subaiee FS (2018) Water resources in the Kingdom of Saudi Arabia. A presentation made on the "Word Water Day" at the college of food and agriculture sciences, King Saud University, Riyadh Saudi Arabia.

Al-Wabel MI, Sallam A, Ahmad M, Elanazi K, Usman ARA (2020) Chapter 25 extent of climate change in Saudi Arabia and its impacts on agriculture: a case study from Qassim Region. In: Fahad S et al (eds) Environment, climate, plant and vegetation growth. Springer Nature Switzerland AG 2020, pp 635–657. https://doi.org/10.1007/978-3-030-49732-3_25

Al-Zaidi AA, Elhag EA, Al-Otaibi SH, Baig MB (2011) Negative effects of pesticides on the environment and the farmers awareness in Saudi Arabia: a case study. J Anim Plant Sci 21(3):605–611

Al-Zaidi AA, Baig MB, Elhag EA, Al-Juhani MBA (2014) Farmers' attitude toward the traditional and modern irrigation methods in Tabuk Region—Kingdom of Saudi Arabia. Chapter 8. In: Behnassi M et al (eds) Science, policy and politics of modern agricultural system. https://doi.org/10.1007/978-94-007-7957-0_8, © Springer Science+Business Media Dordrecht 2014

Al-Zawad FM (2008) Impacts of climate change on water resources in Saudi Arabia. In: The 3rd international conference on water resources and arid environments and the 1st Arab water forum, Riyadh (Saudi Arabia)

APSB (2019) Agricultural Production Survey Bulletin. General Authority for Statistics (GASTAT). Available at: https://www.stats.gov.sa/sites/default/files/Agriculture%20Production%20Survey%202019%20EN.pdf

Arab Environment: Food Security (AFED) (2014) Annual report of the arab forum for environment and development. In: Sadik A, El-Solh M, Saab N (eds) Beirut, Lebanon. Technical Publications

Baig MB, Straquadine GS (2014) Sustainable Agriculture and Rural Development in the Kingdom of Saudi Arabia: Implications for Agricultural Extension and Education. In: Behnassi M, Syomiti M, Gopi Chandran R, Kirit Shelat (Eds.), Climate Change: Toward Sustainable Adaptation Strategies. Springer Science +Business Media B.V., Dordrecht. Netherlands. pp 101–116. https://doi.org/10.1007/978-94-017-8962-2_7.

Baig MB, Gary SS, Aldosari FO (2017) Revisiting extension systems in Saudi Arabia: emerging reasons and realities. J Exp Biol Agric Sci 5(Spl-1- SAFSAW). https://doi.org/10.18006/2017.5 (Spl-1-SAFSAW). S160.S164

Baig MB, Alotibi Y, Gary S. Straquadine., Abed Alataway (2019). Water Resources in the Kingdom of Saudi Arabia: Challenges and Strategies for Improvement. Chapter 7. In: S. Zekri (ed.), Water Policies in MENA Countries, Global Issues in Water Policy. Springer Nature Switzerland AG 2020 23, https://doi.org/10.1007/978-3-030-29274-4_7

Baig MB, Burgess PJ, Fike JH (2020) Agroforestry for healthy ecosystems: constraints, improvement strategies and extension in Pakistan. Agroforest Syst Published on line Jan 01, 2020 by Springer https://doi.org/10.1007/s10457-019-00467-4

Behnassi M, Yaya S (2011) Land resource governance from a sustainability and rural development perspective. In: Behnassi M, Shabbir SA, D'Silva J (eds) Sustainable agricultural development: recent approaches in resources management and environmentally-balanced production enhancement. Springer, DE, Berlin, Heidelberg, pp 3–23

Brandlay P, Marulanda C (2001)Simplified hydroponic to reduce global hunger. Acta Hortic 554:289–296. https://doi.org/10.17660/ActaHortic.2001.554.31 https://doi.org/10.17660/ActaHortic.2001.554.31

Chowdhury S, Al-Zahrani M (2012) Implications of climate change on water resources in Saudi Arabia. Arabian J Sci Eng (AJSE)

Chowdhury S, Al-Zahrani M (2013) Reuse of Treated Wastewater in Saudi Arabia: an assessment framework. J Water Reuse Desalination 3(3):297–314. https://doi.org/10.2166/wrd.2013.082

CTCS (2013) Agri-Food Sector Profile—Riyadh, Saudi Arabia. Produced by the Canadian Trade Commissioner Service, June 2013. Available at: https://www.futuredirections.org.au/wp-content/uploads/2015/07/Agri-Food-Sector-Profile-Saudi-Arabia.pdf.

DeNicola E, Aburizaiza OS, Siddique A, Khwaja H, Carpenter MD (2015) Climate change and water scarcity: the case of Saudi Arabia. Ann Glob Health 81:342–353

Drewes JE, Garduño PRC, and Amy GL (2012) Water reuse in the Kingdom of Saudi Arabia – status, prospects and research needs. Water Supply 12(6):926–936. https://doi.org/10.2166/ws.2012.063

Ecodna. Art (2019) Farming desert evolution

Elhadj E (2004) Camels Don't fly, deserts don't bloom: an assessment of saudi arabia's experiment in desert agriculture. occasional paper No 48. SOAS/KCL Water Research Group. Water Issues Study Group at the School of Oriental and African Studies (SOAS)/King's College London. University of London, UK May 2004

El Mostafa D, Al Assiri A (2010) Response to climate change in the Kingdom of Saudi Arabia. FAO-RNE, Cairo

Embassy of the Kingdom of Saudi Arabia (2020) Agriculture & Water. Washinton, DC USA. Available at: https://www.saudiembassy.net/agriculture-water

FAO (2008) Climate Change: Implications for Agriculture in the Near East [NERC/08/INF/5] Synopsis of the Twenty-Ninth FAO Regional Conference for the Near East held in Cairo, Egypt, 1–5 March 2008. Rome: Food and Agricultural Organization (FAO)

FAO (2011) FAO technical cooperation program, achievements, the technical cooperation between Kingdom of Saudi Arabia and Food and Agriculture Organization of the United Nations (FAO), FAO Office in Riyadh, Kingdom of Saudi Arabia

FAO-GIEWS (2014) Country brief—Saudi Arabia. GIEWS—global information and early warning system. Reference date Oct 27, 2014. Available at: http://www.fao.org/giews/countrybrief/country/SAU/pdf_archive/SAU_Archive.pdf

FAO (2015) Regional Coordination Mechanism (RCM) Issues brief for the arab sustainable development report sustainable agriculture and food security in the arab region. Available at: http://css.escwa.org.lb/SDPD/3572/Goal2.pdf

FAO (2016) Evaluation of FAO's technical cooperation assistance in the Kingdom of Saudi Arabia. Country Programme Evaluation Series. Food and Agriculture Organization of the United Nations. Office of Evaluation October, 2016. Rome Italy

FAO (2017). Country brief—Saudi Arabia. GIEWS—global information and early warning system. Available at: http://www.fao.org/giews/countrybrief/country.jsp?code=SAU&lang=en

FAO-GIEWS (2020) Country brief—Saudi Arabia. GIEWS—global information and early warning system. Available at: http://www.fao.org/giews/countrybrief/country.jsp?code=SAU&lang=en

FAOSTAT (2021) May of the Kingdom of Saudi Arabia. Available at: https://www.fao.org/faostat/en/#country/194

FAO and OECD (2021) Water and agriculture. A note on issues produced for the G20 Presidency of the Kingdom of Saudi Arabia. Rome, FAO. Available at: http://www.fao.org/3/cb2392en/CB2392EN.pdf

Frank Water (2021) Water Resources in KSA. Published on November 3rd, 2021. Available at: https://water.fanack.com/saudi-arabia/water-resources-in-ksa/

GCC Food Industry (2017)Report published by Alpen Capital. Available at: www.alpencapital.com/downloads/reports/2017/GCC-Food-Industry-Report-February-2017.pdf

Global Agriculture (2020) Land Grabbing. Agriculture at a Crossroads. Findings and recommendations for the future farming. Available at: https://www.globalagriculture.org/report-topics/land-grabbing.html

Haque MI, Khan MR (2020) Impact of climate change on food security in Saudi Arabia: a roadmap to agriculture-water sustainability. J Agribus Dev Emerg Econ. Emerald Publishing Limited. https://doi.org/10.1108/JADEE-06-2020-0127. Available at: https://www.emerald.com/insight/2044-0839.htm

Hartmann M, Khalil S, Bernet T, Ruhland F, Al Ghamdi A (2012)Organic Agriculture in Saudi Arabia. Sector Study (2012). Deutsche Gesellschaft für Internationale Zusammenarbeit GIZ (GmbH), Saudi Organic Farming Association (SOFA), Research Institute of Organic Agriculture (FiBL) & Ministry of Agriculture of Saudi Arabia (MoA), Riyadh, KSA 2012

Hasanean H, Almazroui M (2015) Rainfall: Features and variations over Saudi Arabia, a review. Climate 3:578–626. https://doi.org/10.3390/cli3030578

Karam S (2008) Saudi Arabia scraps wheat growing to save water. Available at: https://www.reuters.com/article/instant-article/idUSL08699206

Khan GA (2016) Organic farming becoming more popular with Saudis. Appeared in the Daily Arab News. Feb. 07, 2016. Available at: https://www.arabnews.com/saudi-arabia/news/876586

KSA Water Report (2020) Building a better food and water balance in Saudi Arabia. A report produced by Farrelly & Mitchell Business Consultants Ltd., pp 1–35. Available at: https://farrellymitchell.com/wp-content/uploads/2020/11/MARE11.pdf

Lovelle M (2015) Food and water security in the Kingdom of Saudi Arabia. Global Food and Water Crises Research Programme. Published by Future Directions International Pty Ltd. Australia. Future Direction. 28 July 2015 FDI Team https://www.futuredirections.org.au/publication/food-and-water-security-in-the-Kingdom-of-saudi-arabia/

Mahmoud MSA, Abdalla SMA (2013) Managing infrastructure water and petroleum demand in KSA by GIS. Int J Comput (IJC) 10.1 (2013): 18–41.

Malek C (2019) Ways Saudi Arabia is looking to save water. An article appeared in the Daily Arab News on January 15, 2019. Available at: https://www.arabnews.com/node/1435621/saudi-arabia

Market Research (2020) Available at: https://www.marketresearchsaudi.com/insight/million-riyal-to-support-saudi-farmers-going-organic

MAW (1994) Atlas of land resources of Saudi Arabia. Ministry of Agriculture and Water (MAW). Riyadh, Saudi Arabia

Mestre-Sanchís F, Feijóo-Bello ML (2009) Climate change and its marginalizing effect on agriculture. Ecol Econ 68(3):896–904

MEWA (2017) National environmental strategy: executive summary for the council of economic and development affairs. Ministry of Environment Water and Agriculture. Available at: https://www.mewa.gov.sa/en/Ministry/initiatives/SectorStrategy/Documents/6.%20BAH-MEWA-KSA%20NES-CEDA%20Executive%20Summary%20v3%2020180221%20ENG.pdf.

MEWA (2019) National Agricultural Strategy. Ministry of Environment, Water, & Agriculture. Riyadh, Kingdom of Saudi Arabia

MEWA (2020) Local agricultural production achieves high self-sufficiency in 2020. Ministry of Environment, Water, & Agriculture. Riyadh, Kingdom of Saudi Arabia. Available at: https://mewa.gov.sa/en/MediaCenter/News/Pages/News142010.aspx

MEWA (2020) https://www.arabnews.com/saudi-arabia/news/876586

MOEP (2010) The ninth development plan (2010–2014). The ministry of economy and planning, Saudi Arabia. Legal Deposit No.16/0694 ISSN: 1319–4836

Muhammad F (2014) Saudi Arabia to stop wheat production by 2016. Saudi Gazette. Available at: https://english.alarabiya.net/business/economy/2014/12/11/KSA-to-stop-wheat-production-by-2016

National Environment Strategy (2017)Executive Summary for the council of economic and development Affairs. Ministry of Environment, Water and Agriculture

Nations Encyclopedia (2020) Agriculture—Saudi Arabia. Available at: https://www.nationsencyclopedia.com/economies/Asia-and-the-Pacific/Saudi-Arabia-agriculture.html

Ouda OKM (2014) Impacts of agricultural policy on irrigation water demand: a case study of Saudi Arabia. Int J Water Resour Dev 30(2):282–292. https://doi.org/10.1080/07900627.2013.876330

Parry ML, Swaminathan MS (1993) Effect of climatic ccange on food production. In: Mintzer IM (ed) Confronting climatic change: risks, implications, and responses, Chap. 8. Cambridge University Press

Parry ML, Rosenzweig C, Iglesias A, Fischer G, Livermore M (1999) Climate change and world food security: a new assessment. Global Environ Change 9:S51–S67

Pellikka P, Alshaikh AY (2016) Remote sensing of the decrease of juniper woodlands in the mountains of Southwestern Saudi Arabia—reasons and consequences. Arab J Geosci (2016) 9: 457. https://doi.org/10.1007/s12517-016-2481-z

Presidency of Meteorology and Environment (PME) (2005) Saudi Arabia's "First National Communication" submitted to the United Nations Framework Convention on Climate Change

(UNFCCC). Submitted by the Presidency of Meteorology and Environment (PME). Ministry of Defense and Aviation Kingdom of Saudi Arabia

Ramadan MFA, Abdel-Hamid MMA, Altorgoman MMF, Al-Garamah HA, Alawi MA, Shati AA, Shweeta HA (2020) Evaluation of pesticide residues in vegetables from the Asir Region, Saudi Arabia. Molecules 25(1):205. MDPI AG. Accessed from https://doi.org/10.3390/molecules25010205

Şen Z, Al Alsheikh A, Al-Dakheel AM, Alamoud AI, Alhamid AA, El-Sebaay AS, Abu-Risheh AW (2011) Climate change and Water Harvesting possibilities in arid regions. Int J Global Warming 3(4):355–371

TADCO (2020) Water resources management, community resources and environmental protection. Available at: https://tadco-agri.com/en/corporate/

Taha SM (2014) Kingdom imports 80% of food products. An article based on Report of the World Bank appeared in the Daily Arab News on April 20, 2014. Available at: https://www.arabnews.com/news/558271

US-SABC (2010) Public-private agreements upgrade Saudi water infrastructure. U.S.-Saudi Business Brief. US-Saudi Arabia Business Council, Vienna, pp 9–10

Weatheronline (2020) Climate of the world: Saudi-Arabia. Available at: https://www.weatheronline.co.uk/reports/climate/Saudi-Arabia.htm

Woertz E (2011) Arab food, water, and the big land grab that wasn't. Brown J World Affairs XVIII:119–132

WTO (2020) Saudi Arabia appeals panel report regarding intellectual property measures. World Trade Organization. Available at: https://www.wto.org/english/news_e/news20_e/ds567apl_30jul20_e.htm

Chapter 10
Role of Agriculture Extension in Ensuring Food Security in the Context of Climate Change: State of the Art and Prospects for Reforms in Pakistan

Abdullah Bin Kamal, Muhammad Kamal Sheikh, Bismah Azhar, Muhammad Munir, Mirza Barjees Baig, and Michael R. Reed

Abstract Pakistan is predominantly an agrarian economy with a cropped area of 22 million ha, the largest integrated irrigation system of the Indus Basin, and a 19% share of agriculture in GDP. The agricultural economy is generating employment for 45% of total population and more than 60% directly or indirectly relying on it for livelihoods. This sector has been growing well above the population growth rate. Pakistan has made significant progress in food production over the last several decades. However, the agriculture sector has been facing a number of challenges, resulting in a performance less than its potential, with a growth rate of 3.3% over the last decade. The major factors include a slow rate of technological innovation and inadequate extension services and technology transfer. Due to this low performance, the country is a net importer of agricultural products despite being one of the world's large growers of wheat, rice, cotton, fruits, livestock, and livestock products. Subsequently, food security has remained a key challenge due to high population growth, rapid urbanization, low purchasing power, high price fluctuations, erratic food production, and inefficient food distribution systems. According to available figures, 18% of the population is undernourished with child wasting and stunting. Furthermore, climate change projections indicate that there will be more frequent

---

Present Address:
A. B. Kamal (✉) · B. Azhar
National Defence University, Islamabad, Pakistan

M. K. Sheikh
Planning & Development Division, Pakistan Agricultural Research Council, Islamabad, Pakistan

M. Munir
Pakistan Agricultural Research Council (PARC), Islamabad, Pakistan

M. B. Baig
Prince Sultan Institute for Environmental, Water and Desert Research, King Saud University, Riyadh, Saudi Arabia
e-mail: mbbaig@ksu.edu.sa

M. R. Reed
Emeritus of Economics, University of Kentucky, Lexington, KY, USA
e-mail: mrreed@email.uky.edu

extreme events such as floods and droughts. The agriculture sector will primarily be severely impacted by such changes, thus threatening food security. Hence the importance of coping strategies, including revamping of agricultural extension services given their pivotal role in ensuring food security. Extension services need capacity building to deal with the emerging challenges in agriculture. Sharing of agricultural knowledge and good practices in extension and advisory provision can go a long way in strengthening these capacities and enhancing productivity and rural incomes. The government, therefore, has considered starting mega projects with massive investment for agricultural production enhancement through reviving and upgrading the existing agricultural extension and adaptive research to meet the challenges of national food security.

Keywords Pakistan · Agriculture · Food security · Agricultural extension services · Climate change · Adaptation

1 Introduction

National food security is as important as the defence and security for countries such as Pakistan, with an economy greatly dependent on the performance of its agriculture sector. The livelihood of the majority of its population directly or indirectly depends on farming and off-farm activities. Climate change is posing many threats to the agrarian economy of Pakistan as it is affecting crops, livestock, fisheries, and biodiversity due to changes in temperatures at the critical stages of the crop cycle, changing patterns and uncertainty of rains, and frequent droughts and floods. These phenomena affect millions of people every year, pushing them below the poverty line. The government has given high priority to food security and has adopted various policy measures and started several programs to mitigate related challenges. Most of these programs and actions are taken by provincial governments and have evolved into a very elaborated system of agricultural extension and adaptive research. This has developed a very rich system of outreach, innovation testing and adaptation, staff and farmer training, technology and information dissemination, and input supply. There are also programs aimed at subsidizing inputs and outputs; such programs, which have various forms in different provinces, have employed various models of agricultural innovation and development from around the world with a variable success. Most agricultural innovation and development programs are implemented by provincial agriculture extension departments, which are playing an effective role in achieving domestic food security under climate change.

This chapter gives a brief account of this system of support for agriculture and its role in tackling climate change induced threats for food and nutrition security in Pakistan.

2 Population and Labour Force—The Profile of Pakistan

Pakistan has a population of more than 211 million people, the fifth most populated country in the world, and it is growing at a rate of 1.9% (GoP 2020). It is predominantly an agrarian economy with agriculture accounting for 19.3% of national GDP. The population density stands at 265 per Km^2. The majority of the population falls in the working age group of 15–64 years (61.4%). The total labour force in 2017–2018 was 65.5 million (GoP 2018a).

2.1 Agriculture Sector of Pakistan—An Overview

2.1.1 Agricultural Resource Base

Pakistan consists of five provinces: Punjab, Sindh, Khyber Pakhtoon Khwa (KPK), Balochistan, and Gilgit Baltistan. The total area of the country is 79.8 million ha and the economy is basically agrarian and heavily dependent on irrigation, largely confined to the Indus Plain. The country is part of the sub-continent south of the Himalayan mountains situated between longitude 61° and 76° E and latitude 24° and 37° N. The climate in the country is arid to semi-arid with temperatures ranging between 20 and 50 °C. Three-fourths of the country receive rainfall of less than 250 mm (mm) annually (ADB 2017).

Pakistan is blessed with a wide range of natural resources and has many ecological and climatic zones. Furthermore, arable land and water are among the major natural resources of Pakistan (Azam and Shafique 2017). The country has the world's largest integrated irrigation network serving 14 million ha of contiguous land fed by the Indus river and its tributaries. Of the 80 million ha of mainly arid and semi-arid land, 34 million ha are suitable for agro-forestry use. Approximately 22 million ha are under cultivation (Sheikh 2001).

2.1.2 Major Agricultural Crops and Products

There are two crop seasons, Kharif (summer) and Rabi (winter), with a limited choice of crops according to the weather in these seasons. Major Kharif crops include rice, cotton, maize, and sugarcane; the Rabi crops are mainly wheat and oilseeds. In addition to this, a number of fruits and vegetables (both winter and summer) are grown all over the country. Pakistan has large livestock herds and fisheries, but limited forestry resource (only 4% of land cover). Poultry is a very organized industry providing revenue of Pak Rs.700 billion and serving as a cheap source of meat and eggs. Average crop yields are much lower than the potential yields achieved at experimental stations and progressive farms (Sheikh 2001; GoP 2020).

Pakistan has a coastal line of approximately 1050 km, so fit for deep sea fishing and seaport, and fishing is an important income-generating activity. Even though fisheries account for a miniscule contribution of GDP (1%), they provide a substantial part of national income through exports of good quality seafood to various countries (Azam and Shafique 2017).

2.1.3 Contribution of the Agriculture Sector to Economy

Agriculture is the backbone of Pakistan's economy and it plays a significant role in economic development. Its annual growth rate has been 3% in the last few decades (Azam and Shafique 2017). Agriculture is the 3rd largest income-generating sector contributing more than 19% to the GDP after services (61%) and industry (20%). Within agriculture, the share of livestock is 63% and crops 33% (Fig. 1). It employs 45% of the total labour force and supports directly or indirectly about 68% of the population. Raw and processed agricultural commodities accounts for about 60% of export earnings. The sector provides food, feed, and raw materials for major industries, such as textiles, cotton ginning, sugar, beverages, leather, and dairy accounting for about 50% of the total industrial production (GoP 2020). It is thus evident that welfare of the vast majority of the population is critically dependent on agricultural resources of the country (Sheikh 2001).

3 Current Situation of Agricultural Growth (2019–2020)

During the period 2014 to 2020, national GDP growth was largely influenced by movement in agricultural GDP. In 2014, when agricultural GDP increased by 2.5%, the national GDP recorded a growth rate of 4.1%. During 2016, the agriculture sector

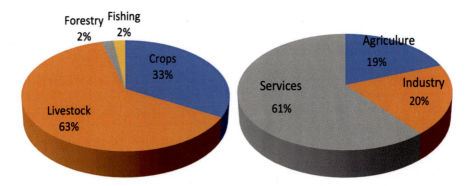

Fig. 1 Overview of Agriculture in Pakistan. *Source* Statistical Supplement of Pakistan Economic Survey 2019–2020 (GoP 2020)

10 Role of Agriculture Extension in Ensuring Food Security ...

suffered a setback and its growth was only 0.15%, mainly due to a negative growth in crops output (−5.27%). However, the services and industrial sectors performed better than expected and compensated for this low growth of agriculture sector. Consequently, the national economy grew by 4.6%. During 2019, the growth rate in agriculture was 0.58%, again due to a drop in crop output (−4.69%), and the economy grew by 1.9%. The growth/decline in crop GDP directly affects the agriculture and national GDP growth (GoP 2020).

Pakistani agriculture is largely dependent on five major subsectors: major crops, minor crops, livestock, fisheries, and forestry. The major crops are wheat, sugarcane, rice and cotton, which typically account for 6.5% of national GDP. Pakistan was the fourth largest producer of cotton, the major non-food crop used as a raw material by the textile industry. Rice, wheat, and sugar cane are among the major food crops; rice is one of the main exports of the country and sugar cane is used for sugar and sugar related products (GoP 2020).

The agriculture sector recorded strong growth of 2.67% in year 2019–2020, considerably higher than 0.58% growth achieved in 2018–2019. For Kharif crops, rice production increased by 2.9% to 7.410 million tonnes, maize production increased by 6.0% to 7.236 million tonnes, cotton production declined by 6.9% to 9.178 million bales, and sugarcane production increased by 0.4% to 66.880 million tonnes. Wheat is the most important Rabi crop, and its production increased by 2.5% to reach 24.946 million tonnes.

Thus, the performance of agriculture during 2019–20 remained remarkable. Overall, the sector recorded strong growth of 2.67% considerably higher than 0.58% growth achieved last year (Table 1).

Minor crops showed growth of 4.57% mainly due to an increase in production of pulses, oilseeds and vegetables. Thus, the crops sector experienced a remarkable growth of 2.98% for 2019–2020 (GoP 2020).

Livestock is an important sub-sector of agriculture and is comprised of buffalos, cows, goats, donkeys, horses, and poultry. Many rural households depend on livestock for daily food consumption and income. Livestock contributes to 11% of GDP with 58% of agricultural value addition. The livestock sector grew by 2.58% in 2019–2020. Pakistan is the fourth largest milk producer in the world after India, China, and

Table 1 Agriculture Growth (Base = 2005–2006) (%)

Sector	FY2014	FY2015	FY 2016	FY2017	FY 2018	FY 2019	FY2020P
National	4.1	4.1	4.6	5.2	5.5	1.9	−0.4
Agriculture	2.5	2.13	0.15	2.18	4.00	0.58	2.67
Crops	2.64	0.16	−5.27	1.22	4.69	−4.69	2.98
Livestock	2.48	3.99	3.36	2.99	3.70	3.82	2.58
Forestry	1.88	−12.45	14.31	−2.33	2.58	7.87	2.29
Fishing	0.98	5.75	3.25	1.23	1.62	0.80	0.60

P: provisional
Source Pakistan Bureau of Statistics/Economic Survey of Pakistan, 2019–2020 (GoP 2020)

the United States. Current milk production is estimated at 35 billion litters with 8 million households producing milk and a total herd size of approximately 50 million animals. In addition, the fishing sector grew by 0.60% while the forestry sector increased by 2.29% during the period 2019–2020 (GoP 2020).

In terms of labour participation, agriculture is the largest sector of economy with the majority of population directly or indirectly dependant on it for livelihood. However, during the last few decades, its contribution to GDP has gradually decreased to 19.3%. Yet there is a great potential for the sector to increase its share in GDP through increased productivity by utilizing the latest agricultural technologies. Being the sector with the largest workforce and providing raw materials to other manufacturing sectors, its development not only contributes in achieving poverty alleviation but also uplifts the socio-economic structure of the population (GoP 2020).

4 Challenges Facing Agriculture Sector in Pakistan

The agriculture sector has faced many major challenges over the last decade. As a result, the performance of this sector has been less than its potential, with low growth of around 3.3% over the last decade. Consequently, agricultural growth has not benefited the rural poor to the extent expected. The major factor behind this underperformance is that the wheat, rice, and sugarcane—being major food crops—were given more attention in previous policies. The sidelining of minor crops and fruits, vegetables, and livestock has resulted in low production of oilseeds, pulses, and quality fruits and vegetables. Other factors include: a slow rate of technological innovation; problems with the quality, quantity, and timeliness of input supply; inadequate extension services and technology transfer; limited investment in construction, road maintenance, and market infrastructure; marketing and trade restrictions; pest and livestock disease problems; feed and fodder shortages; limited credit for agricultural production and processing; and lack of agriculture-specific loan products (GoP 2018b).

Despite its tropical wet and dry climate, rich alluvial soils, and extensive irrigation, agriculture is confronted with stagnate yields, loss of fertile soils due to urbanization, dwindling water resources, land fragmentation, climate change, low pace of diversification, increasing costs of production, energy crisis (frequent and long hours of electricity supply interruptions), and issues related to marketing (Ali 2014). All these challenges are interlinked and have cause and effect relationships with agriculture productivity and growth.

5 The Role of Agriculture in Achieving National Food Security

The agriculture sector has the capacity to not only produce for domestic consumption, but also to have a surplus for export; so it can ensure food security as well as contributing in foreign exchange earnings (GoP 2020). Despite significant progress in Pakistani food production over the last several decades, food security is still a key challenge due to high population growth, rapid urbanization, low purchasing power, high price fluctuations, erratic food production, and inefficient food distribution systems. According to the Food Security Assessment Survey (FSA), 2016, 18% of population in Pakistan is undernourished. The National Institute of Population Studies (NIPS) reported a high level of severe stunting (45%), wasting (11%), and underweight (30%) among children (Table 2). The malnourishment problems are worse in rural areas (46% of rural population) (GoP 2018b).

Food security is a situation that exists "when all people, at all times, have physical, social and economic access to sufficient, safe and nutritious food that meets their dietary needs and food preferences for an active and healthy life" (World Food Summit Declaration 1996). Food security has four main determinants—availability, accessibility, utilization, and stability—which must be improved to ensure adequate nourishment and nutrition for all segments of the population (GoP 2018b). Achieving food and nutritional security for its population has remained one of the core underlying objectives of all policies, programs, and strategies of Pakistan since its independence (Sheikh 2001).

However, future strategies and initiatives in Pakistan are to be undertaken in light of Sustainable Development Goals (SDGs). The Ministry of National Food Security and Research (MNFS&R) prepared a comprehensive National Food Security Policy (2018). The first ever food security policy document, approved by the National Cabinet, manifests the 'Vision' of 'A Food Secure Pakistan'. The 'Mission' set in the document is to ensure a modern and efficient food production and distribution system that can contribute to food security and nutrition, according to its four basic

Table 2 Food and Nutrition Insecurity Situation in Pakistan

Indicator	(%)	Source
Prevalence of Under Nourishment (POU)	22 (41.4 m)	State of Food Insecurity (SOFI) Report (FAO, WFP and IFAD) 2016
Stunting (<5 children)	45	Pakistan Demographic and Health Survey (NIPS, Islamabad) 2012–13
Wasting (<5 children)	11	Pakistan Demographic and Health Survey (NIPS, Islamabad) 2012–13
Under weight (<5 children)	30	Pakistan Demographic and Health Survey (NIPS, Islamabad) 2012–13

Source National Institute of Population Studies (2013)

determinants. The goals of such a food security policy are to: alleviate poverty, eradicate hunger and malnutrition; promote sustainable food production systems (crop, livestock, and fisheries) by achieving an average growth rate of 4% per annum; and make agriculture more productive, profitable, climate resilient and competitive (GoP 2018b).

The major challenges to food security and agricultural development delineated in the policy document are: increase in focus on dietary diversity, nutritious and healthy food; enhancing the level of affordability for nutritious food by the poor segments of the society; improving the quality, quantity, and timing of supply of agricultural inputs; developing infrastructure and technologies for post-harvest management and value addition; improving the rate of diffusion of technological innovations; increasing farm gate prices, reducing price fluctuations and managing declining international prices; addressing market infrastructure requirements and trade restrictions; sustainable use of natural resources (land, water, rangelands, pastures, and forests), capitalizing on the potential of mountain agro-ecological zones; mitigating and adapting to climate change effects on agriculture and livestock; mainstreaming women's contribution in value-added agriculture and family nutrition; enhancing non-farm income opportunities, particularly in the marginalized and remote areas (i.e., mountains and deserts); promoting innovative livelihood practices, i.e., medicinal plants, fisheries, bee-keeping, local food products, seed production, rural poultry, and raising nurseries; improving per unit animal productivity and managing endemic livestock diseases; efficient utilization of land and water resources; securing qualified human resources for food security and food systems analysis; ensuring the placement of qualified persons in food departments; and considering water uncertainty due to Pakistan's status as a low riparian state in the semi-arid region (GoP 2018b).

A number of important policy initiatives have been taken in this direction, which include: The National Zero Hunger Program; the food security assessment survey; the recent commitment of the Government to SDGs, particularly the SDG-1 and SDG-2 about poverty and zero hunger challenges. The Prime Minister's Agriculture Emergency Program to enhance the productivity of crops, reduce hunger, and poverty initiated a number of mega projects involving federal and provincial agricultural R&D and extension departments. These programs allocated huge amounts of resources to implement climate resilience and adaptation strategies all over Pakistan (GoP 2018a). They will be discussed in details in the subsequent sections.

6 Climate Change and Food Security

Studies reveal that increasing temperature and changing pattern of rainfall have substantial impact on food production (Kirby et al. 2016; Janjua et al. 2010; Mahmood et al. 2012). A recent study anticipated that wheat production by 2050 will decrease by 50% in South Asia, almost 7% of the world crop production (CGIAR 2017). Lobell et al. (2011) estimated that the climate change has already reduced global yields of maize and wheat by 3.8% and 5.5%, respectively.

Accordingly, Pakistan needs to build a strong resilient agriculture sector to cope with climate change risks. Climate change projections indicate that there will be greater variability in the weather with more frequent extreme events such as floods and droughts. Much of the impact of these changes will be on agriculture sector, which needs mechanisms to cope and adapt. It is further projected that there will be immense pressure on limited area as well as ground water resources. These challenges could be managed through adopting soil and water conservation technologies, using high efficiency irrigation systems, developing drought resistant varieties, and introducing climate-smart agriculture (GoP 2020). Climate change disrupts food markets and poses threats to food supply. The risks can be reduced by increasing farmers' adaptive capacity and enhancing the resilience and resource-use efficiency in agriculture production system (Lipper et al. 2014).

Akhtar and Olaf (2017), in a study survey of climate change adaptation practices and impact on food security and poverty in Pakistan, found that 80% of farmers were aware of climate change and following three adaptation practices: adjustment in sowing time (22% households); use of drought tolerant crop varieties (15%); and shifting to new crops (25%). Farmers adopting more adaptation practices had higher food security levels (8–13%) and experienced less poverty (3–6%) than those who did not.

7 Impact of Climate Change on Agriculture in Pakistan

According to the Intergovernmental Panel on Climate Change (IPCC) fifth assessment report, agrarian countries, such as Pakistan, are more sensitive to climate change threats primarily because of their unique geography, climate, and distinct demographic features, coupled with socio-economic factors and a lack of adaptive capacity. All these factors taken together determine the country's vulnerability profile and the likelihood of climate change generating a vicious circle of poverty. As per the report, the main threats beyond the average climate change trends that will significantly impact Pakistan are the melting of glaciers and amount of precipitation in monsoon regions. Both will have negative consequences for efficiency and productivity of water dependent sectors, especially in agriculture (ADB 2017).

The most critical climate change threats that will exert significant pressure on agriculture and productivity in Pakistan are the increase in temperature, specifically in arable areas, changing rainfall patterns that are irregular and severe in nature, the increased variation and irregularity in the monsoon seasons, changed availability of irrigation water, and extreme weather events (Stern 2006; IPCC 2007; Karl et al. 2009) such as cyclones, floods, droughts, and heat and cold waves. Small-farm holders constitute 80% of the total farming community and are considered the most vulnerable to the adverse impacts of climate change in the country (Maskrey et al. 2007).

Pakistan is ranked 21st by the Global Climate Risk Index (GCRI) in terms of exposure to extreme weather conditions for the period 1993–2012 (Kreft et al., 2015).

According to the World Bank report, Pakistan was listed as the 12th most highly exposed country to climate change (Nomman and Schmitz 2011).

7.1 Impact of Climate Change on the Yield of Major Crops in Pakistan

Various studies on the impact of climate change on agriculture found that climate change—i.e., minimum and maximum temperatures, sunshine (radiation), irregular patterns of rainfall and humidity levels—directly and adversely affects the yield and production of major food (wheat, maize and rice) and cash crops (sugar cane). Climate change impacts are significantly felt in developing countries due to their poor infrastructure and a lack of coping and mitigation mechanisms. Climate change significantly affects Pakistan because the livelihoods of most of the rural households are dependent on agriculture. Since the impacts will be large, food security has remained a significant focus for policy, with a particular emphasis on the production and yields of rice, wheat, maize, and sugarcane.

Studies have revealed that high temperatures, prolonged winters, and irregular rain patterns have resulted in decreased production of wheat. Similarly, in the case of rice crops, the maximum temperature and less rainfall have negative impacts on yield; however, lower humidity levels played a positive role. Both rainfall and sunshine showed a negative impact on the yield of sugar cane. In contrast, the maximum and minimum temperature did not significantly affect the yield of sugar cane crop but rather had a positive implication.

Several studies on Pakistan show that cereals and other crops are vulnerable to heat stress due to higher temperature. A 1 °C rise in temperature would result in a 5–7% decline in wheat yield (Aggarwal and Sivakumar 2011). Sultana and Ali (2006) found that wheat production in arid, semi-arid, and sub-humid areas decline by 6–9%, while Hussain and Mudassar (2007) find that it can rise in humid areas of Pakistan. Studies further indicate that rice yield can decline by 15% from 2012 to 2039, 25% from 2040 to 2069, and 36% from 2070 to 2099, if the rise in temperature continues (Ahmad et al. 2013). Decreasing rainfall by 6% along with increasing temperature will require a 29% increase in net irrigation water requirements for Pakistan, which would negatively affect 1.3 million farm families as well as the production of most of cereals, fruits, and vegetables (GoP 2003).

Sajjad et al. (2017) studied the impact of climate change using changes in four major predictors: maximum and minimum temperatures, rainfall, sunshine, and humidity on the production/yields of four major crops (wheat, rice, maize and sugarcane). They related the change in production to the food security situation of Pakistan using data from 1989 to 2015 with the Feasible Generalized Least Square (FGLS) and Heteroscedasticity and Autocorrelation (HAC) models. Their results show that the maximum temperature has a significant negative impact whereas the minimum temperature has a positive and significant influence on the yield of wheat. Rainfall had a negative but non-significant effect on yield. Both the relative humidity and sunshine showed non-significant effects on the yield. The results further showed that 30% of the variation in wheat yield was due to changes in climate. In the case of rice crop, 33% of the yield variation is explained by climatic variations. Whereas for the maize crop, 39% and sugarcane crop 25% of exhibited yield variations are due to climatic factors. The study strongly suggested the need for developing heat and drought resistant high yielding crop varieties to mitigate climate change effects and ensure food security of Pakistan. It is believed that the increase in temperature can adversely affect the production due to increased requirements of irrigation water, increased evapotranspiration, and increased heat stress on crops.

Apart from increased temperature, the agriculture sector has experienced severe floods due to climate change, especially since 2010. These gigantic monsoon floods have overwhelming effects on crops, livestock, forestry, fisheries, infrastructure, and equipment. In 2010, floods destroyed almost 13.3 million tons of major crops, standing crops of approximately 2 million hectares, and a loss of 1.2 million livestock units. In 2011, the loss was even worse, with a total damage reaching USD 3.7 billion of major crops, minor crops, and livestock (Ali et al. 2017).

7.2 Impacts of Climate Change on Biodiversity in Pakistan

Climate change has negatively impacted forests, the coastal habitat including the marine life (fish, turtles, and other flora and fauna), mangroves, coastal wetlands, corals, and seagrass. It has also adversely affected biodiversity, resulting in shifts in the avian effect of the coastal habitat, including mangroves, coastal wetlands, corals, and seagrass. Climate change induced several changes in birds, including changes in avian phenology, poleward shifts in avian distributions, modification of migratory distances, direction and activity, and alterations to movement patterns and destinations. It has also produced negative impacts on avian biodiversity of wetlands due to changing water levels as revealed by a study of the Uchalli Wetland Complex (Qureshi and Ali 2011). Due to climate threats it is facing, the Indus is ranked third among the global vulnerable deltas. By 2050, 2.73% of the Indus delta area could be lost from sea-water incursion (ADB 2017).

7.3 Impacts of Climate Change on Water Resources in Pakistan

Climate change results in faster glacier melt and changes in river flows, leading to irrigation water shortages. The irregular, uncertain, and erratic patterns of rainfall affect arid and hyper arid areas. As a result of increased temperature, evapotranspiration also increases resulting in increased water demands of 10–30% for crops.

Melting glaciers in the Himalayas are projected to increase flooding within the next two to three decades. As the glaciers recede, there will be decreased river flow. The glacier lakes and their outburst, known as GLOFs, have already affected the lives and livelihoods of nearby communities (Ullah 2017).

Food security and water availability are highly vulnerable to a rapidly changing climate. All the models expect that summer season rainfall will increase (Mirza 1997; GoB and UNEP 2009). The studies estimate that 75% of Himalayan Glaciers have melted and will disappear by 2035. This glacial melt will cause droughts and floods (Misra 2014). It is expected that the major impact of climate change is going to hit the agriculture sector most due its dependence on water and climatic conditions (Mendelsohn 2001).

Climate change is also impacting river inflows and affect freshwater supply and quality. The Indus river for instance is highly threatened by climate change due to its dependence on glacial water. Originating from Mansarovar Lake in Tibet, the Indus River flows for about 3000 km into the Arabian Sea through the Indus Delta with a drainage area of 950,000 km^2 (including area in India) and an annual flow of 175 million-acre feet. Such a flow fluctuates more seasonally than any other river in Asia. Melting from Himalayan Glaciers account for 70–80% of the Indus' water. The Indus River already faces acute water shortages due to increased extraction for agricultural activities, causing saltwater intrusion into the Delta. The Indus Basin has also lost 90% of its forest cover. Nearly 10% of the water loss is due to evaporation, but 41 million-acre feet of water is lost annually by seepage from canals resulting in water logging and leaving the land unfit for agriculture (Qureshi and Ali 2011).

7.4 Impacts of Climate Change on Livestock in Pakistan

Climate change impacts livestock production particularly through higher temperatures, which result in increased animal stress, irregular conception and reproduction, productivity losses in terms of meat and milk, increased susceptibility to diseases and epidemics, reduction in the production of fodder for livestock, decreased quality of forage, and increased demand of water for animals and fodder crops. Climate change has negatively impacted livestock broadly in two main ways: by affecting the forage production; and by directly impacting livestock kept under various production systems. Furthermore, the losses faced by livestock are not only due to heat stress but also to disasters such as floods and droughts faced over the years (Younas et al. 2012).

It is evident that climate change's impact on livestock production has had further devastating consequences for the overall food security. These include reduced milk production, reproduction problems, and increased animal mortality. It has been observed that milk production and reproduction rates are directly affected by air temperature, humidity, and wind speed. Pakistan has experienced its worst droughts in history over the past 10 years. The persistent drought periods caused a shrinkage in the lactation period of the animals. It is estimated that the livestock production is likely to decrease by 20–30% in the next few decades due to increasing temperatures. This will ultimately increase the prices of meat and dairy products making it more difficult for lower- and middle-class families/households to purchase such items, causing a national food crisis. Climate change, poor infrastructure at the farm level, vulnerability and poor capacity of adopting mitigation strategies are the major challenges confronting dairy farming in Pakistan (Abbas et al. 2019).

Furthermore, extreme weather events, such as floods, droughts, and cyclones caused by climate change, can also adversely affect livestock. The 2010 floods alone inflicted a loss of 1.2 million livestock units; 1.16 million livestock units were killed and another 5.0 million were affected through diseases and dislocation (WFP 2010; NDMA 2011; GoP 2020). Not only were feeding and animal's health negatively affected, but extreme weather conditions also significantly reduced animal productivity. Feeding is considered the most important and expensive part of dairy production. Fodder quantity and quality are severely reduced due to increased temperatures and intensity of pests and diseases. Moreover, the fodder production period also diminishes due to persistent droughts and reduced levels of precipitation. Higher temperature (32–47 °C) and relative humidity regimes (33–75%) are shown to have a significant negative impact on buffaloes and calves (Younas et al. 2012).

8 Response Mechanisms to Face Climate Change in Pakistan

8.1 National Climate Change Policy—An Analysis

The Government of Pakistan has been very keen to understand and take necessary steps to mitigate and adapt to the negative impacts of climate change on various aspects of economy, human life, biodiversity, and natural resources. A Ministry of Climate Change was created in 2012 under the 18th Constitutional Amendment. The Ministry launched the First National Climate Change Policy (NCCP) in 2012, which proposed more than 120 measures to positively deal with climate change. The NCCP provides a framework to deal with current and future climate issues that will be faced by Pakistan. The Task Force on Climate Change and its 2010 report provided the building blocks for the elaboration of the NCCP. Taking into consideration the Pakistan's high vulnerability to climate change, the focus of this policy has been on mitigation and adaptation measures in various sectors, particularly agriculture, livestock, energy, water, forestry, and biodiversity (Mumtaz 2018).

As Pakistan's economy is largely dependent on agriculture, all vulnerable areas are therefore directly or indirectly linked with agriculture. The NCCP provides a complete framework for the development of a national adaptation and mitigation action plan. Moreover, other important components of the policy include measures related to disaster management including preparedness and response, capacity building, strengthening of institutions, introduction of climate change in curricula, ensuring environmental compliance through Initial Environmental Examination (IEE) and Environmental Impact Assessment (EIA), highlighting and resolving the issue of illegal timber trade and deforestation, and promoting Clean Development Mechanisms (CDM) with an overall goal to further enhance the stance of Pakistan regarding climate change on various forums (Mumtaz 2018).

Climate change adaptation strategies for agriculture include: micro level options, such as crop diversification and altering the timing of operations; market response such as income diversification and credit schemes; international changes, mainly government response including subsidies/taxes and improvement in agricultural markets; technological development such as development and promotion of new crop varieties and conservation technologies; many other adjustments at the local level under climate-smart agriculture (Ali and Erenstein 2017). The Agriculture Extension and Adaptive Research System (AE&ARS) in the provinces of Pakistan has remarkably responded to these strategies through new initiatives and begun to coop crate with the Federal Government under the Agricultural Emergency Program. The historical evolution, present structure, spread of the AE&ARS, and various initiatives taken in the area of climate change adaptation to achieve food and nutrition security are explained in the following sections.

8.2 Role of Agricultural Extension Services in Climate Mitigation and Adaptation Within the Agricultural Sector

8.2.1 Evolution of Agricultural Extension and Adaptive Research System

In order to fully understand the role of agricultural extension in achieving food and nutrition security in Pakistan, it is necessary to outline the evolution of the agricultural extension system and its present structure, functions, and workings. Gaps in understanding and mitigating the negative effects of the climate change and preparedness for climate-resilient agriculture and use of climate-smart technologies are also identified.

The extension system in Pakistan is inherited from the administration of British India. The system at partition was very weak and insufficient. Several indigenous and exogenous efforts were made to revamp the agricultural research and extension system. Since its founding, Pakistan started a number of rural and agricultural development projects and programs to increase the productivity and income of farm families. Some were national initiatives while others were supported by various international donor agencies. These include: The Village Cooperative Movement; the Village Agricultural and Industrial Development Programme (V-AID); Agricultural Development Corporations (ADCs); the Integrated Rural Development Program (IRDP); Agricultural Extension Services; and lastly the Training and Visit (T&V)

Program (Sheikh 2001). The evolution, successes, and failures of these extension efforts are analysed below.

The Village Cooperative Movement began immediately after independence under the aegis of the Cooperative Department. All farmers in every village were united under the umbrella of the village cooperative societies to choose their own management committees and find the means to development on a cooperative basis. The primary thrust was to educate member farmers about new technologies and to arrange farm input delivery on soft-term credit. However, the experience suggests that the cooperative movement was not able to achieve a consistent success and it was confronted with a plethora of problems (Agriculture Extension and Adaptive Research Wing, Government of Punjab).

The Village AID program was started in 1952, with substantial help from the US Aid Agency (USAID) and the Ford Foundation. It aimed to bring about all round development of the villages through organizing village councils, building roads, digging wells, constructing schools, and disseminating improved agricultural technology. This program achieved a good level of success in the beginning but became victim of departmental rifts and political change in the country (Sheikh 2001). This Village AID program was phased in 1959; moreover, rural development became a part of the Basic Democracies System (BDS) in 1961. It was designed to bring together the elements of community and political development, especially at the local level. The government administrative and development tiers were organized into five levels. The councils undertook a variety of social and economic development works in their respective areas. The problems that union councils tried to solve were in the realm of education, infrastructure, agriculture, and sanitation. The BDS went a long way in developing awareness and local leadership among rural masses, but met the same fate as its predecessor programs (Extension and Punjab 2020; Sheikh 2001).

In the early 1960s, agricultural development corporations, one each in the West and East (now Bangladesh) Wings of Pakistan were established with the idea to improve the performance of agriculture. Later, in 1970, each province organized its own agricultural supply and development corporation. In the Punjab province, it was named the Punjab Agricultural Development and Supply Corporation (PAD&SC) with the task to promote cooperatives, disseminate farm information, establish seed farms, procure and distribute improved seeds and chemical fertilizers, conserve soil, develop new lands, and rent out farm equipment and machinery to growers. Subsequently, its functions were cut back to the distribution of farm inputs and development of an extensive network of farm input delivery outlets across the country. They were entrusted to distribute information through pamphlets and posters, and to demonstrate improved technology through demonstration and production farms. Though the ADCs established a good input delivery system, they performed poorly in information dissemination (Extension and Punjab 2020).

The new government in 1970 eliminated the BDS and introduced a new rural development approach the 'Integrated Rural Development Program' (IRDP) with the main objective of agriculture development. The IRDP was created as a subsidiary of the Agriculture Department. Its leadership was heavily drawn from this agency, and all frontline workers used to run this program were agricultural graduates. On the other hand, rural development funds were controlled by the Local Government Department (Extension and Punjab 2020). This dichotomy in modus operandi not only resulted in a good deal of tension between the agencies, but also created frustration among the workers of this newly launched program.

The professionally skilled IRDP staff started a campaign to enhance agricultural productivity which had a tremendous impact on crop yields (Sheikh 2001). It integrated the functioning of various line departments and facilitated delivery of services to farmers at one point. However, this model also fell into the hands of hard departmental rigidities. Subsequently, in 1978, the IRDP was turned into a routine bureaucratic agency of the Local Government Department (Extension and Punjab 2020). Afterward, separate departments of Agricultural Extension Services were created under the provincial ministries of agriculture. The structure, organization, functions, and working were almost the same in each province (Sheikh 2001). The example of Punjab province is conveniently used for the elaboration of the system as Punjab has the largest number of private farms (more than 4.0 million) with a good organization in structure and function.

8.2.2 The Current System of Agriculture Extension

Under the World Bank Training and Visit (T&V) program of agricultural extension, the functions of technology transfer are clearly separated from functions of provision of inputs. Technology transfer was kept with agriculture extension in the public sector and the supply of inputs and services were handed over to the private sector. The T&V system could not continue as per program after the completion of the project in 1994–95, because the provincial governments could not arrange the required funds. Thus, the agriculture extension service created under the T&V system has gradually weakened and that resulted in a modified form of T&V. There were no regular staff training and no funds for traveling, thereby limiting the mobility of staff, causing adaptive research to be discontinued, and reducing the morale of extension staff. Hundreds of vacant positions of agriculture officers were never filled. This situation caused complaints from farmers against the extension service, and instead of addressing the debilitating causes, the Government put a further squeeze on the service (Extension and Punjab 2020).

Despite all such odds, the extension staff maintained contacts with farmers and organized field days and seminars. The pesticide private companies supported activities like training of the trainer's program and media extension. Extension in cotton and rice belts of Punjab province was organized differently due to their export potential. The growers were very receptive to improved production and protection practices. The electronic and print media replaced the traditional person-to-person contact of extension staff. The growers were encouraged to visit commodity research institutes and acquire state-of-the-art knowledge and seeds of new varieties (Extension and Punjab 2020).

The training of extension staff was organized on a regular basis employing modern training techniques. A monitoring and evaluation system for the field staff by district and provincial extension managers was developed. Use of information and communication technologies (ICTs) helped sub-district and district extension officers promptly feed the provincial government with the latest information about availability of seed, fertilizer, irrigation water, machinery, and other inputs in addition to observations on crop stand, prevalence of any insect, pest, or disease, anticipated yield/production levels, and marketing of the farm produce (Extension and Punjab 2020).

8.2.3 The Agriculture Extension and Adaptive Research

It was realised that agricultural production is a very complex system. It depends on several interrelated components such as development of appropriate production technology, dissemination of modern technology to end-users, and the formulation of farmer-friendly agricultural policies (Extension and Punjab 2020). Now, the Punjab is following a modified T&V program, where graduate extension agents visit groups of farmers at a specified time and date. The Farmers' Field School approach is also used for fruit and vegetable programs. The Plant-Wise approach of CABI is also used for plant diagnostics through expert contacts. ICTs are also deployed to reach large number of farmers (Ali 2014).

Dissemination of appropriate technology to farmers is highly important if benefits are to be derived from technological advances. Yet, huge gaps exist between research findings and farmers' fields, so farmers need appropriate research and advice on adapting technologies. This task is carried out through agriculture extension system in Punjab province at their ecology-based adaptive research farms.

Adaptive Research aims to devise and showcase site-specific technology packages for increasing agricultural production. It helps adjust the results of experiment station research into suitable form before transmitting it to farmers, keeping in view their local agro-climatic and socio-economic conditions. It bridges the gap between research findings and farmers' achievements and extension (Extension and Punjab 2020). The current system of Agriculture Extension and Adaptive Research is illustrated in Fig. 2.

a. **Functions of Agriculture Extension and Adaptive Research:**

Provincial-Level Functions

To achieve the objectives at the provincial level, the following functions are being performed by department officers at Headquarters (Extension and Punjab 2020):

- Preparation and printing of production technology/plan of crops.
- Fixing of area and production targets.
- Preparation of provincial-level development projects.
- Interaction with other provincial and federal governments.
- Amendments in agriculture laws and provincial-level implementation.

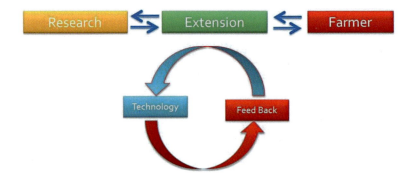

Fig. 2 Current system of Agriculture Extension and Adaptive Research. *Source* Extension and Punjab (2020)

- Monitoring of the district-level activities.
- Planning and testing of new strategies for technology and service transfer at the level of provincial cadres.

District-Level Functions

The district-level functions of the department performed by the field offices are (Extension and Punjab 2020):

- Implementation of production technology and achievement of area and production targets.
- Maintenance of agriculture statistics.
- Implementation of production technologies.
- Conduct village-level farmer trainings.
- Implementation of approved projects.
- Preparation of district-level projects.
- Implementation of agriculture laws.
- Monitoring of agriculture input availability.
- Participation in provincial review meetings.
- Feedback on researchable problems.

The Core Team

The core team of the Punjab Extension System consists of seven key positions, three of which have been vacant for a long time. The charge of these positions is entrusted to other members of the core team. At the end of 2016, the regular Director General of Extension joined the Federal research organization (Pakistan Agricultural Research Council) as a grade 21 officer (Member Plant Sciences), which further weakened the system. Recently, he has re-joined his parent department of Punjab Agricultural Extension and Adaptive Research, which has improved the system's functioning (Fig. 3).

10 Role of Agriculture Extension in Ensuring Food Security …

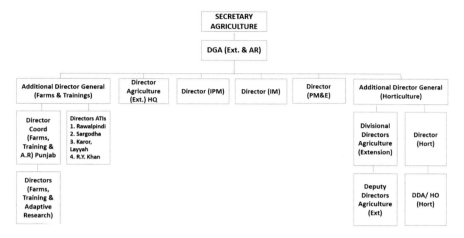

Fig. 3 The Organogram. *Source* Extension and Punjab (2020)

b. **Spread of Agricultural Extension and Adaptive Research (AEAR)**

Provincial Headquarters

The provincial Headquarters is in Lahore with six directorates, each responsible for different tasks. These include: The Director of Agriculture (Extension); the Director of Agriculture Coordination (Farms, Training and Adaptive Research); the Director of Agriculture (Horticulture); the Director of Agriculture (Input Management); and the Director of Agriculture (Project Monitoring and Evaluation), Director of Agriculture (IMP). All directors are housed at Agriculture House in Lahore (Extension and Punjab 2020). They are responsible for the performance of provincial-level functions of the AEAR as stated above.

In-service Agriculture Training Institutes

There are four In-service Agriculture Training Institutes, each headed by a director-level officer. These institutes are located at Rahim Yar Khan, Sargodha, Rawalpindi, and Layyah districts (Extension and Punjab 2020).

Divisional Directors

There are nine divisional-level directors, one each in Lahore, Gujranwala, Faisalabad, Multan, DG Khan, Rawalpindi, Sahiwal, Sargodha and Bahawalpur. These are called Directors of Agriculture Extension (Extension and Punjab 2020).

Adaptive Research Farms

There are seven Farmers' Training and Adaptive Research Farms headed by a director-level officer. These are located at Sheikhupura, Sargodha, Chakwal, Gujranwala, Rahim Yar Khan, Vehari, and Karor (Extension and Punjab 2020).

Deputy Directors of Agriculture Extension and Field Workers

There are 36 Deputy Director-level officers with their facilities and staff in each district. Beside this structure of officers at headquarters and district field stations, there are many field staff like Deputy District Officers Agriculture Extension (one in each of the 136 Tehsils, 3-4 tehsils in each district), and an Agriculture Officer for each Markaz. All the officers are graduates or have higher degrees in agriculture. Each Markaz has 6-7 Union Councils (UC) and each UC comprises 7-10 villages; an Agriculture Officer (AO) covers 45-50 villages. A Field Assistant (FA) works at each UC and has three years of experience and a Diploma in agriculture. The FAs work under the supervision of AOs located at each Tehsil. The FAs are the real front-line force of the agriculture extension system in each province. The success of the extension activities depends on the training and performance of these FAs. The system has a staff of more than 10,000 with all necessary facilities and infrastructure (Extension and Punjab 2020).

Main Services Provided

Main services provided by the Extension department are: Pre-Services and In-Services Trainings of extension staff; Technology Transfer; Production Technology Development; Plant Clinics (Plant Wise); Biological Control Laboratories; and Subsidies on Seeds and Fertilizers (Extension and Punjab 2020).

- *Pre-Services and In-Services Trainings*—A three-year Diploma Course in Agriculture Sciences (DAS) is offered and currently 842 pre-service trainees were enrolled in the Four Agricultural Training Institutes with 11 groups already completing the diploma. In-Service Training for Agriculture Experts is also arranged. Their knowledge in administrative, financial, and technical areas is enhanced (Extension and Punjab 2020).
- *Transfer of Technology*—Training of farmers is a regular activity of Agriculture Extension Services and is carried out in the following way: villages are allocated to every trainer by DOA (Ext.) with specific emphasis on small farmers; FA informs the farmers one day before the activity of the village-level farmer training programme; after which the farmers participate in various activities organised. Village-level trainings include literature distribution and technology demonstration and interactive sessions with strong Q&A (Fig. 4).
- *Production Technology Development*—Agriculture Extension Wing develops the production plans of various crops and vegetables. These plans are constructed and distributed among the field staff throughout the Province. The field staff arranges farmer meetings and guides farmers regarding the production technologies of various crops according to the plans. Copies of publications about the production technologies are also distributed among farmers (Extension and Punjab 2020).
- *Plant Clinics (Plant Wise)*—The Plant Clinic (Plant Wise) is the latest approach launched by Punjab Agriculture Extension under a global programme to enhance food security and improve rural livelihoods by reducing crop losses. This program is developed and led by the Centre for Agricultural Bioscience International (CABI). Working in close partnership with relevant actors, it strengthens national

Fig. 4 Farmers' training session in progress. *Source* Extension and Punjab (2020)

plant health systems to provide farmers with the knowledge they need to 'lose less and feed more'. This is achieved by establishing sustainable networks of local plant clinics, run by trained plant doctors, where farmers can find practical plant health advice. Plant clinics are reinforced by the Plant Wise knowledge bank, a gateway to online and offline actionable plant health information, including diagnostic resources, pest management advice, and front-line pest data for effective global vigilance. Plant clinics are established in the main market near the villages where practical health advice is provided to farmers (Fig. 5).

- *Biological Control Laboratories*—Eleven Biological Control laboratories were established in different districts of the Punjab for mass rearing of beneficial bio-agents (i.e., *Chrysoperla* and *Trichogramma*, to deal with pests through Integrated Pest Management). Suitable pest management techniques and methods are utilized to keep pest population below Economically Injurious Levels. This technique is environmentally safe and compatible with farmers' objectives. *Trichogramma* and *Chrysoperla* Cards are produced and distributed amongst farmers free of cost for installation in their crops. Extension staff also trains farmers on the proper use of these cards (Extension and Punjab 2020)
- *Subsidy on Seeds and Fertilizers*—The Government of Punjab provides massive support to farmers under the Smart Subsidy Program for Seeds and Fertilizers. This subsidy is available by E-Voucher Schemes from Branchless Banking Operators. Farmers register to avail themselves of the subsidy on Seeds and Fertilizers by

Fig. 5 Three Plant Clinics in Progress. *Source* Extension and Punjab (2020)

visiting any extension field office. Seeds of climate-resilient varieties of crops are promoted (Extension and Punjab 2020). Such crop promotion programs are run within special projects funded under the Public Sector Development Program (PSDP) of federal and provincial governments.

c. **Budgetary allocations (2008–2019)**

The budgetary allocations and expenditures have been erratic for both development and non-development budget (recurring expenditures). This suggests a lack of vision for long-term planning, poor focus on agriculture, administrative hurdles, and lacking infrastructure and facilitation. On the contrary, the donor-assisted projects made development budgets to go up manifolds. Furthermore, the matching government grants increased non-development budget under aid agreements in the same period. A record allocation of Rs.4396 million was made in the year 2019–20, but the real picture of expenses against this allocation will appear on 30th June at the end of the

Table 3 Budgetary allocations and expenditures incurred during last 12 years

	Development (PKR in Millions)		Non-Development (PKR in Millions)	
	llocation	Expenditure	Allocation	Expenditure
2019–2020	4396.669	0.047	2330.237	1434.293
2018–2019	239.482	228.535	5504.576	4964.518
2017–2018	448.729	435.590	5620.168	5526.962
2016–2017	1502.994	1293.883	2788.282	2995.453
2015–2016	1210.037	1020.766	745.701	713.290
2014–2015	269.713	268.825	497.848	449.634
2013–2014	54.632	53.980	451.643	552.299
2012–2013	2,123.286	2,119.540	426.802	407.684
2011–2012	295.092	287.497	500.506	491.020
2010–2011	136.915	125.309	1,152.977	1,149.904
2009–2010	919.523	919.417	249.770	250.642
2008–2009	387.132	384.765	398.554	369.255

financial year, when the reconciliations of governments accounts begin (Extension and Punjab 2020) (Table 3).

d. **Collaborating Partners**

The Federal government has a long list of collaborators including: the Food and Agriculture Organization (FAO) and all Consultative Group for International Agricultural Research (CGIAR) system agricultural R&D organizations; International Centre for Agricultural Research in Dry Areas (ICARDA); International Maize and Wheat Improvement Centre—CIMMYT; International Livestock Research Centre (ILRI); Asian Vegetable Research and Development Center (AVRDC); International Centre for Integrated Mountain Development (ICIMOD); International Food Policy Research Institute (IFPRI); International Water Management Institute (IWMI); World Food Program (WFP); and the International Crops Research Institute for the Semi-Arid Tropics (ICRISAT). Six collaborating partners are listed by the Punjab government (Extension and Punjab 2020): CABI; CIMMYT; Punjab Information Technology Board (PITB); USAID; United States Department of Agriculture (USDA); and the University of Agriculture, Faisalabad (UAF).

Private Extension Service Providers

Private sector service providers also help transfer innovative technologies in Punjab and other provinces. These companies provide inputs (seed, fertilizers, pesticides, and machinery) and have their fleet of professional staff working to promote technologies and their products. There are also NGOs, like the Rural Support Program Network, which contribute to the promotion of innovative climate-smart agriculture (CSA) practices (Ali 2014).

e. **Major Initiatives for Productivity Enhancement and Food Security Management**

Federal and Provincial Governments' Initiatives

The following eight mega projects were initiated under the Agricultural Emergency Program for a period of 4–5 years, starting in 2019 with a total cost of Rs.34.667 billion (Extension and Punjab 2020):

- Extension Service 2.0-Farmer Facilitation through Modernized Extension. 4 years (2019–23), cost: Rs. 500 million.
- National Program for Enhancing Profitability through Increasing Productivity of Wheat. 5 years (2019–24), cost: Rs.12535 million.
- National Program for Enhancing Profitability through Increasing Productivity of Rice. 5 years (2019–24), cost: Rs.9988 million.
- National Program for Enhancing Profitability through Increasing Productivity of Sugarcane. 5 years (2019–24), cost: Rs.2049 million.
- National Oilseeds Enhancement Program. 5 years (2019–24), cost: Rs.5490 million.
- Crop Maximization through Cooperative Farming. 2 years (2019–21), cost: Rs.400 million.
- Promotion of Medicinal Plants in Punjab. 4 years (2019–23), cost: 100 million.
- Promotion of Fruit Crop Production through Provision of True to Type/Certified Plants. 4 years (2019–23), cost: Rs.500 million.

National Programs for Enhancing Profitability through Productivity Enhancement of Wheat, Rice, Sugarcane, and oilseeds (under the Agriculture Emergency Program)

The Governments of Pakistan and Punjab have started many mega projects under the Prime Minister's Agriculture Emergency Programme for enhancing productivity of wheat, rice, sugarcane, and oilseed crops. These projects have been prepared after a thorough situation analysis and many discussions by experts, policy makers, and planners to address national food security under the threats of climate change. The Ministry of National Food Security and Pakistan Agricultural Research Council (PARC) took the lead and involved the provincial Departments of Agriculture and their Agriculture Extension and Adaptive Research DGs in the preparation of these projects. The main purpose was to mitigate climate change and adopt the necessary CSA practices for a sustainable national food security.

The projects' purposes are to:

- *Wheat* Enhance the productivity and profitability through: the increase in wheat yields of up-to 7 maunds[1]/acre[2]; the reduction of production cost; ensuring food security; increasing net income of farmers; providing export surplus; and diverting

---

[1] One Maund = 40 kg.

[2] An acre = 0.4046 hectares.

saved area for oilseed and high value crops to minimize import bills on edible oil. Inputs and farm machinery will be provided on subsidized prices.

- *Rice* Enhance rice farmers' profitability through the increase in productivity of Paddy up-to 10 maunds/acre of Basmati and 20 maunds/acre of coarse Rice; the area saved will be diverted for oilseed and high value crops. Enhance international competitiveness through improved quality of seeds. Employment opportunities will be increased through more business development resulting in increased share of rice crop in GDP.
- *Sugarcane* productivity enhancement of sugarcane crop with increase of yield up to 200 maunds/acres through harnessing modern R&D and mechanization technologies. The retrieved area will be diverted to other crops. The relevant CSA practices are also identified for specific regions and their crops are to be tested and adopted through transfer of technology (Extension & A.R, Punjab 2020).
- *Oilseed Crops* Increased productivity of oilseed crops through: increased yields with the objective to make oilseed crops profitable; addressing low yields by enhancing oilseed production both vertically (yield increase) and horizontally (area increase); promoting mechanization, providing quality seeds, and minimizing input bill of oilseeds. Per acre yield of Canola, Sunflower, and Sesame would be increased. In this way, net profit and income of farmers will increase, and area retrieved from other crops (wheat, rice and sugarcane) will be utilized for sowing of oilseed crops (Extension & A.R, Punjab 2020).

One good thing is that other provinces have joined these national projects with separate allocations obligated for a more widespread enhancement of crop productivity, improved livelihoods and income generation for rural people which will help ensure food security.

The Promotion of Fruit Crop Production through Provision of True to Type/Certified Plants, a four-year project (2019–2023) with a total cost of Rs. 500 million, is initiated to enhance productivity and profitability of fruit crops and to improve the socio-economic condition of the fruit growers. This will be accomplished through replacement of old, diseased, and un-economical fruit orchards by true-to-type disease-free, clean, and certified fruit plants for the enhancement of production for domestic consumption and export. The specific objectives of the project are: food security management in a holistic manner with an emphasis on the use of innovative technologies to improve vertical and horizontal crop productivity; and the increase of farmers' income. Better support prices and diversified agriculture will ensure input quality and purity. Development and adoption of new technologies for enhancing crop productivity will also improve the living standard of farming community and reduce poverty. It is envisaged to transform the Punjab's agriculture into a market-driven, diversified and sustainable sector through integrated climate-smart technology (Extension and Punjab 2020).

Another important initiative, Extension Services 2.0—Farmers Facilitation through Modernized Extension, started in 2015 for five years with a total cost of Rs.4105 million. Its major objectives are: technologically transforming the agriculture sector in Punjab by increasing crop productivity; expanding the area under cultivation, and optimizing the crop mix. The project will reach out to millions of farmers across Punjab with localized and specific information and advice to improve crop productivity with the aid of technology; call centres supported by a panel of experts had been established to get technical support and advise according to local conditions, and to integrate with the Pakistan Meteorological Department for accurate, area-specific, and prompt weather information for agriculture related activities. The project makes heavy use of ICTs in its different forms to modernize the extension system (Extension and Punjab 2020).

These programs and projects have included all climate change mitigation and adaptation micro-level options, such as: crop diversification and altering the timing of operations; market response—income diversification and credit schemes; international changes—ainly government response—concerning subsidies/taxes and improvement in agricultural markets; technological development—development and promotions of short duration, early maturity drought and heat tolerant high yielding crop varieties, and resource conservation technologies; and many other adjustments at the local level under CSA (GoP 2020).

9 CSA Practices in Pakistan

Climate-smart technologies and practices offer opportunities to address climate change challenges, economic growth, and the development of agriculture sector. These practices enhance food security and lead to mitigation and adaptation. Hundreds of practices around the world fall under CSA. For Pakistan, these include, but are not limited to, addressing drought, heat, and flood tolerance; improved crop and livestock varieties; integrated pest management; soil and manure management; renewable energy technologies; and water management strategies. The focus is also placed on improved dissemination of under-utilized, low-barrier adaptation and mitigation technologies. In addition to the adjustment of crop calendars and planting dates, the use of agronomic practices, agroforestry, intercropping, and crop rotation are increasingly important CSA practices. High efficiency irrigation and alternate renewable energy are also popular CSA practices (CIAT; World Bank 2017).

The FAO introduced the system of rice intensification in Sindh province. The farmers trained in the newly established farmers field schools adopted the Direct Seeded Rice (DSR) to directly sow rice in the fields, which had water saving of 30%, reduced labour requirements and production costs, thus enabling farmers to earn additional income by selling extra seedlings to neighbouring farmers. They experienced good yields without nursery raising and laborious transplanting in standing waters. The additional income was used for livestock, better family nutrition, and health and education of children (CIAT; World Bank 2017).

10 Current Challenges to Agricultural Extension Services in Pakistan

There are more than 4 million private farms in the Punjab, so the task for extension services is heavy. There is one graduate extension worker for every 9500 farm families. An extension officer can only reach a typical village every 40–45 days. Moreover, the extension service has weakened in Pakistan due to decentralization, to district government under devolution. Extension is a lower priority under the district government, so funds are less available. The extension must currently deal with many new challenges including post-harvest losses, declining soil and water health, adaptation to climate change, and linking small farmers to markets. In order to address these challenges, extension staff needs regular training on technical backstopping from national and international organizations. At the regional level, there is a lack of any kind of professional and technical collaboration for the sharing of viable, economical, adaptable, and sustainable technologies across the countries (Ali 2014).

Given such a situation, actions should be undertaken to overcome current challenges. The following recommendations maybe useful:

- Reforms are needed in agriculture extension and Adaptive Research System. This could be facilitated by the network of Agriculture Extension in South Asia (AESA) and help change the fate of small farmers.
- Continued support is needed in terms of finance and staff training. New and pragmatic approaches need to be identified, piloted, scaled up, and scaled out for the use of many farm households. Extension services in Pakistan need capacity building to deal with emerging challenges in agriculture.
- New programs or project preparations through effective planning are needed for continued and vibrant agriculture extension system support.
- More emphasis on monitoring and evaluation of activities for feedback and fine tuning.
- High priority in government policies with strong vision for the introduction of new extension approaches based on regional and international experiences.
- Sharing of agricultural knowledge and good practices in extension and advisory provision in the region can go a long way in strengthening these capacities and enhancing productivity and rural incomes (Ali 2014). Increased collaboration at the regional level for sharing of professional and technical experiences.
- Private sector, international donors, and NGO/GO-led CSA activities may be encouraged to reach out to the largest number of farm households, especially in the areas most affected by climate change.
- The cost of climate change adaptation needs to be borne by donors, local and federal governments, NGOs, farmers, and private sector service providers. The estimated annual cost of implementing climate change adaptation actions is US$7–14 billion (CIAT and the World Bank 2017). A larger part of financial resources has to be channelled through agriculture extension and Adaptive

Research System for implementation of climate change mitigation and adaptation strategies to enhance crop productivity for a sustainable national food and nutrition security.

11 Conclusion

Several initiatives have been undertaken by federal and Punjab governments under an emergency program to enhance the productivity of various crops. There are a number of technical, administrative, financial, and policy actions proposed and being implemented for climate change mitigation and adaptation as highlighted in various studies and reports. These initiatives will provide quality seeds, fertilizers, pesticides, machinery, and other inputs at a subsidised cost for timely completion of farm operations; resulting in less pre and postharvest losses, savings in farm inputs like water, fertilizers, and other crop management practices, and increased crop yields. There will be price subsidies for many crops, improved fruit plants nurseries, and better infrastructure with a regulated fruit plants nursery production system.

The wheat, rice, sugarcane, oilseed crops, and fruits and vegetables projects will provide climate-resilient varieties. The focus is on the supply of improved production technology and innovative knowhow, timely provision of market and weather information, farmer and extension staff training, and demonstration of innovative conservation technologies. These projects fill many gaps in extension service facilities and infrastructure (transport, office, field and training equipment, machinery, and operational funds) that will revamp and modernize the system.

The efforts made will provide climate-smart and innovative technologies to address food security management through the improvement of income and livelihood of rural and urban poor by providing them income-generating value addition activities. Efforts are being also made to produce more crops per unit of input, lower the cost of production, and make the abundant nutritious food accessible to a growing population; therefore contributing in fighting hunger. However, these actions are always prone to internal and external shocks, hence regular and detailed M&E across all provinces is required for feedback and fine tuning of policies, approaches, and activities.

References

Abbass Q, Han J, Adeel A, Ullah S (2019) Dairy Production Under Climatic Risks: Perception, Perceived Impacts and Adaptations in Punjab, Pakistan. Int J Environ Res Public Health 16:4036

Asian Development Bank (ADB) (2017) Climate Change Profile of Pakistan. Asian Development Bank. Manila, Philippines

Aggarwal P, Sivakumar MV (2011) Global climate change and food security in South Asia: An adaptation and mitigation framework. In: Climate change and food security in South Asia. Springer, Berlin, pp 253–275

Ahmad M, Iqbal M, Khan M (2013) Climate change, agriculture and food security in Pakistan: adaptation options and strategies. Brief, Pakistan Institute of Development Economics, Islamabad

Akhter A, Olaf EA (2017) Assessing farmer use of climate change adaptation practices and impacts on food security and poverty in Pakistan. Climate Risk Management. Elsevier

Ali MA (2014) Face to Face. Agriculture Extension in South Asia (aesa). http://www.aesa-gfras.net

Qureshi N, Ali Z (2011) Climate Change, Biodiversity Pakistan's Scenario. J Plants Anim Sci 21:358–363

Ali S, Liu Y, Ishaq M, Shah T, Abdullah., Ilyas, A., and Ud Din, I. (2017) Climate change and its impact on the yield of major food crops: evidence from Pakistan. Foods 6(6):39

Azam A, Shafique M (2017) Agriculture in Pakistan and its impact on economy. Int J Adv Sci Technol 103:47–50

CIAT and the World Bank (2017) Climate-smart agriculture in Pakistan. CSA country profiles for Asia Series. In: International Center for Tropical Agriculture (CIAT), The World Bank. Washington, D.C. p 28

CGIAR(2017) CGIAR System Annual Performance Report 2017

Directorate General Agriculture (Extension & A.R) Punjab (2020) retrieved from: Extension & Adaptive Research, agripunjab.gov.pk

Facing Climate Change by Securing Water for Food, Livelihoods and Ecosystems. Available online: http://www.iwmi.cgiar.org/wp-content/uploads/2013/02/sp11.pdf. Accessed 21 May 2017

Gamty D, Henry K, Hottle R, Jackson L, Jarvis A, Kossam F, Mann W, McCarthy N, Meybeck A, Neufeldt H, Remington T, Thi Sen P, Sessa R, Shula R, Tibu A, Torquebiau EF (2014) Climate smart agriculture for food security. Nat Clim Change (4)

Government of Bangladesh (GoB) United Nations Development Program (UNDP) (2009) The probable impacts of climate change on poverty and economic growth and options of coping with adverse effects of climate change in Bangladesh. Policy Study; Dhaka, Bangladesh

Government of Pakistan (2003) Establishment Division, Government of Pakistan, Islamabad

Government of Pakistan (GoP) (2018a) Labour Force Survey 2017–2018 (Annual Report). Pakistan Bureau of Statistics, Islamabad (34)

Government of Pakistan (GoP) (2018b) National Food Security Policy 2018. Ministry of National Food Security and Research, Islamabad

Government of Pakistan (GoP) (2020) Pakistan Economic Survey 2019–2020. Ministry of Finance, Islamabad Government of Punjab, Agriculture Extension and Adaptive Research Wing. Accessed from: http://ext.agripunjab.gov.pk/

Hussain SS, Mudasser M (2007) Prospects for wheat production under changing climate in mountain areas of Pakistan–an econometric analysis. Agric Syst 94(2007):494–501

IPCC (2007) IPCC Climate change impacts, adaptation and vulnerability-working group II contribution to the intergovernmental panel on climate change: summary for policymakers. IPCC Secretariat, Geneva, Switzerland (2007)

Janjua PZ, Samad G, Khan NU, Nasir M (2010) Impact of climate change on wheat production: a case study of Pakistan. Pak Dev Rev 2010(49):799–822

Karl T, Melillo J, Peterson T, Hassol S (2009) Global climate change impacts in the United States. Cambridge University Press, New York, NY

Kirby JM., Mainuddin M, Mpelasoka F, Ahmad MD, Palash W, Quadir ME, Shah-Newaz SM, Hossain MM (2016) The impact of climate change on regional water balances in Bangladesh. Clim Chang 2016(135):481–491. https://doi.org/10.1007/s10584-016-1597-1

Kreft S, Eckstein D, Junghan L, Kerestan C, Hagen U (2015) Global climate risk index 2015: Who suffers most from extreme weather events? Weather-Related loss events in 2013 and 1994 to 2013. German Watch. Bonn, Germany

Lipper L, Thornton P, Campbell BM, Baedeker T, Braimoh A, Bwalya M, Caron P, Cattaneo A, Gamty D, Henry K, Hottle R, Jackson L, Jarvis A, Kossam F, Mann W, McCarthy N, Meybeck A, Neufeldt H, Remington T, Thi Sen P, Sessa R, Shula R, Tibu A, Torquebiau EF (2014) Climate smart agriculture for food security. Nat Clim Change 4

Lobell D, Schlenker W, Costa-Roberts J (2011) climate trends and global crop production since 1980. Science 333:616–620

Mahmood N, Ahmad B, Hassen S, Baskh K (2012) Impact of temperature and precipitation on rice productivity in rice-wheat cropping system of Punjab province. J Anim Plant Sci 2012(22):993–997

Maskrey A, Buescher G, Peduzzi P, Schaerpf C (2007) Disaster risk reduction: 2007. In: Global review. Consultation edition. Prepared for the global platform for disaster risk reduction first session. Geneva, Switzerland, pp 5–7

Mendelsohn R (2001) Global warming and the American economy. In: Oates WE, Folmer H (eds) New horizons in environmental economics. Edward Elga, Broadheath, UK

Mirza MMQ (1997) PhD Thesis. Modeling the effects of climate change on flooding in Bangladesh. International Global Change Institute (IGCI). University of Waikato, Hamilton, New Zealand

Misra AK (2014) Climate change and challenges of water and food security. Int J Sustain Built Environ 2014(3):153–165. https://doi.org/10.1016/j.ijsbe.2014.04.006

Mumtaz N (2018) The national climate change policy of Pakistan: An evaluation of its impact on institutional change. Springer. Earth's Syst Environ. https://doi.org/10.1007/s41748-018-0062-x

National Disaster Management Authority (NDMA). (2011). Flood rapid response plan. NDMA; Islamabad, Pakistan.

National Institute of Population Studies [Pakistan] and ICF International (2013) Pakistan Demographic and Health Survey 2012–2013. Calverton, Maryland, USA: National Institute of Population Studies and ICF International

Nomman AM, Schmitz M (2011) Economic assessment of the impact of climate change on the agriculture of Pakistan. Bus Econ Horiz 2011(04):1–12. https://doi.org/10.15208/beh.2011

Sajjad A, Ying L, Tariq S, Aasir I, Izhar U (2017) Climate change and its impact on the yield of major crops: evidence from Pakistan. Accessed from: www.mdpi.com/journals/foods

Shahzad U (2015) Global warming: causes, effects and solutions. Durreesamin J 1:4

Sheikh MK (2001) Agricultural research and extension systems in Pakistan. In: Goraya MM, Razzaque MA, Abdullah M (eds) Agricultural research and extension systems in SAARC countries. SAARC Agricultural Information Centre (SAIC), Dhaka, pp 87–106

Stern N (2006) What is the economics of climate change? World Econ Henley Thames 7(2):1

Sultana H, Ali N (2006) Vulnerability of wheat production in different climatic zones of Pakistan under climate change scenarios using CSM-CERES-wheat model. In: Second international young scientists' global change conference. Beijing, pp 7–9

Ullah S (2017) Climate change impact on agriculture of Pakistan - A leading agent to food security. Int J Environ Sci Nat Resour 6:3

World Food Programme (WFP) (2010) Pakistan flood impact assessment. WFP, Rome, Italy

World Food Summit Declaration (1996) Rome declaration and plan of action. Accessed from: http://www.fao.org/3/w3613e/w3613e00.htm

Younas M, Ishaq K, Ali I (2012) Effect of climate change on livestock production in Pakistan. In: Jakarta: proceeding of the 2nd international seminar on animal industry

Chapter 11
Role of Agricultural Extension in Building Climate Resilience of Food Security in Ethiopia

Burhan Ozkan, Ahmed Kasim Dube, and Michael R. Reed

Abstract Climate change, related disasters and extreme events are currently a leading cause of food insecurity, affecting all its dimensions. Globally, these factors increase the food insecurity risk at unprecedented levels, thus undermining current efforts to eradicate hunger and undernutrition. Regionally, despite the fact that agriculture is the backbone of many sub-Saharan African countries' economy, food production has failed to keep up with fast rising food demands. Challenges are expected to intensify given the extreme climate vulnerability of agriculture. Reduced rainfall, more frequent extreme weather events, and hotter and shorter growing seasons are affecting the productivity of crops, livestock, fishing, and forestry across the region, resulting in serious implications for food and nutrition security. In this context, Ethiopia has been identified as one of the most vulnerable countries since the variability of rainfall and increasing temperature cause frequent droughts and famines, with disastrous impacts on peoples' livelihood. To counter such trends, there is a need to build climate resilience in a way ensuring that a growing population can be fed sustainably without further depletion of natural resources. This requires helping people cope with current changes, adapt their livelihoods, and improve governance systems so they are better able to manage risks in the future. It also requires long-term strategies and interventions that build on agro-ecological knowledge to enable smallholder farmers to counter the impacts of environmental degradation and climate change. Consequently, for development to be climate-resilient, farmers will require greater access to technologies, markets, information, and credit for investment to adapt their production systems and practices. In this case, agricultural extension services may provide an opportunity for strengthening the resilience of rural and farming households. In such a perspective, this chapter assesses how agricultural extension services may play a critical role in promoting agricultural and rural development and improving the resilience of the sector as a whole in Ethiopia.

B. Ozkan (✉) · A. K. Dube
Faculty of Agriculture, Department of Agricultural Economics, Akdeniz University, Antalya, Turkey
e-mail: bozkan@akdeniz.edu.tr

M. R. Reed
Agricultural Economics, University of Kentucky, Lexington, KY, USA
e-mail: Michael.Reed@uky.edu

Keywords Agricultural extension services · Climate change · Food security · Climate resilience · Ethiopia

1 Introduction

Agriculture remains the economic mainstay of many countries in sub-Saharan Africa, employing about 60% of the workforce and contributing approximately 30% of gross domestic product (GDP) (Thornton et al. 2011). In East Africa, the agricultural sector provides livelihoods to 80% of the population, and contributes about 40% of the GDP. Agriculture remains one of the most important sectors in the Ethiopian economy for the following reasons: it directly supports about 83% of the population in terms of employment and livelihood; it contributes over 40% of the country's gross domestic product (GDP); it generates about 85% of export earnings, and it supplies around 73% of the raw material requirements of agro-based domestic industries, such as biofuels (AfDB 2011). Thus, agriculture contributes a remarkable proportion to the Ethiopian national economy, and the GDP is highly correlated to the performance of agriculture sector (FAO 2016).

Agriculture in Ethiopia includes crops, livestock, forestry, fisheries, and apiculture. The existence of diverse agro-ecological conditions enables Ethiopia to grow a large variety of crops, which include cereals, pulses, oil seeds, and different types of fruits and vegetables (CSA 2017b). Crop production is estimated to contribute about 60% of the total agricultural value. Since 2006/07, the share of crop agriculture from GDP exceeded 30% and its share from agricultural gross domestic product (AGDP) exceeded 65% (NBE 2017). With an estimated 59.5 million cattle, 31 million sheep, 30 million goats, 2.16 million horses, 8.44 million donkeys, 0.41 million mules, and nearly one million camels, Ethiopia's livestock has also a lot of potential (CSA 2017a). It is the source of many social and economic values such as food, draught power, fuel, cash income, security, and investment in both highlands and lowlands as well as pastoral farming systems. As in the case of crops, this sector also makes a significant contribution to GDP and is a major source of foreign exchange (FAO 2016). The shares of the livestock, hunting subsector, and forestry were 12.3% and 3.9% from GDP and 26.5% and 8.4% from AGDP, respectively (Table 1) (NBE 2017).

The government of Ethiopia has given top priority to the agricultural sector and has taken a number of steps to increase its productivity. Over the last four decades, the country's agricultural and rural development policies and strategies have changed to keep pace with the economic development and rural transformation goals of the regimes.

However, agriculture in Ethiopia is almost exclusively rain fed (1% to 3% of cultivated land is irrigated) and dominated by small-scale subsistence farmers (about 8 million households) practicing traditional methods. Over 95% of the annual gross total agricultural output of the country is said to be generated from smallholder farmers. The contribution of medium to large-scale commercial farms to gross

Table 1 Shares of crop, livestock, forestry, and fishery in AGDP (percent)

Sector		2012/2003	2013/2014	2014/2015	2015/2016	2016/2017
Growth rate	Crop	8.2	6.6	7.2	3.4	8.1
	Animal farming and hunting	5.2	2.1	4.7	-1.5	4.5
	Forestry	3.3	4.2	3.5	2.2	3.5
	Fishing	19.4	32.5	30.6	0.1	0.5
Share in agriculture	Crop	69.8	70.6	71.1	71.9	65.3
	Animal farming and hunting	21.3	20.6	20.3	19.5	25.3
	Forestry	8.8	8.7	8.4	8.4	8.9
	Fishing	0.1	0.2	0.2	0.2	0.2

Source: NBE (2017)

total agricultural output is only about 5% (CSA 2017b). Many farmers grow slow maturing, high yielding 'long cycle' crops for which they need both the Belg and the Kiremt season (February-September). These crops are highly vulnerable to changes in seasonal rainfall (Gizachew and Shimelis 2014; USGS et al. 2012). In addition, as land is further divided by each generation, most plots are less than 0.5 hectare—insufficient for household food security or adequate income. These small plots reduce the households' capacity to adapt to climate change or to invest in improved farming methods (Yirgu et al. 2013). Rainfall unpredictability is moreover proving a great disincentive to invest in agricultural improvements, suggesting that it has a double adverse effect on productivity (Cipryk 2009). Other contributing factors include limited use of improved seed and fertilizers and inadequately resourced agricultural extension systems. Moreover, the dry lowlands experience erratic rainfall at times with very severe droughts, the impact of which, together with land degradation, human population growth and climate change, has greatly impaired the country's economic and social development and its food security status (FAO 2016).

2 Vulnerability of Ethiopia to Climate Change

Ethiopia's history is associated with major natural and man-made disasters that have affected the population from time to time. The country is also highly vulnerable to climate change as it is ranked the 22nd most vulnerable and the 31st least ready country. An integrated vulnerability assessment of climate change impacts in Ethiopia's regional states found that the top four vulnerable states are Afar, Somali, Oromia, and Tigray, which are all heavily agriculture-dependent (Aragie 2013) and are among the poorest ones in the country. A staggering 90% of all people in Tigray and Afar and 60–70% of those in Oromia and Somali live with less than USD 2 per day (Admassu et al. 2013).

The most common climate-related hazards in Ethiopia include drought, floods, heavy rains, strong winds, frost, and heat waves (NMA 2007). These climate-related disasters, particularly drought, have significant negative impacts on agriculture, rural livelihoods, food security, and economic development. In 2011, Ethiopia was ranked 5th out of 184 countries in in terms of drought risk (Swarup et al. 2011). Between 1900 and 2010, twelve extreme droughts were recorded—killing over 400,000 people and affecting over 54 million (You and Ringler 2010)—of which seven occurred since 1980 (World Bank 2010). The majority of these events resulted in famines: for instance, from the year 2000 up to 2005, 3.8 million to 12.2 million people were estimated to be in need of food assistance, and the number remained around 8 million until 2010/2011(Adem and Bewket 2011). Among these people, 2 million children were malnourished due to drought-induced food shortages. Studies indicate that crop yields have declined and will continue to decline causing food insecurity. Climate change-induced losses in yield, which lead to cereal price increases in international markets, also lead to high cereal prices nationally (Mahoo et al. 2013).

Furthermore, the severe drought of 2015–2016 was worsened by the strongest El Nino in decades, causing successive harvest failures and widespread livestock deaths in some regions. Apart from these major or extreme droughts, there have been dozens of local droughts with equally devastating effects. The country has experienced even more major floods in different parts of the country; 47 major floods since 1900, of which six since 1980 (World Bank 2010) killed almost 2,000 people and affected 2.2 million. Ethiopia was ranked 34th out of 162 countries in terms of flooding risk, and 5th out of 162 in terms of landslide risk (You and Ringler 2010).

Climate change is having and will continue to have significant economic effects on the Ethiopian economy, and especially the agricultural sector, because of many factors such as the loss of arable land due to shifting agro-ecological zones, altered growing cycles that delay planting, and increased incidence of pests and diseases. Ethiopia's livestock sector will be increasingly affected by drought and degradation of land. It has been estimated that climate change will affect the country's GDP growth by 0.5–2.5% per year in the near future with the potential reduction of GDP up to 10% by 2045, primarily through its impact on agricultural productivity (Eshetu et al. 2014). In addition, the study found that about USD 2 billion will be lost in the agricultural sector in the next few years due to rainfall variability alone equal to 32.5% of current real AGDP.

You and Ringler (2010) analysed the potential impacts of climate change on Ethiopia's economy and found that the major impact will result from more frequent occurrences of extreme hydrologic events, which cause losses in agricultural and non-agricultural sectors. If further irrigation development is not undertaken, the country will lose between US$28 billion and US$32 billion by the year 2050. This is about 40% of the GDP.

The projected impacts of climate change on agricultural productivity differ per region and per agricultural activity. Typically, mixed farming or crop farming is practiced in the highlands while the lowlands are characterized by nomadic pastoralism (Gebreegzabher et al. 2011). A study comparing climate change effects in moisture-sufficient highland areas with those in drought-prone highland areas found that the moisture-sufficient highlands are likely to benefit from climate change effects until 2030, but will suffer from productivity loss after that. For drought-prone highlands, no 'grace period' is expected; they will suffer from continuously declining productivity throughout the study period, and much of this decline will occur relatively soon (USGS et al. 2012). It will be crucial for Ethiopia's food security to focus agricultural development in areas that will maintain a moist climate in order to offset the expected productivity losses in other areas (Admassu et al. 2013).

Ethiopia is also socio-economically vulnerable to climate change. Rapid population growth and expansion of agriculture in potentially drier and certainly warmer climates could dramatically increase the number of people at risk (USGS et al. 2012). Rural livelihoods, crop cultivation, pastoralism, and agro-pastoralism are highly sensitive to climatic conditions. The increase in drought and desertification from land use pressures have resulted in significant losses of arable land and increased dependency on food aid (USAID 2016).

3 Climate Change Impacts in Ethiopia

3.1 Impacts on Food Security

Climate change has strong links with poverty and hunger. Not only does climate change increase poverty and hunger through its adverse impacts on food security and economic development, but poverty and hunger also decrease people's resilience and adaptive capacity to climate change (Aragie 2013). Poor smallholder farmers are overall the most vulnerable groups to climate change; others are the rural landless, the urban poor in flood-prone areas, the elderly and sick, and women and children left behind as male adults migrate for employment.

Food insecurity is not a recent occurrence. Ethiopia has historically been vulnerable to food insecurity and has often relied on food aid in recent decades. Production of key crops is not increasing fast enough to keep up with the country's high population growth: per capita cereal production has been estimated to decline by 28% between 2009 and 2025 (USGS et al. 2012). Food insecurity is most pronounced before harvesting, when food stocks are depleted (Admassu et al. 2013). While most of the country has one hunger season per year (June–September), the eastern pastoralists have two (February-April and September–October).

The trends in number of food insecure people due to climate-related calamities are increasing and reached a peak of 13 million in 2002/2003. Over 200,000 people died in the drought of 1973/1974. In 1974, about 1.5 million people (5% of the population) required food aid and the number grew to 7 million (17.4%) during the 1990s. In addition, about 14.5 million (22% of the total population) were food insecure in 2003. According to the 2011World Bank climate risk fact sheet, between 1999 and 2004 more than half of all households in Ethiopia experienced at least one major drought shock. Furthermore, between 1900 and 2009, 47 major floods occurred, killing 1,957 people, affecting 2.2 million people, and costing the country about US$16.5 million in damage (You and Ringler 2010). These shocks are a major cause of transient poverty. Correlations have been also found between rainfall, temperature variations, and stunting and underweight populations in Ethiopia (USGS et al. 2012), suggesting that climate change will exacerbate the situation. In 2005, Ethiopia established a Productive Safety Net Program (PSNP) where millions of people receive assistance (Keller 2009). However, this program may not be sufficient if productivity is further limited in the future. Thus, the number of people affected by drought and famine and people dependent on food aid has been increasing ever since. There has never been a year in which the Ethiopian people were food secure for many decades.

Ethiopia's GDP is heavily dependent on agriculture and it is highly correlated with rainfall variability (World Bank 2006). Therefore, rainfall variability and associated yield reductions are estimated to cost the economy about 38% of its potential growth rate and increase poverty by 25% (World Bank 2006). Climate change could reduce GDP by 3–10% by 2025 (Evans 2012). A marginal impact analysis indicated that a 1 °C increase in annual temperature will lead to a statistically significant change in net revenue of −694.15 Birr from total agriculture, inclusive of crop and livestock (Gebreegziabher et al. 2014). Through its strong negative effect on the economy in general and by reducing crop yields, increasing land degradation, and lowering water availability in particular, climate change poses more pressure on the food security of millions of people in Ethiopia. A bio-economic analysis using maize as a case study indicates that the number of food insecure people in Ethiopia would increase by up to 2.4 million people by 2050 as a result of climate change, not only from domestic production, but also on global agricultural trade and prices (Tesfaye et al. 2016).

3.2 Impacts on Livestock

The livestock sector is the other important area of agriculture and food security in Ethiopia that will be impacted by climate change. The country has the largest livestock population in Africa and the 10th largest in the world (Admassu et al. 2013). Cattle and other livestock are a key source of subsistence for many households in the country. However, due to factors that are directly or indirectly related to climate change, the livestock sector is under great stress. These factors include rainfall variability, temperature increases, invasive species, conflicts, migration to lowlands due to increased population density in highlands, and overgrazing (Yirgu et al. 2013). Climate change impacts differ by species, with income decreases up to 19% for beef cattle, 22% for sheep, and 30% for chicken under a temperature increase of 2.5 °C (World Bank 2010). The other study suggests that goats are better adapted to higher temperatures than other livestock, which may be a reason why many pastoralists are already shifting from cattle to goats (Eriksen and Marin 2011).

3.3 Impacts on Labour Productivity and Health

Climate change will also have a significant impact on labour productivity and health. Due to the increase in temperatures (+4 °C), the WHO predicts that there will be a 2% average daily loss of work hours for those in heavy labour (e.g., agriculture and construction) and an increase in heat related deaths among elderly people (65 +) to over 65 deaths per 100,000 by 2080, compared to the estimated baseline of three deaths per 100,000 in 1990 (WHO and UN 2015).

3.4 Impacts on Gender

Another factor of specific importance in climate change is gender vulnerability. In poor areas, women often have more household responsibilities, so they are often disproportionately affected by climate change impacts. The collection of water for drinking, cooking, and washing, the collection of fuelwoods, and the small-scale cultivation of subsistence crops will be more difficult with higher temperatures and more frequent droughts. An additional indirect impact of climate change is that women and girls have been found to be more vulnerable to sexual abuse since they have to travel to more remote sources of water (Swarup et al. 2011; World Bank 2010). Despite the strong impacts of climate change on their lives, the abilities of women and girls to cope with these impacts are often significantly lower than men because of their reduced access to information, markets, mobility, alternative income sources, and decision-making mechanisms.

4 Adaptation and Mitigation Strategies in Ethiopia

The impacts of climate change are expected to undermine the ability of people and ecosystems to cope with and recover from extreme climate events and other natural hazards. The degree to which societies will experience the negative environmental and socioeconomic impacts of climate change depends in large part on their vulnerability (Mahoo et al. 2013). Vulnerability is the propensity of a system to be adversely affected (IPCC 2014). It can be measured by looking at: the extent to which societies are directly dependent on natural resources and ecosystem services; the extent to which those resources and services are sensitive to climate change; and the capacity of societies to adapt to changes in these resources and services (Mahoo et al. 2013). Thus, adequate and relevant agricultural adaptation strategies are the only solution for the poor to reduce the effect of climate change (Akponikpè et al. 2010). According to IPCC (2014), adaptive capacity is the ability of systems, institutions, humans, and other organisms to adjust to potential damage, to take advantage of opportunities, or to respond to consequences. Those societies that can respond to change quickly and successfully have a high adaptive capacity (Smit and Wandel 2006). Consequently, developing the agricultural sector's adaptive capacity to deal with the adverse effects of climate change is imperative to protect the livelihoods of the poor and to ensure food security. However, the capacity of farming households to cope with climate change (*adaptive capacity*) depends on the households and local resources (*asset base*), social networks, and institutional support (*extension services*) (Adger 2009). Thus, climate change adaptation programs will fail unless their implementation considers farmers' social networks and their asset bases—natural, human, financial and physical capital (Connolly-Boutin and Smit 2015).

The adverse effects of climate change on food productivity and food security in Ethiopia are also significant as small-scale and subsistence farmers depend on rain-fed agriculture with limited irrigation coverage (NAPA 2007). The vulnerability to climate change in Ethiopia is highly related to poverty and the lack of adaptive capacity in most regions. Meanwhile, reducing the vulnerability of farm households and communities to climate change is highly related to increasing food security. According to Devereux (2000), low productivity and food insecurity in Ethiopia are associated with long-term climatic changes related to scarce and erratic rainfall. Moreover, adaptation enables farmers in the country to mitigate climate hazards and increase food productivity and security even under high climate vulnerability (Di Falco et al. 2011). Ethiopia is vulnerable to climatic variability due to its low adaptive capacity, low level of socioeconomic development, high population growth, inadequate infrastructure, lack of institutional capacity, and high dependence on climate-sensitive, natural resource-based activities (Georgis 2010; NMA 2007). Consequently, adaptation capacity depends on the existing policy and institutional arrangements like land tenure, social safety net programs, and extension services (Abegaz and Wims 2015; Weldegebriel and Prowse 2013). Thus, adaptation to climate change is key for food productivity and security in Ethiopia (Di Falco et al. 2011).

Farmers in the country have been implementing a variety of adaptation measures in the agricultural sector including: new crop varieties that are tolerant/resistant to drought, salinity, insects/pests; improved water supply and irrigation systems, and the application of new technologies; new land management techniques; and efficient water-use techniques (Hadgu et al. 2014; Mowo et al. 2013; World Bank 2010). However, farmers' adaptation and coping strategies are mainly short-term, reactive responses to climate shocks, and vary in time and space (Bryan et al. 2009). In addition, Ethiopia is often cited as an extreme case where the impacts of present and future climatic changes could lead to disasters similar to the 1980s famines (Conway and Schipper 2010). Consequently, the negative impacts of climate change are projected to be more severe in Ethiopia where farmers are dependent on rain-fed agriculture and natural resource. Furthermore, adaptive capacity is location specific and varies over time, space, and communities depending on human and social capital levels and the quality of governance and institutional services (World Bank 2010). Consequently, planned long-term and local-level strategic actions are important for implementing effective adaptation plans and policies (Bryan et al. 2009).

5 Climate-Smart Agriculture

Agriculture in Ethiopia must undergo a significant transformation in order to meet the related challenges of achieving food security and responding to climate change through adaptation and mitigation (Tesfaye et al. 2016). The role of future agricultural research is to create productive and more resilient agriculture under a variable and changing climate. This requires a major shift in the way land, water, soil nutrients, and genetic resources are managed to ensure that these resources are used more efficiently. Making this shift requires considerable changes in agricultural technologies, national and local policies, financial mechanisms, and other socio-economic enabling environments. In this regard, an agricultural system based on the Climate-Smart Agriculture (CSA) framework is very important. CSA is not a new agricultural system, nor is it a set of practices; it is a way to guide the changes needed in agricultural systems, given the necessity to jointly address food security and climate risks. CSA also aims to get existing technologies on the shelf and into the hands of farmers, as well as to develop new technologies such as drought-tolerant or flood-tolerant crops to meet the demands induced by a changing climate.

CSA is an integrative approach to addressing the interlinked challenges of food security and climate change through the three dimensions of sustainability: *economic*, including sustainably increasing agricultural productivity to support equitable increases in farm incomes, food security and development; *social* by adapting and building the resilience of agriculture and food security systems to climate change at multiple levels; and *environmental* by reducing greenhouse gas (GHG) emissions from agriculture (including crops, livestock and fisheries) (FAO 2016). In addition to promoting land management practices, such as conservation agriculture, CSA can further strengthen farmers' adaptive capacity by adding early warning systems for climate shocks, promoting crops which exhibit greater resilience to certain shocks, and working with both governments and the private sector to improve social safety nets and risk insurance for protection against crop losses. It also involves innovative practices such as improved weather forecasting, early-warning systems, and climate-risk insurance. Thus, the CSA practices are viable and effective response measures, whereby increased support, through adoption of innovative technologies, policies and strategies to address the barriers and widen the adoption scope of CSA, is important (Adera and Pauline 2017).

To make Ethiopian agriculture climate smart, explicit attention needs to be paid as to how interventions in agriculture and food systems affect each of the three key outcomes: food security, climate change adaptation and mitigation. CSA technologies need also to be context-specific and prioritized according to different farming systems. Ethiopia has implemented a variety of practical CSA techniques (Table 2).

The study conducted by Adera and Pauline (2017) to find out farmers' preferences for, and barriers to, adopting CSA practices, found that high and moderate climate resilience and high crop yield agricultural practices were the most preferred combination of CSA when responding to climate change. Contrary to this, high GHG emissions, low-climate resilience, and low-crop yield were the least preferred

Table 2 Summary of some common CSA practices in Ethiopia

CSA practice	Components	Why it is climate smart
Conservation agriculture	• Reduced tillage • Crop residue management –mulching, intercropping • Crop rotation/intercropping with cereals and legumes	• Carbon sequestration • Reduce existing emissions • Resilience to dry and hot spells
Integrated soil fertility management	• Compost and manure management, including green manuring • Efficient fertilizer application techniques (time, method, amount)	• Reduced emission of nitrous oxide and CH_4 • Improved soil productivity
Small-scale irrigation	• Year-round cropping • Efficient water utilization	• Creating carbon sink • Improved yields • Improved food security
Agroforestry	• Tree-based conservation agriculture • Practiced both traditionally and as improved practice • Farmer-managed natural regeneration	• Trees store large quantities of CO_2 • Can support resilience and improved productivity of agriculture
Crop diversification	• Popularization of new crops and crop varieties • Pest resistance, high yielding, tolerant to drought, short season	• Ensuring food security • Resilience to weather variability • Alternative livelihoods and improved incomes
Improved livestock feed and feeding practices	• Reduced open grazing/zero grazing • Forage development and rangeland management • Feed improvement • Livestock breed improvement and diversification	• Improved livestock productivity • GHG reduction • CH_4 reduction
Other	• Water conservation/harvesting • Early-warning systems and improved weather information • Support to alternative energy-fuel efficient stoves, biofuels • Crop and livestock insurance • Livelihood diversification (apiculture, aquaculture) • Post-harvest technologies (agro-processing, storage)	• Resilience of agriculture • Improved incomes • Reduced emissions • Reduced deforestation • Reduced climate risk

Source FAO (2016)

combination of CSA when responding to climate change. The study found that crop yield and resilience are the most important factors influencing farmers' preferences for CSA practices. The main barriers limiting wide adoption and practicing of CSA include lack of incentives, inadequate and unreliable extension and weather information (Adera and Pauline 2017). Globally, as the role of extension agents in scaling up the utilization of CSA by farmers remains very crucial, extension agents should be trained on the wide range of CSA to enhance sustainable agricultural productivity and food security for all (Olorunfemi et al. 2020).

6 Ethiopia's Climate Resilient Green Economy (CRGE) Strategy

Ethiopia's contribution to global GHG emissions is very low. However, if current practices prevail, GHG emissions in Ethiopia will more than double from 150 Mt CO_2e to 400 Mt CO_2e in 2030. On a per capita basis, emissions are set to increase by more than 50% to 3.0 t CO_2e—and will thus exceed the global target to keep per capita emissions between 1 and 2t per capita in order to limit the evolution of climate change and its negative impacts. In absolute terms, the highest increase adding around 110 Mt CO_2e in GHG emissions—will come from agriculture, followed by industry at 65 Mt and forestry at 35 Mt. In relative terms, the emerging industrialization will manifest itself in an annual emission increase of more than 15% from the industrial sector and around 11% from transport. Ethiopian government plans to reduce GHG emissions and the vulnerability of the population and the economy to the impacts of climate change through the Climate-Resilient Green Economy (CRGE) initiative which began in 2011 (FDRE 2011).

Consequently, to achieve food security the Ethiopian government mainstreamed the CRGE strategy into its Growth and Transformation Plan (FDRE 2015b). In this plan, adaptation and mitigation programmed were prioritized to achieve sustainable economic growth (and achieving lower-middle income status) without net increases in GHG emissions relative to 2010 levels. The CRGE Strategy is considered fairly unique in terms of its integration of economic and climate change goals.

Building a green economy will require increasing the productivity of farmland and livestock rather than increasing the land area cultivated or cattle headcount. In order to offer a viable alternative to the conventional development path without foregoing growth in the short term and significant advantages thereafter, a set of initiatives has been identified that can provide the required increase in agricultural productivity and resource efficiency. Therefore, the CRGE targets include higher yields and food security, carbon sequestration, and soil carbon restoration for adaptation and mitigation, respectively (FDRE 2015b).

The CRGE Strategy consists of two components: Climate Resilience (CR) and Green Economy (GE). The CR component focuses on adaptation to climate change impacts in two sectors: agriculture/forestry and water/energy. These climate resilience strategies focus on three main work (FDRE 2015b): identification of the impact of current and projected climatic changes; identification of options to build and reduce the impact of weather variability and climate change; and mapping of steps necessary to fund and implement measures for climate resilience.

The climate resilience strategy for agriculture and forestry identifies 9 themes and 41 key adaptation options to address the negative impacts of climate change. The expected costs of these 41 options are USD236 million by 2030. The identified 9 themes are: capacity building and institutional coordination; information and awareness; crop and water management on-farm; livestock; value chain and market development; sustainable agriculture and land management; natural resources conservation and management; disaster risk reduction; and social protection for high priority groups, including women and children (FDRE 2015b).

The GE component was launched in parallel with the launching of the overall CRGE vision in 2011. The mitigation measures of the green economy component are built on four pillars that target agriculture (livestock and soil), forestry, transport, electric power, industry, and buildings:

- Improving crop and livestock production practices for higher food security and farmer incomes, whilst reducing carbon emissions;
- Protecting and re-establishing forests for their economic and ecosystem services (including as carbon stocks);
- Expanding electricity generation from renewable sources of energy for domestic and regional markets;
- leapfrogging to modern and energy-efficient technologies in transport, industry, and construction (FDRE 2015b).

The adaptation measures are built on three pillars (FDRE 2015a):

- **Droughts**: (1) increase agricultural productivity, minimize food insecurity and increase incomes; (2) protect humans, wildlife and domestic animals from extreme droughts; (3) improve and diversify economic opportunities from agroforestry and sustainable afforestation of degraded forests; (4) enhance irrigation systems through rainwater harvesting and conservation of water; (5) ensure uninterrupted availability of water services in urban areas; (6) improve traditional methods that scientifically prevent deterioration of food and feed in storage facilities; (7) create biodiversity movement corridors; (8) enhance ecosystem health through ecological farming, sustainable land management practices and improve livestock production; and (9) expand electric power generation from geothermal, wind and solar sources.
- **Floods**: (1) enhance adaptive capacity of ecosystems, communities and infrastructure in the highlands of Ethiopia; (2) build additional dams and power stations for energy generation potential in rivers; and (3) develop and implement climate change compatible building/construction codes.
- **Crosscutting interventions**: (1) develop insurance systems for citizens, farmers and pastoralists, to rebuild economic life following exposure to disasters; (2)

reduce incidence and impact of fires and pests through integrated pest management, early warning systems, harvesting adjustments, thinning, patrols, public participation; (3) effective early warning system and disaster risk management policies; (4) strengthening capacity to deal with expansion of humans, animal and crop diseases; and (5) strengthening and increasing capacity for breeding and distributing disease resistant crops and fodder varieties.

7 Role of Agricultural Extension Services in Building Climate Resilience

Agricultural production and productivity from smallholder farming have been very low in Ethiopia and insufficient to feed the growing food demand. Consequently, the country remained one of the poorest countries in Africa, with a significant proportion of its population still relying on traditional farming practices, which prevent rapid agricultural transformation and structural changes from taking place. Attempts to increases agricultural production and productivity in Ethiopia were associated with expansion in cultivated area of land and increased use of chemical fertilizers. However, given the diminishing access to additional cultivable land due to increasing population pressure, achieving further productivity increases will be growingly difficult without additional investments to improve the quality—not just the quantity—of the extension system. Recognizing the prevailing issues, the government considers improving agricultural production and productivity through expanding and reinforcing agricultural extension as the best way to reduce poverty, ensure food security, and sustainably manage natural resources (Bachewe et al. 2017a).

Since the late 1960s, the country has been actively pursuing agricultural extension services (AES) as a key means of agricultural and rural development as well as economic transformation. Within Africa, Ethiopia is probably the country with the greatest state involvement in the agricultural sector (Lefort, 2012). In recent decades, the country has allocated massive resources to its AES. In fact, agriculture has attracted more investments in the current regime compared to the earlier two regimes: the imperial regime (1930–1974) and the *Derg* military regime (1974–1991) (Spielman et al., 2012). Ethiopia has the largest number of local agricultural extension workers, known as DAs, in Africa and the fourth largest in the world, after China, India, and Indonesia (Swanson and Davis, 2014). By 2010, Ethiopia's DA-farmer ratio was estimated at one DA per 476 farmers—that is, 21 DAs per 10,000 farmers. In comparison, figures for Tanzania stood at one DA per 2,500 farmers—that is, 4 DAs per 10,000 farmers. More recent data from the Ministry of Agriculture and Natural Resource show that more than 72,000 DAs reportedly served about 16.7 million smallholder farmers in 2016/2017—that is, one DA per 230 farmers or 43 DAs per 10,000 farmers (Berhane et al. 2018). One study showed that about 80% of farm households have been reached by extension services (Bachewe et al. 2017b).

The country has also invested heavily in agricultural infrastructure such as Agricultural Technical and Vocational Training (ATVET) colleges and Farmer Training

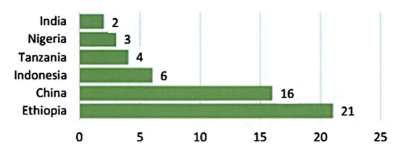

Fig. 1 Number of agricultural extension agents per 10,000 farmers in selected countries. *Source* Davis et al. (2010)

Centres (FTCs), among others. In the early 2000s, 25 ATVET colleges were set up throughout the country (Davis et al. 2010), and a total of 62,303 diploma graduates were trained in these colleges up to 2011. Over 10,000 FTCs have been constructed in the country over the years (Breen 2014). The future of the Ethiopian extension services relies heavily on the use of these FTCs (Gebremedhin et al. 2006) (Fig. 1).

Over the years, many public policies and strategies were introduced to support the implementation of agricultural extension in the country. Agricultural extension is seen as an important component of the Sustainable Development and Poverty Reduction Program (SDPRP) and the Plan for Accelerated and Sustainable Development to End Poverty (PASDEP), which were developed and implemented during 2002–2005 and 2006–2010, respectively (MoFED 2006). SDPRP helped farmers enhance their production capacity by providing agricultural extension services and assigning three DAs to each *kebele*[1] in the country. The key objective of PASDEP was to accelerate the transformation of smallholder agriculture from subsistence to commercial purposes by strengthening extension services through increasing technical and vocational trainings (MoFED 2006). The Growth and Transformation Plan (GTP) is the current state-based development strategy, which is in its second phase of implementation (2016–2020). The plan aims to maintain agriculture as the main source of economic growth and as the foundation of the structural transformation towards industrial growth in the long run (MoFED 2010). Consequently, the GTP also aims to strengthen agricultural extension efforts (NPC 2016).

Agricultural extension services have traditionally played many roles for agricultural development, increased productivity and food security. In Ethiopia, the state and its agricultural extension service system are engaged in the facilitation of credit services through multiple channels. Farmer access to credit services in rural Ethiopia is vital for agricultural development, increased productivity, and food security. Credit services were provided mainly by state-owned microfinance organizations. The newly emerged private microfinance institutes' capacities were limited and their requirements and interest rates were higher than those of the state-owned

---

[1] *Kebeles* are the lowest administrative unit.

microfinance organizations (Leta et al. 2017). Consequently, the agricultural credit service sector is controlled by the government and it was strongly linked to the AES.

Access to improved seeds is vital to improve smallholders' crop production and productivity. In Ethiopia, the AES is also mainly responsible for facilitating farmer's access to packages of technologies and agricultural inputs such as inorganic fertilizers, improved seeds, and agrochemicals. The input delivery is largely organized by state-formed input distribution centres associated with farmers' primary cooperatives. However, many primary cooperatives do not have the needed facilities and technical and personnel capabilities for an effective inputs distribution. As a result, Farmers Training Centres (FTCs) and *kebele* offices often serve as temporary centres for inputs stocking and distribution (Leta et al. 2017).

The extension service also contributed to agricultural development by enhancing the transfer of technology and knowledge. Agricultural knowledge and technologies generated by universities, national research systems, and bureaus of agriculture are transferred to farmers through extension services. Moreover, locally generated knowledge is also scaled up through media and the extension systems (Gebretsadik 2012). Thus, extension systems served as an important service component in the knowledge system as well as in agricultural development processes (Rivera and Sulaiman 2009). Agricultural knowledge is widely transferred through farmers' training and on-farm technology demonstrations by various research and development actors through the AES. However, in Ethiopia, technologies and knowledge transfer are largely delivered by state-planned public mobilization drives and campaigns (Leta et al. 2017).

Yet there are also problems in technology and knowledge transfer. The state media promotes the dissemination and adoption of technologies or best practices in every corner of the country without educating farmers about its compatibility to local agro-ecological conditions or farming systems (Leta et al. 2017). To ensure effective technology dissemination and adoption, stakeholders should be involved in the testing of the technologies in their respective agro-ecological conditions (Hornidge et al. 2009).

Despite its importance, the operation of the agricultural extension system in Ethiopia is plagued with large-scale ineffectiveness. The root cause is the centralized top-down state control. Davis et al. (2010) linked the shortfalls to the trade-offs between the twin objectives of the state extension policy: (a) to improve production and ensuring food security; and (b) to win and maintain the support and loyalty of farming communities. Also, the fact that AES covers vast and scattered geographic areas implies that it is poorly resourced and has weak links with knowledge centres; thus hampering the quality of extension services delivered (Bachewe et al. 2017a). In addition, DAs lack the skills required for functions important to farmers, such as agricultural marketing (value chain) and agricultural intensification and diversification (Davis et al. 2010). Furthermore, a serious challenge to the system is that the policies and focus of agricultural extension and rural development are constantly revised (Leta et al. 2017). In this regard, the role of extension in increasing agricultural productivity has remained limited in Ethiopia, as in many other African countries (Davis et al. 2010). Thus, although agricultural extension is targeted at ensuring food security, it has never resulted in a breakthrough in the Ethiopian agricultural sector, particularly in the interest of smallholder farmers. It is less clear whether or not productivity gains have been achieved and poverty has been reduced in consequence (Bachewe et al. 2017b).

In addition to their traditional function of promoting production technologies and natural resource management practices, agricultural extension services are expected to have expanded roles in building farmers' social networks and supporting climate change adaptation strategies (Swanson and Davis 2014). Institutions play an important role in adaptation at multiple scales—from inception to transfer of technologies and access to agricultural innovations by vulnerable smallholders (Agrawal 2008). The influence of institutions in adaptation and climate vulnerability is critical in the following ways: first, they shape impacts and vulnerability in social and ecological contexts; second, institutions mediate between individual and collective responses to climate impacts, and thereby influence the outcomes of adaptation; and finally they act as the means of delivery of external resources to facilitate adaptation, and thus govern access to resources such as information, technology, funds, and leadership (Agrawal 2008).

Agricultural extension is expected to facilitate adaptation through training and educating farmers to anticipate and change their knowledge, attitudes and adaptive capabilities in response to climate change. Extension services should be farmer oriented (Davis et al. 2010). However, most of the extension systems in sub-Saharan Africa are not responsive to farmers technology and information needs (Bingen and Simpson 2015). Thus, extreme events associated with climate change are putting additional pressure on farmers in Africa whose vulnerability is already affected by weak social networks and poor extension supports (Connolly-Boutin and Smit 2015). In other words, climate change and weak social and institutional support are believed to have overwhelming consequences for food security in sub-Saharan Africa.

Furthermore, the current uniform approach pursued in Ethiopia does not fit well with the diverse agro-ecologies and extension challenges in the country. While the number of extension agents were increased substantially, they still lack skills, incentives, and resources which affect their work motivation and job performance. Moreover, the planning, monitoring, and evaluation system is not effective in regularly assessing what has been achieved at the training centres and what remains to be done in the future. Similarly, there is room for improvement of partnerships and linkages among actors, especially by including key actors that are currently missing (Tesfaye et al. 2016).

At this time, the extension system is not geared towards addressing different challenges; a fact which calls for aligning the extension reforms to the different local farming systems (Tesfaye et al. 2016). In *dega* and *woyna-daga* agro-ecologies, the main challenges are getting information on climate change related hazards (rainfall and temperature), commercial marketing (cooperative development, price, and new markets), and postharvest handling (drying and storage technique). In the *kolla* agro-ecologies, the major problems are mainly related to the lack of dry land farming methods (contour plowing, mulching, strip farming, summer fallow, seedbed preparation, and planning in rows). Thus, extension reforms should consider current agricultural challenges, especially climate change (Tesfaye et al. 2016).

8 Conclusions

It is known that extension services have the potential to influence farmers' decision to change their farming practices in response to climate change (Di Falco and Marcella 2010). However, extension services have never resulted in a breakthrough in the Ethiopian agricultural sector, particularly in the interest of smallholder farmers. Agricultural extension is constrained by multiple challenges such as: high input and low output prices; knowledge and skill gaps among development agents and model farmers; non-inclusive extension services; ambitious top-down allocation of plans; and actors' involvement in non-extension activities. As a result, the majority of farmers resort to social learning and local networks for interactions and knowledge acquisition. Also, involving the private and non-governmental organizations extension services to operate alongside the pubic system may help address the prevailing gaps of inadequate capacity and skills, shortage of inputs (such as improved seeds), and price escalation (Tesfaye et al. 2016).

Consequently, significant reforms of the AES system are critical to Ethiopia's agricultural transformation. Reforms will need to extricate the system away from single-minded, top-down, package approaches of cereal intensification to more dynamic, responsive, and knowledge-based service provision. This will obviously require some major policy choices between a system that covers a wider area of the country with thinly spread resources (given Ethiopia's topography and limited resources) and a more focused but better-resourced system that is capable of addressing critical knowledge bottlenecks to proceed with transformation.

More importantly, institutional innovations require the channelling of new knowledge to agricultural extension agents, with a strong link between AES and research that is country specific. Given the complexities associated with the size of farm communities to be served, and the physical and infrastructural constraints of Ethiopia, recasting the AES system as one that will be responsive to farmers' demands and to knowledge sources is the most pressing agenda for policy makers (Leta et al. 2017).

As AES is very much tied to access to productive inputs, there is a need not only to continue public investment to promote fertilizer, seed, credit, and AES, but also to support private sector development. The key actors, such as the private sector and NGOs, are missing from effective provision of extension services (Leta et al. 2017). However, being more flexible in how inputs and services are provided will ensure a greater degree of choice for smallholders and pave the way for the emergence of new market and technological opportunities in Ethiopia's agriculture sector. The AES may have a comparative advantage in activities such as provision of improved seeds, fertilizers, pesticides, vaccination, deworming, and artificial inseminations. Their engagement in these areas will also allow the regional government to free up and reallocate funds to its broader extension strategies such as development of new incentive schemes, education and training, technical advisory services, sustainable natural resource management practices, and organizing farmers to link them with new markets. Thus, the roles of NGOs and the private sector in the provision of extension services should be enhanced.

Furthermore, though adaptation is a critical factor in building social resilience from individual to institutional levels in Ethiopia, there is no clear sign of mainstreaming climate change in ATVET curriculums (Tesfaye et al. 2016). Consequently, it should be investigated and reconsidered as pre-service climate change and adaptation education is vital to prepare AES to face rising technical and methodological challenges.

References

Abegaz DM, Wims P (2015) Awareness of climate change in Ethiopia. J Agric Educ Extension Agents 37–41. https://doi.org/10.1080/1389224X.2014.946936

Adem A, Bewket W (2011) A climate change country assessment report for Ethiopia. Addis Ababa: Epsilon International R&D

Adera W, Pauline N (2017) Evaluating smallholder farmers preferences for climate smart agricultural practices in Tehuledere District. Northeastern Ethiopia 39:300–316. https://doi.org/10.1111/sjtg.12240

Adger WN (2009) Social capital, collective action, and adaptation to climate change. Econ Geograph 79(4):387–404

Admassu H, Getinet M, Thomas TS, Waithaka M, Kyotalimye M (2013) Chap. 6., pp 149–182

AfDB (African) D (2011) Federal democratic republic of Ethiopia: Country Strategy Paper 2011–2015

Agrawal A (2008) The role of local institutions in adaptation to climate change. In: Paper prepared for the social dimensions of climate change, social development department. The World Bank, Washington, DC, March 5–6.

Akponikpè PBI, Johnston P, Agbossou EK (2010) Farmers' perception of climate change and adaptation strategies in sub-saharan West-Africa. Submitted to 2nd International 14 D.M. Abegaz and P. Wims Downloaded by [University College Dublin] at 03:19 03 February 2015 Conference on Climate, Sustainabilit

Aragie A (2013) Climate change, growth, and poverty in Ethiopia

Bachewe FN, Berhane G, Minten B, Taffesse AS (2017a) Agricultural transformation in Africa? assessing the evidence in Ethiopia. World Development (forthcoming)

Bachewe FN, Berhane G, Minten B, Taffesse AS (2017b) Agricultural transformation in Africa? assessing the evidence in Ethiopia. World Development (forthcoming)

Berhane G, Ragasa C, Abate GT, Assefa TW (2018) The state of agricultural extension services in Ethiopia and their contribution to agricultural productivity

Bingen R, Simpson B (2015) Farmer organizations and modernizing extension and extension services : a framework and reflection on cases from Sub-Saharan Africa. MEAS, Discussion Paper

Breen M (2014) Agricultural training in Ethiopia. Consultant report, Ministry of Agriculture, Addis Ababa, Ethiopia

Bryan E, Deressa TT, Gbetibouo GA (2009) Adaptation to climate change in Ethiopia and South Africa: options and constraints. Environ Sci Policy 12(4):413–426

Cipryk R (2009) Impacts of climate change on livelihoods: what are the implications for social protection?.CDG Working Paper No. 1. CDG, Brighton

Connolly-Boutin L, Smit B (2015) Climate change, food security, and livelihoods in sub-Saharan Africa. Regional Environ Chang

Conway D, Schipper L (2010) Adaptation to climate change in Africa: challenges and opportunities identifed from Ethiopia. Global Environ Chang 21(1):227–237

CSA (2017a) The federal democratic republic of ethiopia agricultural sample survey 2016/17 [2009 e.c.] volume ii report on livestock and livestock characteristics (private peasant holdings) addis ababa ethiopia. ii(april)

CSA (2017b) The federal democratic republic of ethiopia agricultural sample survey 2016/2017 (2009 e.c.) volume i report on area and production of major crops (private peasant holdings, meher season, addis ababa ethiopia. i

Davis K, Swanson B, Amudavi D, Mekonnen DA, Flohrs A, Riese J et al (2010) In-depth assessment of the public agricultural extension system of Ethiopia and recommendations for improvement. International Food Policy Research Institute (IFPRI) Discussion Paper 1041, IFPRI, Washington, DC

Devereux S (2000) Food insecurity in Ethiopia: a discussion paper for DFID. IDS Sussex

Di Falco S, Mahmud Y, Gunnar K, Ringler C (2011) Estimating the impact of climate change on agriculture in low-income countries : household level evidence from the Nile Basin. Ethiopia. https://doi.org/10.1007/s10640-011-9538-y

Di Falco S, Marcella V (2010) Does adaptation to climate change provide food security ? Micro-Perspective Ethiopia 22:1–33

Eriksen S, Marin A (2011) Pastoral iPathways: climate change adaptation lessons from Ethiopia. Norwegian University of Life Sciences

Eshetu Z, Simane B, Tebeje G, Negatu W (2014) Climate finance in Ethiopia 113

Evans A (2012) Resources, risk and resilience: scarcity and climate change in Ethiopia. New York University, Center on International Cooperation

FAO (2016) Ethiopia climate-smart agriculture scoping study. In: Jirata M, Grey S, Kilawe E (eds) Addis Ababa, Ethiopia

FDRE (2011) Ethiopia's climate-resilient green economy:green economy strategy (1)

FDRE (2015a) Ethiopia's climate-resilient green economy climate resilience strategy: water and energy

FDRE (2015b) Ethiopia's climate resilient green economy: climate resilience strategy agriculture and forestry

Gebreegzabher Z, Stage J, Mekonnen A, Alemu A (2011) Environment for development climate change and the Ethiopian economy

Gebreegziabher B, Ahmed S, Gebre-ab N (2014) Ethiopia's vision for a climate resilient green economy

Gebremedhin B, Hoekstra D, Tegegne A (2006) Commercialization of Ethiopian agriculture: extension service from input supplier to knowledge broker and facilitator. Addis Ababa: International Livestock Research Institute

Gebretsadik M (2012) The impact of climate change and adaptation through agroecological farming practices: a case study of the konso area in Ethiopia. Thesis, Swedish University of Agricultural Sciences, MSc

Georgis K (2010) Agricultural based livelihood systems in drylands in the context of climate change; inventory of adaptation practices and technologies of Ethiopia. Food and Agriculture Organization of the United Nations, Rome

Gizachew L, Shimelis A (2014) Analysis and mapping of climate change risk and vulnerability in central rift valley of Ethiopia Climate change is one of the current issues that severely impact all climate sensitive sectors like agriculture. Manifestation Clim Change Such 22(Vi):807–818

Hadgu G, Tesfaye K, Mamo G, Kassa B (2014) Analysis of farmers' perception and adaptation methods to climate variability/change in tigray region, northern Ethiopia. Res J Agric Environ Sci 1(1):15–25

Hornidge A-K, Hassan MU, Mollinga P (2009) Follow the innovation. A joint experimentation and learning approach to transdisciplinary innovation research. ZEF Working paper series 39. Bonn, Germany

IPCC (2014) Summary for policymakers In: Climate change 2014: impacts, adaptation, and vulnerability. In: Field CB et al (eds) Part A: global and sectoral aspects. contribution of working Group II to the fifth assessment report of the intergovernmental panel on climate change. https://doi.org/10.1017/CBO9781107415324

Keller M (2009) Climate risks and development projects

Lefort R (2012) Free market economy, developmental state and party-state hegemony in Ethiopia: the case of the model farmers. J Mod Afr Stud 50:681–706

Leta G, Kelboro G, Stellmacher T, Hornidge AK (2017) The Agricultural Extension System in Ethiopia—operational setup, challenges and opportunities. Working paper 158, ZEF working paper series

Mahoo H, Raden M, Kinyangi J, Cramer L (2013) Climate change vulnerability and risk assessment of agriculture and food security in ethiopia CGIAR research program on climate change. Agriculture and Food Security (CCAFS) Edited

MoFED (2006) Ethiopia : Building on Progress A Plan for Accelerated and Sustained Development to End Poverty (PASDEP) vol I. Main Text

MoFED (2010) Growth and transformation plan 2010/ 11 -20 14/ 15

Mowo J, Bishaw B, Abdelkadir A (2013) Farmers' strategies for adapting to and mitigating climate variability and change through agroforestry in Ethiopia and Kenya. Oregon State University, Corvallis, Oregon, Forestry Communications Group

NAPA (2007) Climate change National Adaptation Programme of Action (NAPA) of Ethiopia. Report of the Federal Democratic Republic of Ethiopia, Ministry of Water Resources, National Meteoro-logical Services Agency

National Bank of Ethiopia (NBE) (2017) Quarterly bulletin third quarter 2016/17 fiscal year series. Addis Ababa 33(3)

NMA (2007) The federal democratic republic of ethiopia climate change national adaptation programme of action (napa) of ethiopia climate change national adaptation programme of action (napa) of Ethiopia

NPC (2016) Growth and transformation plan II (GTP II) (2015/16–2019/20): Main Text. Addis Ababa: NPC ((National Planning Commission))

Olorunfemi TO, Olorunfemi OD, Oladele OI (2020) Determinants of the involvement of extension agents in disseminating climate smart agricultural initiatives : implication for scaling up. J Saudi Society Agric Sci 19(4):285–292. https://doi.org/10.1016/j.jssas.2019.03.003

Rivera W, Sulaiman V (2009) Extension: object of reform, engine for innovation. Outlook Agric 38(3):267–273

Smit B, Wandel J (2006) Adaptation, adaptive capacity and vulnerability. Global Environ Change 16:282–292. Department of Geography, University of Guelph, Canada

Spielman D, Kelemework D, Alemu D (2012) Seed, fertilizer and agricultural extension in Ethiopia. In: Dorosh PA, Rashid S (eds) (2012) Food and agriculture in Ethiopia: progress and policy challenges. IFPRI Publishing, University of Pennsylvania press USA, pp 84–122

Swanson B, Davis K (2014) Status of agricultural extension and rural extension services worldwide. Global Forum for Rural Extension services (GFRAS)

Swarup A, Dankelman I, Ahluwalia K (2011) Weathering the storm: Adolescent girls and climate change

Tesfaye K, Seid J, Getnet M, Mamo G (2016) Agriculture under a Changing Climate in Ethiopia: challenges and opportunities agriculture under a changing climate in Ethiopia: challenges and opportunities for research

Thornton P, Herrero M, Freeman A, Mwai O, Rege E, Jones P, Mcdermott J (2011) Mapping climate vulnerability and poverty in Africa: impacts on livestock and livelihoods. 4(1):1–23

USAID. (2016), Resilience at USAID 2016 Progress report. Retrieved from https://www.usaid.gov/sites/default/files/documents/1867/082816_Resilience_FinalB.PDF

USGS, USAID, and FEWS-NET (2012) A climate trend analysis of Ethiopia

Weldegebriel ZB, Prowse M (2013) Climate-change adaptation in Ethiopia: to what extent does social protection influence livelihood diversification? Dev Policy Rev 31(S2):35–56

WHO and UN (2015) Climate and health country profile—2015 Ethiopia.

World Bank (2006) Ethiopia: managing water resources to maximize sustainable growth. Country water resources assistance strategy. Report No. 36000-ET. Washington, DC: World Bank

World Bank (2010) Economics of adaptation to climate change

Yirgu L, Nicol A, Srinivasan S (2013) Warming to change? climate policy and agricultural development in Ethiopia

You GJ, Ringler C (2010). Hydro-economic modeling of climate change impacts in Ethiopia

Chapter 12
The Nexus of Climate Change, Food Security, and Agricultural Extension in Islamic Republic of Iran

Peyman Falsafi, Mirza Barjees Baig, Michael R. Reed, and Mohamed Behnassi

Abstract Iran is a water-scarce country with an average rainfall of about 240 mm per annum. Agriculture is one of the most important sectors, meeting domestic food and nutrition needs, and providing employment opportunities to a large segment of the population. Undoubtedly, wise planning for the development of this sector can play a very important role in improving the overall country's economic situation. However, this sector is highly dependent on climatic and environmental conditions. Like many other countries, especially in the Global South, climate change will likely have several negative impacts on agriculture by changing environmental parameters such as rainfall and temperature. Such impacts are also expected to result in extreme events, mainly destructive floods and long-term droughts. A cluster analysis of weather station data in Iran shows that the ten currently defined agro-environment zones will be reduced to eight by 2025 and to seven by 2050. The climate change impacts on agriculture, and subsequently on the development of this sector, vary depending on the spatial and temporal scale. Due to its special geographical location and topographic characteristics, the country has a different climate in each region. Because of severe fluctuations in rainfall throughout the country, droughts have had

P. Falsafi (✉)
Agricultural Extension Education, Agricultural Education and Extension Institution, Agricultural Research, Education and Extension Organisation, Ministry of Agricultural Jahad, Tehran, Islamic Republic of Iran
e-mail: p.falsafi@areo.ir

M. B. Baig
Prince Sultan Institute for Environmental, Water and Desert Research - King Saud University, Riyadh, Kingdom of Saudi Arabia
e-mail: mbbaig@ksu.edu.sa

M. R. Reed
Emeritus of Agricultural Economics, University of Kentucky, Lexington, KY, USA
e-mail: mrreed@email.uky.edu

M. Behnassi
International Politics of Environment and Human Security, College of Law, Economics and Social Sciences, Ibn Zohr University of Agadir, Agadir, Morocco
e-mail: m.behnassi@uiz.ac.ma

Center for Environment, Human Security and Governance (CERES), Agadir, Morocco

© The Author(s), under exclusive license to Springer Nature Switzerland AG 2022
M. Behnassi et al. (eds.), *Food Security and Climate-Smart Food Systems*,
https://doi.org/10.1007/978-3-030-92738-7_12

several detrimental effects on agriculture and the economy. However, planning and mobilization of facilities and resources focusing on agricultural extension can help ensure food security through an enhanced supply of basic foodstuffs and commodities. Currently, the agricultural extension system is experiencing a radical change to make it more effective in helping farmers to be more productive, profitable, and climate resilient. This chapter aims at presenting the current changes in climate, examining the food security situation, and outlining workable and viable strategies by redefining the vibrant role of an agricultural extension system, particularly addressing climate change challenges and ensuring food security of the country.

Keywords Climate change · Food security · Agricultural extension · Extension system · Iran

1 Introduction

The Islamic Republic of Iran lies in western Asia, situated in the south-west of Asia, and covers an area of 1,648,000 km^2. In the north, it is littoral to the Caspian Sea and borders Azerbaijan and Turkmenistan. It is contiguous with Turkey and Iraq to the West. In the South, the country is littoral to the Persian Gulf and the Sea of Oman and abuts Pakistan and Afghanistan to the East (Fig. 1).

The principal and official language is Farsi (Persian). The population in 1994 was about 57.7 million and now estimated at 80.0 million. Nearly 25% of the total population live in rural areas and account for 4.3 million farm-units. The country is divided into 31 provinces, 252 districts, 680 sub-districts and more than 60,000 villages/settlements. About 50% of the active rural populations of age 10 and above are engaged in agriculture, with the rest in the industry (27%) and services (22%) (Falsafi 2005) (Fig. 2).

Iran has a diverse climate: (1) dry and semi-dry in large parts of internal lands and the southern border of the country; (2) mountainous subdivided into two categories,

Fig. 1 Map of Iran (left) and a picture on the climate diversity in Iran (right)

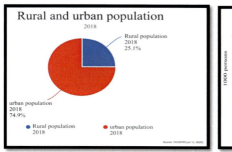
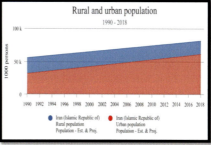

Fig. 2 Rural and urban population of Iran, 1990–2018. *Source* FAOSTAT (2019)

cold and moderate; and (3) Caspian in a narrow and small area between the Caspian Sea and Alborz Mountain Belt with 600 2000 mm of annual rain. The area coverage of different types of climate in Iran is 35.5% hyper-arid, 29.2% arid, 20.1% semi-arid, 5% Mediterranean, and 10% wet (of the cold mountainous type). Thus, more than 80% of Iran's territory is in the arid and semi-arid zone of the world. The average rainfall in the country is about 250 mm, which is less than 1/3 of the world average rainfall (860 mm) (Amiri and Eslamian 2010).

Currently, Iran's average annual temperature is 17 °C and this average has shown an increasing trend in the past half-century as a direct result of climate change (FAOSTAT 2019). It is expected that the upward temperature trend will continue with an average increase of up to 2.2 °C for the coming century (Fig. 3).

In 2013, Carbon Dioxide Information Analysis Center (CDIAC) ranked Iran as the 7th largest CO_2 emitter at 168,251 thousand metric tons due to its fossil-fuel burning, cement production, and gas flaring (Akbarieh 2017).

The agricultural sector is highly dependent on climatic and environmental conditions. Like many other countries, especially in the global South, climate change is expected to have several negative impacts on agriculture by changing environmental parameters such as rainfall and temperature patterns. Other negative impacts

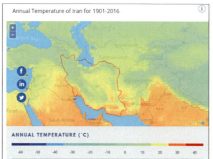

Fig. 3 Annual temperature of Iran, 1901–2016. *Source* FAOSTAT (2019)

are more destructive floods and long-term droughts. A cluster analysis of weather station data shows that 10 currently defined agro-environment zones will be reduced to 8 by 2025 and to 7 by 2050 (Koocheki et al. 2006). Due to its special geographical location and topographic characteristics, the country has a different climate in each region. Because of severe fluctuations in rainfall throughout the country, droughts have had several detrimental effects on agriculture and the economy, but the mobilization of facilities and resources focusing on agricultural extension can help ensure food security through an enhanced supply of basic foodstuffs and commodities.

Currently, agricultural research, education, and extension in Iran are interlinked systems of the Agricultural Research, Education and Extension Organization as one of the key departments of the Ministry of Agriculture. Recently, the agricultural extension system is experiencing a radical change to make it more effective in helping farmers to be more productive, profitable, and sustainable (Falsafi and Shahpasand 2014).

Extension programs in agriculture aim to increase farm productivity, decrease the consumption of basic resources including water, seeds, pesticides and fertilizers, strengthen market-led production, improve food safety, and reduce food security.

Accordingly, innovation delivery is the process which helps agricultural extension system identify, select, organize, and disseminate important innovations including knowledge, information and experiences to the farmers with a specific focus on small ones. This enables farmers to utilize innovations, solve their problems in the field, and increase the farm productivity. In the Iranian agriculture extension system, innovations are managed through codifying and integrating knowledge created by researchers, experts, farmers' indigenous knowledge and experiences and other existing sources. The agriculture extension system is officially affiliated to the public research sector from one side, and is also connected to the two other important sectors including the education system and farmers. Agricultural extension transfers existing knowledge to recipients through different communication means, including training packages, CDs, handbooks, posters, guidelines etc. Farmers can obtain detailed information through participation in training courses, farmer field schools, field visits, and workshops (Falsafi and Shahpasand 2014).

2 Role of Agriculture in Iran's Economy and Food Security

Agriculture is one of the most important sectors of Iran's economy, meeting domestic food and nutrition needs, and providing employment opportunities to a large segment of population. Therefore, this sector may play a very important role in improving the overall economic situation. Indeed, it accounts for 10 to 13% of GDP, employs 18 to 20% of the workforce, constitutes 30% of non-oil export (excluding gas condensates), provides 95% of the food consumption needs of the country, produces approximately 95–120 million tons of products (crops, horticulture, livestock, and fisheries), and ranks among the top ten producers for 20 agricultural products (Falsafi and Shahpasand 2014).

12 The Nexus of Climate Change, Food Security, and Agricultural ...

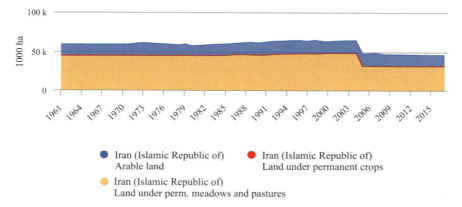

Fig. 4 Iran's arable lands, under permanent crop and pastures lands, 1961–2015. *Source*: FAOSTAT (2019)

The total land area of Iran is 164.82 million hectares (ha), from which 86 million ha (52.4%) are rangelands, 14.2 million ha (8.6%) forests, and 32 million ha (19.5%) deserts, including bare salty lands. Approximately 18.5 million ha (11%) are under cultivation, of which 8.5 million ha are irrigated and the rest rainfed. Approximately 51 million ha of the total land area is characterized by a high or medium degree of cultivability and is considered potentially arable land (Fig. 4).

Moreover, Iran is ranked in the top-10 producing countries for 63 products. The country produced more than 120.000.000 tons of different agricultural products in 2014. The main horticultural products are pistachios (60% of the world total), dates (19% of the world total), pomegranates, grapes, almonds, walnuts, barberry, berries, apricots, cherries, cantaloupe, apples, oranges, and peaches. The major farming crops are saffron, wheat, barley, rice, corn, canola, potato, peas, tomatoes, tea, and melons. Animal husbandry and dairy products include sheep and beef meat, milk, butter and cheese, wool, chicken and eggs, honey, and fish products (Falsafi and Shahpasand 2014).

3 National Policies and Objectives in Agriculture, Food, and Nutrition Security

The main policy frameworks governing agriculture and economic development in Iran are the following:

- The Vision 2025, adopted in January 2009, which is the overall framework that defines long-term policy objectives in all areas;
- Broad Policies for Agriculture, adopted in July 2005;
- The 4th Five-Year National Economic, Social and Cultural Development Plan (FYNDP) 2005–2009, which has been extended to March 2011; and

- The 5th Five-Year National Economic, Social and Cultural Development Plan (FYNDP) for the financial years 2011–2016.

These policy frameworks share the same objectives regarding the agricultural and rural sector, which are:

- Enhancing the role of agriculture in the national economy by improving agricultural productivity and its contribution to poverty reduction;
- Achieving national food security through higher domestic productivity and self-sufficiency in staple crops, improving food safety, and upgrading food consumption patterns through increasing the share of animal protein intake; and
- Focusing on commercialization, sustainable development, disaster and risk management, and private sector participation in agriculture.

The Government has emphasized producer-support measures through its self-sufficiency strategy, especially by providing high support prices for farmers on several commodities. This policy requires a lot of budgetary support at a time when the medium-term outlook for economic growth is negative. However, it is not clear how long Iran can sustain the high cost of its food self-sufficiency strategy (FADPA 2014).

After the Iran's 1979 Islamic Revolution, public participation was the highest priority of rural development policies. To this end, various rural development projects have allowed for public participation to be institutionalized and rural people to be trained and empowered, thus encouraging their active participation in the development process.

4 Case Studies on Impacts of Climate Change on Agro-Climatic Indicators in Iran

The contemporary empirical literature regarding the estimation of climate change-induced impacts on farming systems is rooted in three predominant approaches: crop simulation models; agronomic statistical models; and hedonic price models (Jacoby et al. 2011; Hertel and Rosch 2010; Zhai et al. 2009; Schlenker and Roberts 2009, cited in Heshmati et al. 2015). Li et al. (2012) say crop simulation models use quantitative descriptions of ecophysiological processes to predict plant growth and development as influenced by environmental conditions and crop management, which are specified for the model as input data (Hodson and White 2010). Crop modelling has been used primarily as a decision-making tool for crop management, but crop modelling, coupled with crop physiology and molecular biology, also could be useful in breeding programs (Slafer 2003). Agronomic statistical models estimate statistical relationships between crop yields, on the one hand, and climatic parameters, especially temperature and precipitation, on the other, using relatively less calibrated data. This is readily implemented for large geographic areas (Hertel and Rosch 2010; Lobell and Burke 2010, cited in Heshmati et al. 2015). Finally, the hedonic approach (Jacoby et al. 2011, cited in Heshmati et al. 2015) is known as the

Ricardian Model that focuses on the impact of climate on land values, not yields. This technique draws heavily on the underlying observation by Ricardo (1817) that, under competition, the rental value of land reflects net productivity/profit from the land in its highest use. The main advantage of this approach is that it automatically takes the farm-level adaptations into consideration, while assessing the direct effect of climate on crop performance.

Nasiri Mahalati et al. (2006) investigated the effects of climate change on climatic indicators of Iranian agriculture. The purpose of this study was to calculate the climatic indicators of agriculture under future climatic conditions, compare those indicators with the current conditions, and predict agricultural production with the future climate of Iran. These indicators were calculated in the current conditions as well as for 2025 and 2050. The results showed that the date of occurrence of the first autumn frost in the studied stations for 2025 and 2050 will be delayed by 5–9 days and 8–15 days, respectively, and the intensity of change will increase from north to south and from west to east of the country. Predictions from public circulation models indicate a delay in the date of occurrence of the last spring frost of 4–8 days and 7–12 days, respectively, for the years 2025 and 2050. Due to the delay in the date of occurrence of the first autumn frost and the advance of the date of the last spring frost, the length of the growing season in all studied stations increased by 5–23 days and 16–42 days for 2025 and 2050, respectively.

Maleki Morsht (2017) studied the effects of climate change on Iran's agricultural development using a Positive Mathematical Planning (PMP) technique. The results showed that climate change has the greatest effect on the studied indicators. Based on these results, serious and immediate planning is needed in water management, and changes in macro-plans of agricultural management and agricultural policies are among the recommendations to adapt to climate change and manage its negative effects. Climate change enhances the risks of agricultural production and marketing, acting as a threat multiplier, particularly regarding the availability of water and the changes in temperatures. In many places, climate change expresses itself through higher variations in moisture and increasing periods when the weather is too dry or too wet.

5 Climate Change Direct Impact Assessment in Iran

5.1 Water Resources

Research on climate change impacts on hydrology and water resources in Iran has been undertaken for several rivers and lake basins by using historical hydrometeorological data and runoff models in combination with climate change scenarios. The results of historical runoff data surveys collected at 398 hydrometric stations show that flood indicators have changed in 47% of them. In addition, of 600 climatologically stations studied, 68 indicate **climate change**s during the period 1990–2000.

The long-term runoff model applied to 30 basins shows that the temperature rise increases the runoff volume during winter and decreases it during spring as rising temperature melts snowfall into rain and hastens the time of snow melt. It also indicates that the temperature increase affects runoff of basins and decreases the amount of runoff variation of rainfall (Amiri and Eslamian 2010).

5.2 Agriculture

The predicted increase in temperature due to global warming may lead to spike sterility in rice, loss of pollen viability in maize, reversal of vernalization in wheat, and reduced formation of tuber bulking in the potato (Amiri et al. 2009). The changing climate will affect wheat, which is the main staple crop. The historical data indicates that wheat production will be sharply reduced due to drought and decreased rainfall (Fig. 5). Losses inflicted by the 1998–1999 droughts on wheat production nationwide were estimated at about 1,050,000 tons of irrigated wheat and 2,543,000 tons of rainfed wheat. The values indicate that agricultural areas are highly vulnerable to **climate change**.

Accordingly, there is a large uncertainty about the future of agricultural sector in the context of climate change. The existing studies show that Iran's total crop yield is projected to decrease in all scenarios. However, the extent of changes in yield depends on the crop type, assumptions related to the CO_2 fertilization effect, climate scenarios, and adaptation capacities. Also, it is predicted that climate change will contribute to reduced water resources in various regions of Iran. Moreover, decreased precipitation and rising temperatures greatly increase the irrigation water requirements. Economic studies reveal that as water availability declines, the welfare of farm families also reduces. The results of a study by Asmare et al. (2019) indicate heterogeneous effects of climate variables on farm income between adopters and non-adopters of crop diversification (CD). The study also confirms the win–win effect of adopting CD with a positive and significant effect on farm income and a

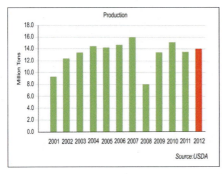
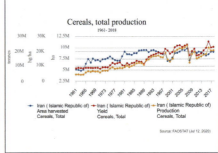

Fig. 5 Iran historical wheat production, 2001–2012. *Source*: FAOSTAT (2019)

reduction in demand for on-farm labour. The results suggest that adoption of CD helps improve the well-being of farm households and build a resilient agricultural system. Investment in water storage and irrigation is particularly import for rural incomes. An extreme decline in crop yields in arid and semi-arid areas globally has caused food shortages and a manifold increase in food inflation. Countries in Africa, Middle East, Arab region, and Asia have close economic ties with natural resource and climate-dependent sectors such as forestry, agriculture, water, and fisheries (Misra 2014). Changes in the pattern and timing of precipitation under future climate scenarios have the potential to seriously disrupt water availability worldwide (Molle and Mollinga 2003). Case studies in Asia and the Middle East have shown that farmers react to water scarcity by challenging the water allocation environment, including tampering with infrastructure, colluding with water officials, organizing protests, and lobbying political connections (Molle et al. 2009).

Despite a growing literature on climate change and its impacts on Iran's agricultural sector, there are still several research gaps that need to be addressed in the future. *First*, most studies analysing climate change impacts on crop yields have focused on wheat; therefore, it is imperative to cover other crops, too. *Second*, most studies have assessed the impacts of climate change on crop yields and water resources, but the economic impacts of climate change on rural communities are rarely covered; therefore, the economic analysis of climate change impacts should be promoted. *Third*, agricultural production and demand for irrigation water are heavily influenced by socio-economic factors. As a result, they cannot reflect farmers' adaptive responses appropriately. Consideration of alternative adaptation strategies on agriculture production is imperative. *Finally*, the CO_2 fertilization effect is neglected in almost all studies. This may lead to overestimating the importance of climate change impacts when they are considered in isolation from CO_2 fertilization effects (Karimi et al. 2018).

5.3 Livestock

Climate change directly and indirectly affects the livestock sector. The main direct effect of drought is through fodder deficiencies and their impact on animal health. This, in turn, has a significant effect on livestock production and the economic value of the entire livestock herd. With the onset of drought and the drying up of rivers, their natural habitat is destroyed and it is difficult to replace it. The drought brings about a sharp decline in the number of livestock and livestock production due to problems with drinking water supply, weakening livestock immune systems, and increasing losses due to diseases, abortions, and internal and external parasites. Some suggested solutions to reduce the effects of drought on livestock include: providing subsidized therapeutic supplements to address some of the vitamin and mineral deficiencies in livestock; providing subsidized vaccines to combat animal diseases such as *enterotoxemia* or *agalaxia*; and introducing and promoting breeds that work well under drought conditions (Zahedi Bidgoli et al. 2018).

The results of a study by Mirjalili and Zarekia (2016) showed that the reduction of both the number of livestock from 2011 to 2017 and the percentage of vegetation had caused a severe reduction in forage. As a result, the population of livestock fell drastically and farmers and the animal husbandry industry have faced huge problems in 2009–2010. Accordingly, the farmers had to sell their livestock and abandon this profession due to the excess pressure on their fields and pastures.

5.4 Forestry and Rangeland

Climate change has a profound impact on forestry sector; this includes changing the habitat location of forest species, especially the less tolerant ones, and the extinction of low tolerant species. The natural regeneration regime of forest plants is upset and results in the reduction of wood and non-wood products. Forests are witnessing pests and plant disease infestation and an intensification of land erosion, particularly in arid and **semi-arid** zones. Sea-based mangrove forests are degraded and sometimes destroyed because of the rise in sea level in the Persian Gulf and the Sea of Oman. Environmental conditions for wildlife in forest areas decline sharply as does forage production in rangeland, which can in some cases signal the onset of desertification. **Soil erosion** is the natural result of the decline of vegetative cover, and all such conditions are exacerbated by **high temperature** and aridity. One social consequence of this environmental downgrading and resource shortage is human displacement.

Monitoring the rangeland forage production at specified spatial and temporal scales is necessary for grazing management and implementation of rehabilitation projects in rangelands. Jaberalansar et al. (2017) collected data from 115 monitored records of forage production over 16 rangeland sites during the period 1998–2007 in Isfahan Province, Central Iran. Present and future forage production maps (for 2030 and 2080) were estimated. Rangeland forage production exhibited strong correlations with environmental factors, such as slope, elevation, aspect, and annual temperature. Under climate change, the annual temperature was predicted to increase, and the annual precipitation was predicted to decrease. Future maps of forage production indicated that areas with current low forage production (0–100 kg/hm^2) will increase while the areas with moderate, moderately high and high levels of forage production ($\geq$100 kg/hm^2) will decrease in 2030 and 2080. It was predicted that forage production of rangelands will decrease in the next couple of decades, especially in the western and southern parts of Isfahan Province. These changes were more pronounced in elevations between 2200 and 2900 m. This may be attributable to the increasing annual temperature and decreasing annual precipitation. Therefore, rangeland managers must cope with these changes by holistic management approaches through mitigation and human adaptations.

5.5 Recent Floods and the Disruption of Favourable Conditions for Winter Crop Development

The harvesting of the 2019 winter season barley crop started in March whereas the wheat harvest started in June. Up to March, the conditions for the development of the winter cereal crop were exceptionally favourable, as rains started on time and rainfall amounts were abundant throughout the whole season, promising an above average harvest. However, torrential rains at the end of March led to major flooding across the country, affecting more than 20 provinces, with Mazandaran, Golestan, Lorestan and Khouzestan being the most affected areas, resulting in casualties, economic losses, and severe infrastructure damage (Fig. 6).

The worst affected areas of Golestan Province received 70% of the average annual rain in the first 24 h of the downpour. According to the Ministry of Agriculture, Golestan Province produces about 10% of the total wheat output. Across the other heavily affected areas, rainfall totals for the two days at the height of the downpour were above the average rainfall normally received throughout the whole month. Losses caused by floods were estimated to be between USD3.5 to USD4.1 billion, including USD1.5 billion in the agriculture sector. In the current season, about 5.85 million hectares of wheat were planted, including two million hectares with irrigation. Following the floods, national authorities downgraded the national cereal harvest from excellent to good, like last year. The final production figures will depend on the impact of floods. The Government intends to locally purchase about 10 million tonnes of the 2019 wheat harvest with the aim to eliminate the reliance on imports for domestic consumption (FAO 2019).

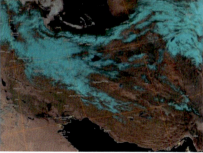

Fig. 6 A picture of Khuzestan flood in southern Iran in 2019 (left) and Map of flood-prone areas of Iran (right)

5.6 Links Between Influx of Desert Locusts to Iran and Climate Change

The influx of desert locusts to the coastal provinces of Iran has caused the FAO to change the situation in Iran from yellow to orange. Recently, the news that large herds of desert locusts have flown from the Arabian Peninsula to Iran and pose a threat to the country's food security have raised many concerns in the country. The FAO has also warned of locusts entering the six southern provinces of Iran. Statistics show that these desert locusts are growing on both sides of the Red Sea. According to FAO, a group of locusts moved to Saudi Arabia and laid eggs in March, while in Southwest Asia there are larger groups and a number of herds off the southern coast of Iran that hatched at the end of Marc (Zist Online News Site 2019).

The FAO has warned that all countries invaded must take serious action against desert locusts (Fig. 7). According to the FAO's report, desert locusts are a major pest for agriculture and can endanger a country's food security. In recent years, this type of locust has attacked Iran's farming system several times. These locusts destroy all plants and trees wherever they pass because they eat fodder, grains, summer crops and weeds up to several times their own weight daily (Zist Online News Site 2019).

Accordingly, one of the reasons for the widespread prevalence of locusts is certainly the climatic conditions, particularly the increased rainfall necessary for rapid multiplication of this pest. Indeed, it seems that the torrential rains that occurred recently in Iran have improved conditions for the presence of these insects.

Locust surveillance and control operations in infected countries are mainly carried out by ministries of agriculture with the public participation in addition to the involvement of several regional organizations. During times of outbreak, international cooperation and the assistance of volunteer organizations are often needed (Dadashi 2020).

Fig. 7 FAO map of the prevalence of desert migratory locusts in different countries, 2019. *Source*: Dadashi (2020)

6 Green Climate Fund Readiness and Preparatory Program

Iran's economy heavily depends on extraction and processing of fossil fuels. The country's energy use intensity and per capita CO_2 emission levels are among the highest in the world. Iran's CO_2 emissions have an increasing trend; surging from 337,325 Gg CO_2 equivalent in 1994 to about 832,043 Gg in 2010, and it is projected to increase if not addressed. While the lack of holistic adaptation and mitigation plans is still among the challenges that the country confronts, increasing frequency and intensity of extreme climatic events—which are driven by carbon emissions—contribute to reduced vegetation cover, soil erosion, desertification, water shrinkage, and loss of livelihoods across the country.

The Expected Results Green Climate Fund (GCF) Readiness and Preparatory Programme supports the Government of Iran in strengthening its capacities to access and manage climate financing with the goal to fulfil its mitigation commitments and resilience building. This Readiness Programme aims at strengthening the Department of Environment as the GCF National Designated Authority (NDA), improving the country's coordinated climate change investment, planning and decision-making mechanisms, and fostering stakeholders' engagement in all mitigation and adaptation projects. More specifically, the Programme's objectives include:

- Establishing a coordination mechanism to develop, appraise, and finance climate-change related projects;
- Strengthening NDA's institutional, functional, and technical capacities;
- Developing operational guidelines for the NDA;
- Training NDA staff on climate finance and fiduciary standards, country-ownership, and accountability;
- Developing stakeholders' engagement framework for taking part in GCF portfolio development and decision-making; and
- Supporting stakeholders in identifying adaptation and mitigation priorities of the country.

7 Priorities for Agricultural Research and Innovation

The Agricultural Research, Education and Extension Organization (AREEO) has: 19 nationwide research institutes (including multidisciplinary and crop-based research centres); 34 provincial research and education centres; 360 research stations, farms and bases; 34 agricultural extension directorates; 1213 district extension centres; 23 research incubation centres; and 142 knowledge-based enterprises. It is considered the largest national agricultural research system (NARS) in the Middle East with over 10,000 employees, including nearly 3000 scientists (Figs. 8 and 9).

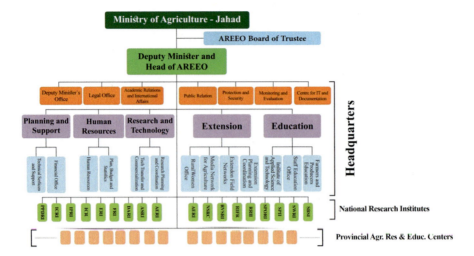

Fig. 8 Organizational chart of the Agricultural Research, Education, and Extension Organization (AREEO)

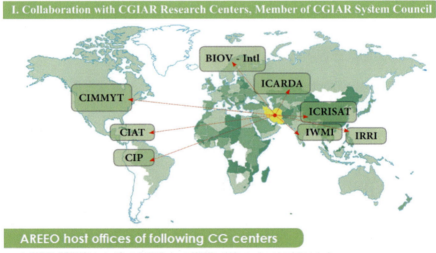

Fig. 9 International cooperation of AREEO

As Iranian agriculture encompasses various production systems, the priorities for agricultural research include many disciplines and hundreds of research topics. Some of the main conventional areas are listed below:

Research priorities:

- Water management and irrigation
- Mechanization and agricultural engineering
- Soil conservation and watershed management
- Soil, water, and crop relations
- Plant nutrition
- Postharvest, food science and technology
- Food safety
- Plant protection
- Dry-land agriculture
- Horticulture
- Salinity tolerance
- Plant genetics
- Conservation and use of genetic resources
- Crop production and breeding
- Seed production
- Animal production and breeding
- Silk production (sericulture)
- Honeybee production and breeding
- Fisheries and aquaculture
- Vaccine and serum production
- Veterinary and animal diseases
- Forest management and wood production
- Rangeland management
- Production and use of medicinal plants
- Agriculture economy and food security
- Conservation agriculture
- Integrated farming system
- Agricultural production value chains

Frontier areas:

- Agricultural biotechnology
- Support system decisions
- Precision and smart agriculture
- Adaptation to climate change

8 Extension Approaches for Climate Change in Iran

Efforts to reduce the impacts of climate change have been undertaken by many developing countries, especially those which are highly exposed to changing climatic conditions and weather extremes. Many attempts have been made to adapt agriculture to climate change because it is the main source of income and livelihoods of many people in these countries. Extension services have been at the centre of governments' efforts to build farmers' adaptation capacities (Kalimba and Culas 2020, cited in Venkatramanan et al. 2020).

In this section, we observe that climate change extension delivery builds upon the evolution of extension approaches over a span of decades. The agricultural extension system has functions that have been explored in different dimensions, conditions, and times with respect to its rationale. These functions are manifested in roles. Although the pivots of agricultural extension have been unchanged or just slightly changed over time, the roles assumed for the extension service may vary at different times depending on the conditions governing their development and goals (Falsafi 2011). Ponniah et al. (2008) stated that the agricultural extension has mainly three functions, and this system in the future should focus on: farmers' guidance in problem-solving; the transfer of knowledge and skill through technical recommendations; and the supply of inputs and other services to farmers. Accordingly, agricultural extension has a critical role in helping farmers adapt to climate change.

The promotion of best practices that assist smallholder communities in developing countries to be resilient and cope with the risk of weather extremities have their roots in participatory extension approaches. NGOs have been at the forefront of farmer-centred extension in developing countries that incorporate problem analysis, prioritizing needs, and choosing technologies that are socially, culturally, and economically viable (Prokopy et al. 2017).

Currently, the agricultural extension system in Iran is experiencing a radical change to make it more effective in helping farmers improve their performance and becoming productive, profitable and sustainable. Serious revision and restructuring started in 1996 and fostered after the amalgamation of two ex-ministries: the Ministry of Agriculture and the Ministry of Jihad Sazandegi. In this perspective, the basic principles for reinventing agricultural extension are:

- Focusing on small-holding farmers and enhancing their access to public services in agriculture;
- Starting from the grass-roots level and focusing on people/farmer's needs/issues and their demands (demand-oriented);
- Learning from past experiences and developing their own methodology and model of extension (grounded methodology); and
- Benefiting from emerging paradigms, the latest theories, and experiences to inform collective action (thinking globally, acting locally).

Eight strategic and master plans for agricultural extension have been prepared with all activities people-oriented and based on participation and privatization. Moreover, application of state-of-the-art communication technology to organize media is a core strategy. The following master strategies have been formulated to reinforce the changes in Iran's extension system:

- Restructuring the extension system to establish new and efficient modalities, approaches and institutional structure;
- Renovating all extension sites for optimal exploitation and management of facilities and capacities;

- Human resource development through promotion of training and employment of skilled staff;
- Reorganization of extension media aimed at improving its context and connection to the clientele;
- Networking agricultural knowledge and information systems;
- Restructuring and developing production cooperatives to overcome bottlenecks and formulate new policies for improvement;
- Studying and assessing farming and agricultural knowledge and information systems; and
- Reorganizing and expanding non-governmental institutions in the agriculture sector.

In addition to the above, other actions have been undertaken to re-invent the agricultural extension to tackle climate change. The plan for the New Agricultural Extension System has been implemented in some provinces of the country since 2014. Better agricultural extension services require the formulation of policies in this sector in line with future threats and opportunities. Iran is the pioneer in the pilot implementation of this new system in five provinces of Fars, Kermanshah, Ardabil, Golestan, and East Azerbaijan (Falsafi and Shahpasand 2014). The system includes the following:

- The enhancement of the accessibility to extension system and services for small farmers by: establishing a decentralized agricultural extension administration; and the restructuring of 1213 Agricultural Services Centres (ASCs) located at the district level, which have been equipped with necessary equipment like modern surveillance, meteorological systems, and other ICT-based facilitations (Fig. 10);
- The organization of small-holding farmers in different forms of institutions, including the National Confederation of Farmers and the Elite Farmers Association;
- The rearrangement of human resources to serve small farmers through: the relocation of extension agents and SMS to ASCs; the organization of 40,000 elected

Fig. 10 Applying ICT-based instruments in farmers' lands

farmers (male and female) as group leaders to serve as the bridge between smallholding farmers and ASCs; and the assignment of 2500 agricultural graduates in ASCs to fulfil their military service; and
- The introduction of new participative extension methods and techniques through the establishment of Integrated Participatory Crop Management (IPCM), Comprehensive Production and Extension Model Sites (CPEM), and Learning Focus Sites (LFS). The extension administration tries to deliver innovation in all activities and empowers the small-holding farmers to increase agricultural productivity (Falsafi and Shahpasand 2014).

9 Conclusions

Climate change is not just a future scenario, it is already happening in many regions with increased exposure to droughts, floods, and storms destroying opportunities and reinforcing inequality. Meanwhile, there is overwhelming scientific evidence that we are getting closer to an irreversible ecological catastrophe. According to many IPCC reports (i.e., 1996, 2007), this could lead to an unprecedented reversal in current human development and acute risks for future generations.

In Iran, climate change has affected the agricultural sector and human development indicators. Although the average production of agricultural products has increased over the last four decades, in some years, due to successive droughts, frost, and heat the quantity and quality of agricultural products have decreased. However, the government's policies in support of agriculture have made this sector more stable than other sectors, even in the most difficult economic conditions.

The agricultural extension system plays an important role in the development process, especially rural development. A look at the starting point of extension activities in Iran shows that the agricultural extension system is not a fledgling system and is nearly 70 years old. Iran has the largest network of agricultural research, education, and extension in the Middle East region and it has support programs dedicated to empower farmers and help them benefit from new innovations and initiatives. These programs have had favourable effects on controlling and managing climate change implications in the country.

The use of new approaches to agricultural extension with a focus on farmers, as well as government supportive policies, such as targeted subsidies and the implementation of some regional projects in cooperation with international organizations, have been effective in counteracting the adverse effects of climate change. However, some events, such as locust infestations, are partly related to climate change and have intensified in the past two years due to drought or sporadic and asymmetric rainfall in Persian Gulf countries. Iran is a country that has always faced drought problems and water shortages, but climate change has imposed new conditions on the agricultural sector. However, the government's supportive policies and farmers' efforts have made it one of the leading countries in the region in dealing with climate change.

To conclude, the following recommendations may guide future developments: (1) strengthening the capability for agrometeorological networking and monitoring to improve advisory services for agricultural and sustainable development in Iran. There is a window of opportunity to avoid the most damaging impacts of climate change, but that window is closing. The world has less than a decade to change course; (2) the new extension models should focus on strategies for assisting agricultural innovation through pluralism in service providers and the development of demand-driven extension services; (3) enhancing more integrated development of agriculture and coordination between the extension, research, and market sectors; and finally, (4) expediting the shift from capital and resource intensive (land, water, energy, inputs) to restorative, regenerative, and vibrant agriculture and food systems (Mendoza et al. 2020, cited in Venkatramanan et al. 2020).

References

Akbarieh A (2017) Climate change impacts in Iran, University of Luxembourg. http://www.researchgate.net/publication /31661111.

Amiri MJ, Eslamian SS (2010) Assessment of direct adverse impacts of climate change in Iran. J Environ Sci Technol 3:208–216

Amiri MJ, Ebrahimizadeh A, Amiri S, Radi M, Niakousari M (2009) Comparative evaluation of physicochemical properties of corn flours through different water qualities and irrigation methods. J Appl Sci 9:938–943

Asmare F, Teklewold H, Mekonnen A (2019) The effect of climate change adaptation strategy on farm households welfare in the Nile basin of Ethiopia. Is there synergy or trade-offs? Int J Clim Change Strategies Manag 11(4):518–535. Emerald Publishing Limited. https://doi.org/10.1108/IJCCSM-10-2017-0192

Dadashi D (2020. Investigation of the prevalence and control of desert locusts in Iran and Saudi Arabia in 2019, Ministry of Agriculture-Jahad

Falsafi P (2005) Integrated management of agricultural extension activities in Javadabad Pilot Project, the country report for the training course on agricultural extension planning and management. Tokyo-Japan

Falsafi P (2011) Farmers attitude towards the role of agricultural extension system in paving the way for their institutionalized participation in realizing agricultural sector's priorities (A case study in Buein Zahra District). J Agric Econ Dev 19(75):4–32

Falsafi P, Shahpasand MR (2014) Towards innovative extension services in national agricultural innovation system in Iran. In: Country paper for the workshop on innovative extension services to agricultural productivity. Manila, Philippines

FAO (2019) GIEWS country brief Iran (Islamic Republic of). Reference Date: 17-April-2019, Food and agriculture organization of the United Nations

FAOSTAT (2019) Iran (Islamic Republic of), Food and Agricultural Organization

FAPDA (2014) Country fact sheet on food and agriculture policy trends-IRAN, food and agriculture policy decision analysis

Hertel T, Rosch S (2010) Climate change agriculture and poverty. Apply Econ Prospect Policy 32(3):355–385

Hertel T, Rosch S (2010) Climate change, agriculture and poverty, Policy Research Working Paper, 5468, the World Bank Development Research Group Agriculture and Rural Development Team

Heshmati A, Maasoumi E, Guanghua W (2015) Climate change, agricultural production and poverty in India. In: Book: poverty reduction policies and practices in developing Asia, Chap. 4. Springer, Singapore. https://doi.org/10.1007/978-981-287-420-7_4

Hodson D, White J (2010) GIS and crop simulation modelling applications in climate change research. In: Reynolds MP (ed) Climate change and crop production. CABI, Oxfordshire, UK, pp 245–262

Intergovernmental Panel on Climate Change (1996) Climate change, 1995: the Science of Climate Change: contribution of working group. In: Houghton E (ed) (1995). Cambridge University Press, Amazon.com, Barnes&Noble.com, Books-A-Million, IndieBound

Intergovernmental Panel on Climate Change (2007) Climate change, 2007—Impacts, adaptation and vulnerability. Working group 2

Jaberalansar Z, Tarkesh M, Bassiri M, Pourmanafi S (2017) Modeling the impact of climate change on rangeland forage production using a generalized regression neural network: a case study in Isfahan Province, Central Iran. J Arid Land 9:489–503

Jacoby H, Rabassa M, Skoufias E (2011) On the distributional implications of climate change: the case of India. policy research working paper, World Bank, Washington

Kalimba U, Culas R (2020) Climate change and farmers' adaptation: Extension and capacity building of smallholder farmers in Sub-Saharan Africa. In: Venkatramanan V, Shah S, Prasad R (eds) Global climate change and environmental policy. Springer Nature Singapore Pte Ltd. https://doi.org/10.1007/978-981-13-9570-3_13

Kar S, Das N (2015) Climate change, agricultural production and poverty in India. https://www.researchgate.net/publication/281232641

Karimi V, Karami E, Keshavarz M (2018) Climate changes and agriculture: Impacts and adaptive responses in Iran. J Integr Agric 17(1):1–15. Available online at www.sciencedirect.com

Kouchaki A, Nasiri Mahallati M, Jafari L (2015) Investigating the impact of climate change on Iranian agriculture. In: Predicting the agro-climatic situation. Iranian agricultural research. Winter, vol 13(4):651 664

Koocheki A, Nasiri M, Kamali GA, Shahandeh H (2006) Potential impacts of climate change on agro climatic indicators in Iran. J Arid Land Res Manag 20(3):245–259

Li X, Yu J, Zhu C, Wang J (2012) Computer simulation in plant breeding. Adv Agron 116. Elsevier Inc, ISSN 0065-2113. https://doi.org/10.1016/B978-0-12-394277-7.00006-3

Lobell DB (2010) The use of statistical models to predict crop yield responses to climate change. Agric Meteorol 150(11):1443–1452

Maleki Morsht R (2017) Assessing the effects of climate change on water and agricultural resources of Iran. In: The first international and the fifth national conference on organic agriculture versus with conventional agriculture. Faculty of agriculture and natural resources. Mohaghegh Ardabili University

Mendoza T, Furoc-Paelmo R, Makahiya H, Mendoza B (2020) Strategies for scaling up the adoption of organic farming towards building climate change resilient communities. In: Venkatramanan V, Shah S, Prasad R (eds) Global climate change and environmental policy. Springer Nature Singapore Pte Ltd. https://doi.org/10.1007/978-981-13-9570-3_13

Mirjalili AB, Zarekia S (2016) A study of management strategies in drought seed for steppe animal husbandry farmers (Case Study of Tang Chenar rangelands in Yazd province). In: 7th National conference on rangeland and rangeland management of Iran, May 2016, pp 18–19

Misra AK (2014) Climate change and challenges of water and food security, the review article, gulf organization for research and development. Int J Sustain Built Environ 3:153–165. Production and hosting by Elsevier BV

Molle F, Mollinga P (2003) Water poverty indicators: Conceptual problems and policy issues. Water Policy 5(5):529–544

Molle F, Venot J-P, Lannerstad M, Hoogesteger J (2009) Villains or heroes? Farmers' adjustments to water scarcity. Irrig Drain n/a-n/a. https://doi.org/10.1002/ird.500

Nasiri Mahallati M, Kouchaki A, Kamali GA, Marashi SH (2006) Investigating the effects of climate change on agricultural climatic indicators of Iran. Agric Sci Technol 20(7):71–82

Ponniah A, Puskur R, Workneh S, Hoekstra D (2008) Concepts and practices in agricultural extension in developing countries: a source book. International Livestock Research Institute, pp 1–274

Prokopy L, Burniske G, Power R (2017) Agricultural extension and climate change communication, book chapter: Oxford Research Encyclopedia of Climate Science. Oxford University Press USA.https://doi.org/10.1093/acre fore/9780190228620.013.429

Rezaei Moghadam K, Fatemi M (2020) Strategies for improvement of agricultural extension new approach of Iran (Text in Persian). J Iran Agric Ext Educ Sci 15(2):223–251

Ricardo D (1817) The principles of political economy and taxation. Murray J, London Sanghi A, Mendelsohn R (eds) (2008) The impacts of global warming on farmers in Brazil and India. Glob Environ Change 18:655–665

Schlenker W, Roberts MJ (2009) Nonlinear temperature effects indicate severe damages to U.S. crop yields under climate change. Proc Nat Acad Sci USA (PNAS). 106(37):15594–15598. http:/www.pnas.org/content/106/37/15594

Slafer GA (2003) Genetic basis of yield as viewed from a crop physiologist's perspective. Ann Appl 142(2):117–128

Venkatramanan V, Shah S, Prasad R (2020) Global climate change and environmental policy agriculture perspectives. Springer Nature Singapore Pty Ltd. ISBN 978-981-13-9569-7ISBN 978-981-13-9570-3(eBook), https://doi.org/10.1007/978-981-13-9570-3

UNDP (2019) Human development report, inequalities in human development in the 21st century, briefing note for countries on the 2019, human development report (Islamic Republic of Iran)

Zahedi Bidgoli A, Dehghani F, Abdi Darkeh S, Boland Morteba M, Tehrani Sharif S (2018) Drought and its impact on livestock. J Lab Res 10(2) of the Twelfth Iranian Veterinary Congress, Semnan University, Spring and Summer 2018, pp 112–112. https://doi.org/10.22075/jvlr.2018.3185

Zhai F, Lin T, Byambadorj E (2009) A general equilibrium analysis of the impact of climate change on agriculture in the People's Republic of China. Asian Dev Rev 26(1):206–225

Zist Online News Site (2019) https://www.zistonline.com/news/.

Chapter 13
Better Crop-Livestock Integration for Enhanced Agricultural System Resilience and Food Security in the Changing Climate: Case Study from Low-Rainfall Areas of North Africa

Mina Devkota, Aymen Frija, Boubaker Dhehibi, Udo Rudiger, Veronique Alary, Hatem Cheikh M'hamed, Nasreddine Louahdi, Zied Idoudi, and Mourad Rekik

Abstract Increasingly frequent droughts, declining soil fertility, and poor plant-animal-atmosphere interactions are threatening the sustainability of integrated crop-livestock systems in the rainfed drylands of North Africa. Previous research from

---

M. Devkota (✉)
International Center for Agricultural Research in the Dry Areas (ICARDA), Rabat, Morocco
e-mail: m.devkota@cgiar.org

A. Frija · B. Dhehibi · U. Rudiger · Z. Idoudi · M. Rekik
International Center for Agricultural Research in the Dry Areas (ICARDA), Tunis, Tunisia
e-mail: a.frija@cgiar.org

B. Dhehibi
e-mail: b.dhehibi@cgiar.org

U. Rudiger
e-mail: u.rudiger@cgiar.org

Z. Idoudi
e-mail: z.idoudi@cgiar.org

M. Rekik
e-mail: m.rekik@cgiar.org

V. Alary
International Center for Agricultural Research in the Dry Areas (ICARDA), International Centre of Agronomic Recherche for Development (CIRAD), Tunis, Tunisia
e-mail: veronique.alary@cirad.fr

H. C. M'hamed
Agronomy Laboratory (LR16INRAT05), Carthage University, National Institute for Agricultural Research of Tunisia (INRAT), Tunis, Tunisia

N. Louahdi
Technical Institute of Field Crops (ITGC), Setif, Algeria

© The Author(s), under exclusive license to Springer Nature Switzerland AG 2022
M. Behnassi et al. (eds.), *Food Security and Climate-Smart Food Systems*,
https://doi.org/10.1007/978-3-030-92738-7_13

around the globe has verified that better integration of crop and livestock activities within agricultural production systems is promising in boosting food productivity, soil health, and overall farm profitability. This is especially relevant for rainfed drylands, particularly in areas with low rainfall where livestock production is predominant. Although integrated crop-livestock farming already exists in these regions, the decreasing integration between the two activities—induced by a variety of factors during the previous three decades—resulted in perennial depletion of soil fertility and an overall decrease of relative farm incomes. The North African region owes its sustainable intensification benefits to numerous synergistic interactions. This chapter, therefore, aims to highlight options for better integration of the crop-livestock system into the region's long-existing cereal-based livestock farming system, in order to help boost food and nutrition security, farmers' income and soil health. The chapter looks at case studies from Algeria and Tunisia. In particular, it considers as the key integrating factors for crop-livestock system: diversifying cereal monocropping by introducing of food and forage legumes; integrating alternative grazing/feeding systems; integrating tree-crops and livestock; adopting Conservation Agriculture practices in order to effectively address the crop residue tradeoff between providing feed for livestock and leaving residues as mulch; improving the management of herd health and increasing the availability of scale-appropriate mechanization. A combination of all is considered as the key integrating factor for the crop-livestock system. The combination of all or a few of these components helps improve overall farm incomes, crop productivity, and soil health, increases the efficiency of input use, provide healthy protein for human's diet and fodder for livestock, and also has the potential as a sustainable intensification strategy. For the wider adoption of these alternative options by smallholder farmers, it is important to consider different approaches; for example, participatory evaluation, field visits, farmers field schools and the use of information and communications technology, along with improving farmers' capacity to access and use these tools.

Keywords Conservation agriculture · Crop-livestock integration · North Africa · Diversification · Resilience

1 Introduction

North Africa, often considered as a "climate change hotspot" (Diffenbaugh and Giorgi 2012), is characterized by high rates of population growth, rapidly increasing food deficits, and limited natural resources, particularly arable land and water (FSI 2015). Drought has become more severe and frequent in the region, and climate change is projected to continue this trend in the future. Agriculture production is, therefore, highly affected by climate change, as more than 90% of agriculture depends on rainfall across North Africa, except Egypt (Mrabet et al. 2021). The agriculture sector is critical to the region's economy and represents about 15% of its gross domestic product (FAOSTAT 2021). Most of the region is located within arid and semi-arid climates characterized by high temperatures and low levels of precipitation with significant inter and intra-annual variability for rainfall and extreme events (drought and heat). An average annual rainfall mostly ranges from 200 to 500 mm, and this varies widely from year to year and within seasons. Climate change has and will impact agricultural productivity through an increase in mean temperatures, alterations in rainfall patterns, an increasingly frequent occurrence of weather extremes (including high temperatures and levels of precipitation, and drought, and heat), and altered patterns of pest pressure (Mrabet et al. 2021).

Most agricultural soils in the dry climates of the region contain small amounts of organic matter (less than 1%) with poor soil aggregate structure, and the predominant land-use practices of tillage, overgrazing and exposing bare soils continue to worsen the situation. In the long run, this can only lead to severe land degradation, and finally to desertification, as can be observed already in many parts of the region. Mixed crop and livestock systems are the dominant production systems, especially for smallholders and in areas with low-rainfall (Dixon et al. 2001). Half of the population in the region live in rural areas, and two out of three countryside dwellers work in farming. Therefore, when severe drought strikes, it can have a major impact on food and water security, as well as on productive assets (trees and livestock flocks), livelihoods, and health.

2 Characterization of North Africa's Crop Production System and Crop-Livestock Integration

Most farming systems in North Africa are dominated by traditional subsistence agriculture and smallholder farms run by multigenerational families. Livestock plays a key role and contributes a significant share to food security and incomes in the region. Smallholder farmers also consider livestock as a primary asset that can be easily converted into cash in dry years. Livestock farming is characteristically interrelated with cropping systems through weedy fallows, residue, and stubble grazing and the use of woodlands and rangelands (Magnan 2015; Moujahed et al. 2015). Hence, the food, nutrition, and livelihood security of rural populations in the region largely depend on both crops (mostly cereals and trees) and livestock, with preference given to livestock due to its high and multiple contributions to incomes and assets.

Most of the poorer smallholder farmers of North Africa live in drylands where their crop-livestock systems are rainfed making them more vulnerable to limited and increasingly unpredictable and variable rainfall. Again, in such a production environment, crop productivity and biomass are typically low, due mainly to drought stress and low levels of inputs. Poor land management, namely, continuous cereal monocropping, intensive soil tillage, and overgrazing have further degraded land through soil erosion and loss of soil organic matter. Moreover, an increased frequency of droughts, floods, and other climatic risks in recent years has further exacerbated abiotic stresses in rainfed drylands. Indeed, environmental threats related to climate change and water scarcity, together with biased socio-economic and market policies and other demographic and technological drivers are constraining agricultural productivity and lead to continuous degradation of natural resources, mostly soil fertility and water.

Studies are showing that crop-livestock integration in the region can reverse this degradation dynamic and help to sustainably intensify agricultural production in mixed systems (Ezeaku et al. 2015; Murendo et al. 2019). Crop-livestock integration further helps reduce food waste through increased use of by-products and recycling of nutrients both of which are key measures for optimizing the efficiency of nutrient use across the full food chain. This chapter highlights options for better integration of crop-livestock production systems as well as pathways for their sustainable upscaling across the region. Highlights provided in this chapter are based on pilot research for development projects adapted to farmers' preferences and capacities.

3 The Framework of Crop-Livestock Integration

Integrated crop-livestock systems (ICLS) are based on temporal and spatial interactions between animal and crop activities. Since the 1990s, many studies have highlighted the multiple benefits of these systems in better exploiting the complementary resources of specific biophysical conditions (Moraine et al. 2014), and the emergent properties that result from soil–plant-animal-atmosphere interactions (Silva and Lambers 2020). Among these benefits are an improvement of nutrient cycling and soil fertility at field and farm levels achieved through animal-waste recycling and by including grasslands in field-crop systems (Ryschawy et al. 2012). From a technical point of view, these integrated systems induce an increase in the efficiency of land resource and machinery use, a decrease in plant diseases and weed incidence, and increased diversity, profitability, and incomes (Bell et al. 2021; Ryschawy et al. 2012), with overall greenhouse gas mitigation. Synergies between cropping and livestock husbandry also offer many opportunities for sustainably increasing production, both for households and regions (Herrero et al. 2010). One of the critical elements for a sustainable crop-livestock system is developing sustainable intensification methods that improve efficiency and help produce more food using less land, water, or other resources (Doré et al. 2011; Matson et al. 1997).

Despite these advantages, mixed crop-livestock systems have experienced a significant decline globally (Moraine et al. 2014). In North African countries, this decline has been most prevalent in more favorable environments where irrigation has been possible—farmers have been able to extend cereal crops with high yields and develop vegetable crops with high economic returns (Alary et al. 2019). In more vulnerable zones characterized by erratic rainfall, policies have encouraged tree planting of mainly olive or fig trees, which has also contributed to a reduction of grazing lands. In this new configuration, farmers have significantly reduced flock sizes. Livestock activity based on grazing land has, therefore, relocated to non-cultivable mountainous and arid zones. In the present context, competing pressures on the land, water and nutrients used to produce biomass to feed animals raise many challenges that can affect the overall sustainable development of livestock systems (Dixon et al. 2010).

3.1 Opportunities and Challenges Relating to the Sustainable Intensification of Crop-Livestock Production Systems

Despite their overall decline, mixed crop-livestock farming systems, one of the key dynamics in the process of agricultural intensification, are highly prevalent in the North Africa region, offering a range of benefits and challenges for farmers. They are, however, extremely impacted by climate variability and associated reduced availability of feed production, forage biomass, and vegetation in North Africa (Tibbo and van de Steeg 2013). Several studies conducted globally as well as in the region clearly indicate that integrated, rather than individual crop and livestock system are more resilient in term of mitigating risk and enhancing food and nutrition security (Bell et al. 2021; Ezeaku et al. 2015; Murendo et al. 2019). Also, better integration of crop-livestock systems can lead to the reclamation of degraded lands: crop residues can restore soil fertility and quality; and crop revenues can fund further system improvement (Costa et al. 2014).

Crop residues and forage crops can be can be considered as a source of feeding. Indeed, crop residues (i.e., stubble and straw) are a valuable forage resource for livestock. Residues are easily available, cheap to purchase, and can be stored for long periods of time to meet future feed shortages (Ates et al. 2018). Grazing of stubble by livestock, especially during the summer period, is also a traditional and common practice in the North Africa region. With increased numbers of livestock (mainly small ruminants) and an expansion of cereal monocropping at the expense of food, the livestock feed gap has become more acute in the rainfed Mediterranean (Ryan et al. 2008). Therefore, with increasing monocropping, cereal stubbles and straw have become the primary source of feed for small ruminants in the region (Salem and Smith 2008). But stubble grazing can often be intensive with high stocking rates, leaving soil barren. These intensive grazing practices also damage soil structure, create compaction, and lower the soil's capacity to capture rainwater (Bell et al. 2021). In addition, the dry matter intake and nutritional quality of the stubble decrease linearly with an increased number of grazing days and stocking rates (Moujahed et al. 2015). Although, leaves and some grains, which are considered high-quality plant parts, left immediately after crop harvest, have been found to benefit grazing livestock in live weight gain, the feeding value of stubble declines overtime. Hence, it is important to provide supplements alongside low-quality stubble.

The growing forage crops for direct grazing, conservation or as cover crops have an important role in agronomic sustainability and livestock production in a mixed crop-livestock system. The inclusion of forages in a cereal-based livestock system improves livestock production, enhances soil health and biodiversity, increases carbon sequestration, and minimizes disease infestation and economic risk through diversifying the cropping system (Christiansen et al. 2015). However, the lack of quality seed of improved forage varieties, poorly functioning seed and fodder markets, and biased national policy towards the production of strategic food grains and crops constrain the economic incentives for sustainable forage production and marketing. Additional non-technical solutions are needed to facilitate farmers' access to forage seed, technical skills and other related services that allow them to enhance their crop rotations and satisfy their flock feeding requirements. Redirecting subsidies on forage crops, for example, has provided successful outcomes for smallholder crop-livestock producers in various Mediterranean countries (Demir and Yavuz 2010; Lloveras et al. 2004). Developing farmers capacity for seed multiplication in addition to related skills, and providing machinery for seed cleaning and packaging were also found to be effective for upscaling forage integration in Tunisia (Rekik et al. 2019). Thus, public funds should be invested in supporting forage-based crop-livestock farming and related research, including seed systems, rather than cereal monocropping or industrial-scale meat and milk production.

3.2 Conservation Agriculture: An Integrated Approach for Crop-Livestock System Intensification

Crop-livestock integration also needs to be considered within a broader framework of sustainable intensification and resource conservation. This will help guide the development and selection of appropriate technical solutions. Conservation Agriculture (CA) offers an innovative framework for integrating the system into this broader framework. CA principles, including minimum mechanical soil disturbance or no-tillage, permanent soil cover with crop residues and/or cover crops, and crop diversification through varied crop rotation, sequences and associations have proved to be key interventions for enhancing crop productivity and improving resource-use efficiency and soil health (Corsi and Muminjanov 2019; Devkota et al. 2021; Moussadek et al. 2014; Mrabet et al. 2012).

CA is also widely promoted to reduce soil degradation and develop systems that are more resilient to climate change (Bahri et al. 2019; Kassam et al. 2019). Under CA systems, residues retained on the soil surface lead to soil water retention, by decreasing soil evaporation and increasing the infiltration and deep percolation that lead to increased yields and water-use efficiency (Moussadek et al. 2011). In fact, improved soil moisture and nutrient availability under the CA system can improve crop yields by 20-120% in the dry Mediterranean climates of different continents (Fernández-Ugalde et al. 2009; Mrabet et al. 2012). Additionally, studies found that cropping systems under CA sequestered soil organic carbon, especially in the topsoil layer (Moussadek et al. 2014).

Fig. 1 Sheep grazing stubble in Fernana region—North West Tunisia. *Photo* Zied Idoudi (ICARDA)

Most of the areas under CA in North Africa are located in semi-arid regions under rainfed conditions. Production systems in these regions are mainly based on field crops and especially cereals (wheat, barley, and oat) combined with ruminant livestock. In these regions, CA farmers usually practice both cereal crop and livestock activities and give preference to their livestock as these are usually a primary source of income and risk management. These farmers rarely practice all three CA principles together, and no-tillage is the most adopted of the three. The farmers face major constraints in terms of permanent land cover since stubble grazing is the usual practice. They also tend to overgraze crop residues due to a lack of summer sources of feed for their animals (Fig. 1).

Therefore, under the CA system, the practice of retaining crop residues creates a conflict of interest between mulch for that can cover the soil surface and stubble grazing for livestock, especially during the summer period. Trade-offs between the use of stubble for livestock feeding or for covering the soil must therefore be resolved, particularly in drylands where fodder potential is low. In order to inform and promote better integration of crop-livestock systems, while adopting conservation agriculture with national partners in the public and private sector, International Center for Agricultural Research in the Dry Areas (ICARDA) set up the project Crop-Livestock under Conservation Agriculture (CLCA). This initiative funded by the International Fund for Agricultural Development was initiated in North Africa and other similar regions in 2015 to assist in the sustainable intensification of the crop-livestock system. Under the framework of the CLCA project, ICARDA developed a stubble grazing model (30:30 model) in Tunisia in order to give farmers adopting CA some solutions for reasonable stubble grazing during the summer period (Guesmi et al. 2019;

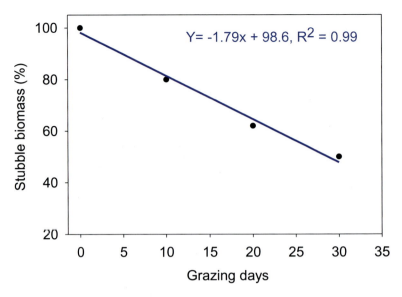

Fig. 2 Relationship between biomass of residues (%) on soil surface and grazing duration (day). (Author's own work)

Moujahed et al. 2015). The 30:30 model was developed based on a stocking rate of 30 animals per hectare, during a 30-day stubble grazing period (Fig. 2). This model allows for the retention of adequate crop residues (mulch) in the soil surface (more than 0.4 t ha^{-1} of residue in the soil surface or 40% of the initial biomass of residues on the soil surface) and at the same time helps preserve the health body conditions of animals (Guesmi et al. 2019; Moujahed et al. 2015).

Another requirement of for better crop-livestock integration under CA is the combination of diversified crop rotation with controlled and improved grazing. This can be effective at preserving or even enhancing soil function and health. Crop diversification is also an issue under CA, and it has been reported that several crops can be grown successfully in sequence without tillage and seedbed preparation in various contexts of North Africa. Among such crops are oilseed crops (sunflower and sesame), food legumes (chickpea, bean, lentil, fababean, and soybean), forage crops (vetch), and most cereals (corn, wheat, barley, sorghum, and oat) (Ben-Hammouda et al. 2007; Devkota et al. 2021; Harb et al. 2015; Mrabet 2011). In this context, to improve the integration of livestock in cereal-based systems, new species of forage crops (vetch, triticale, etc.) and crop mixtures (triticale + vetch, oats + vetch) were included in crop rotations and were introduced as highly suitable alternatives in marginal wheat-based systems (Cheikh M'hamed et al. 2016). In fact, increasing forage production reduces grazing pressure on residues during the summer. For example, when several alternative forage species (*Vicia sativa* (Vetch), *Medicago sativa* (Lucerne), *Hedysarum citinarium* (Sulla) and forage crops mixtures (triticale

40% + vetch 60%; oat 30% + vetch 70%; triticale 30% + vetch 70%) were introduced and disseminated among farmers adopting CA in Northern Tunisia (Abidi et al. 2020, 2019; Cheikh M'hamed et al. 2018; Rekik et al. 2019). Results showed that the yields of forage crops and forage crop mixtures introduced under CA ranged from 4 to 12 t ha^{-1} depending on the bioclimatic zone. In addition, the fodder was of high nutritional value, adequate for maintaining an intensive production system for dairy products and small ruminants. Indeed, for vetch crops, the crude protein content was an average of 14% (Abidi et al. 2019; Rekik et al. 2019).

3.3 Good Agriculture Practice for Reducing Crop Yield and Feed Gaps

Most of the currently cultivated field crops in the small mixed crop-livestock systems of North Africa are cereal forages such as barley, oat, and triticale. Similar to other cereal crops, forage crop yields are exceptionally low compared to what can potentially be achieved (Rekik et al. 2019). The existence of large attainable yield gaps for major food crops indicates an opportunity to increase the average farmer's yield through the adoption of good agricultural practices (Devkota and Yigezu 2020). The adoption of good agriculture practices not only increases the crop yield but also increases the quantity and quality of grains and biomass produced. This will, in turn, help satisfy the feeding quality and requirements of livestock in these mixed systems. The use of improved varieties, quality seed, balanced fertilizer application, timely weed management, and legumes in cereal monocropping is considered a major determinant for closing the yield gaps in rainfed wheat in Morocco (Devkota and Yigezu 2020), for example. However, it is important to find better ways for scaling up those innovations in ways that consider the biophysical and socio-economic environment and reach to the end user.

4 Integration of Crops, Livestock, and Trees: An Adapted Production System in Low-Rainfall Areas of North Africa

4.1 Conservation Agriculture Based Crop-Livestock Integration

Work initiated in Algeria and Tunisia as part of the CLCA initiative aims to promote CA in a semi-arid environment through the integration of crop–livestock systems (Rekik et al. 2019). It is known that CA faces significant constraints in North Africa with regards to integrating crop-livestock systems because of critical tradeoffs related to biomass and soil resource use across the agricultural production system (ICARDA 2017). Crop-livestock farmers live with an acute shortage of biomass and would be impeded by grazing solely on crop residues and stubble (a central pillar for CA). Additionally, farmers prefer to use land for wheat as the market is secured for this commodity; this leaves little room for including forage in their system. Faced with these challenges, the CLCA initiative has been promoting a set of innovative livestock and crop management practices to optimize climate-resilient and integrated crop-livestock systems under CA in the fragile livestock-cereal belt of semi-arid North Africa. The promoted options for crop-livestock integration (CLI) in these dry areas encompass, among others: the inclusion of forages in the cereal rotation system (Abidi et al. 2020); the improvement of crop management to enhance grain and straw yield (Souissi et al. 2020); the use of dual-purpose crops and varieties, and forage combinations that provide the flexibility with regard to livestock (Ayeb et al. 2013); the provision of stubble management for mulching (Guesmi et al. 2019), cover crops for feed and soil, feed alternatives during the summer gap, mechanization of alternative feed production, and specific health interventions to reduce the parasite burden and to prevent feed-borne diseases (Idoudi 2020). These options range from pure productivism to rather conservative practices and from the farm to the landscape level (Fig. 3) in order to favor their context-specific adaptation.

Filtering of these various CLI options based on their agroecological attributes, in addition to their short-term impact on farmers' livelihoods, is needed to ensure a successful transition within the sustainable intensification pathway. Some of the more agroecological CLI options are relevant for the farm household level. In contrast, others might be more relevant at a landscape or community level, which involve more collective action and broader ecosystem services. A combination of options at the landscape level is, therefore, required to promote sustainable intensification.

Fig. 3 Clustering Crop-Livestock Integration (CLI) options based on the scale of implementation and resource-orientations. Graphic Design: Zied Idoudi and Mourad Rekik (ICARDA)

This also means that interventions to strengthen farmers' organizations are sometimes crucial for the successful implementation of some of these agroecological CLI options towards sustainable intensification.

4.2 Small-Scale Mechanization for Better Crop-Livestock Integration

Small-scale crop-livestock systems in low- and middle-income Countries, such as those in North Africa, are characterized by low levels of mechanization (Fig. 4), which undermine their productivity and sustainability. The machinery promoted in local markets is made and imported mostly from industrial countries where farm systems are on a larger scale (Rudiger et al. 2020). These machines are unaffordable for farmers with small to medium-sized landholdings who remain incapable of upgrading and modernizing their farming operations. Through the CLCA initiative and the CGIAR Research Program on Livestock (Feed and Forage Flagship), ICARDA and its national partners in Tunisia have been working on developing machinery well adapted to small-scale farming systems and able to contribute to crop rotation by allowing for the inclusion of forage crops, enhancing the quality

Fig. 4 Traditional and manual seed cleaning by woman farmer. Photo: Zied Idoudi (ICARDA)

Fig. 5 Mobile seed cleaning and treatment unit. *Photo* Zied Idoudi (ICARDA)

of animal feed, and reducing the impact of livestock grazing on soil covers. These machines include 'small mobile seed-cleaning and treatment units' (Fig. 5) and 'small mobile feed-grinder machines' (Fig. 6), and are locally manufactured at a low cost. The partnership has also developed business models to deploy these machines to small- and medium-scale farmers' cooperatives and other enterprises with the potential to deliver machinery service, thus contributing to agricultural diversification and the enhancement of their respective incomes.

The conventional national seed system in Tunisia and other North African countries does not provide enough quality forage seeds. Production of forage seed for crops like barley, vetch, or alfalfa is mainly undertaken by large seed-producing cooperatives through subcontracts with individual farmers. In Tunisia, one private seed enterprise, COTUGRAIN, and Tunisia's Livestock and Pasture Office (OEP) are equally engaged in forage seed production.

Due to an insufficient supply of forage seed, but also to save costs, many small-scale, mixed farmers prefer using their own saved seed. The quality of these farm seeds is generally low as they are normally cleaned manually, so the final product still contains some unproductive seeds (broken seeds or small-sized seeds). In addition, these seeds are sometimes attacked by pests and diseases as they are not treated. The results of using these poor-quality farm seeds are low forage yields and quality and reduced farm incomes.

Fig. 6 Training in use of mobile feed grinder. *Photo* Udo Rudiger (ICARDA)

To tackle this constraint, the CLCA project promoted the use of innovative locally produced seed cleaning and treatment units. These were designed to provide business development for lead farmers and farmer cooperatives around forage seed production. The cooperatives can provide seed cleaning and treatment services for their members. This provides additional income for the cooperative and enables members to engage in forage seed production. Also, the seeds are used by the members themselves, thereby improving production and overall farm incomes. For example, introducing four seed cleaning and treatment units with a capacity of 800 kg/hr each has increased the availability of quality seed for major food and forage crops by three-fold (240 t in 2019 to 691t in 2020) (Rudiger et al. 2020).

The cost of feed supplies is another constraint for small-scale livestock farmers in Northern, Central, and Southern Tunisia, in particular during summer. A solution is to grind locally available feed, in order to increase intake, improve digestibility, and gain productivity (Rudiger et al. 2020). The introduction of locally produced mobile grinders helped farmers develop quality feed at a local level while at the same time reducing feed waste. Such mobile grinders can serve for feed mash production as well as simple grinding of bulky feed like straw and hay and a range of other agro-industrial feed resources to reduce feed wastage and provide alternative diets for the summer feeding of flocks. More than 3,000 beneficiaries including young farmers and women are now benefiting from this equipment. It is an ideal tool for enabling smallholder farmers to improve their incomes and represents an opportunity for improved livelihoods in traditional small-scale farming. The use of these tools also reduces the labor time spent on feed-farming operations, especially for women farmers (Rudiger et al. 2020).

5 Impacts of Adopting Agronomic Innovations

Diversifying cereal monocropping by including legume food or forage crops in the rotation, adopting conservation agriculture practices, using improved crop varieties, and applying timely fertilizer and weed management are the major agronomic innovations implemented in Algeria and Tunisia under the CLCA project. The implementation of single or multiple innovations has enhanced several sustainability indicators in the two countries, as presented in Table 1. The inclusion of forage or food legumes in existing cereal monocropping helped diversify the rotation system. Although we have not quantified the effect of rotations in different situations, a previous study confirms that the inclusion of legume in cereal monocropping enhances crop yields (Devkota and Yigezu 2020), increases farm profits (Yigezu et al. 2019), minimizes problems of insect pests and reduce fertilizer requirements for the following cereal crops. Similarly, switching from more water-demanding wheat crops to less water-demanding and more drought-resistant crops, for example lentil, vetch or chickpea, can be an important adaptation strategy for food and nutrition security in a changing climate. The adoption of CA under the CLCA initiative produced a similar or higher yield of wheat, barley, and forage crops with a more stable yield compared to the current farmers' practice. The adoption of CA also improved water use efficiency and the soil organic matter compared to farmers' practice in both Algeria and Tunisia (Table 1). The production costs have reduced with the adoption of CA technologies, mainly due to a reduction in land preparation costs, and the CA system avoided soil tillage before seeding. Similarly, the adoption of CA produced a higher net return and total benefit–cost ratio than did existing farmers' practice. These findings suggest that the adoption of a number of CLCA conservation practices has the potential to provide a relative advantage (particularly in economic terms in the case in Algeria) or a range of accommodations facilities that smallholders farmers may have in trailing them (easy to test and implement).

6 Innovation Management for Technology Scaling at Scale

Scaling up the CA concept and practices among rural farming communities in North Africa remains a challenge (Kassem et al. 2019). CA technologies and principles have been disseminated and piloted by ICARDA and International Maize and Wheat Improvement Center (CIMMYT) with significant impact and success for the farmers under the CLCA project. The practices enable smallholders to integrate livestock

Table 1 Summary of the impact of different agronomic innovations (technology from the Crop–Livestock under Conservation Agriculture [CLCA] initiative) on improving sustainability indicators in Algeria and Tunisia under rainfed conditions

A. Tunisia				
S. No	Crop and practices	With CLCA innovations	Farmers' practice	Type of innovation
1	Cropping system	Cereal- legume	Cereal- cereal rotation	Crop diversification
2	Crop yield (t/ha)			Conservation Agriculture, improved variety, crop diversification
	Wheat	2.57 ± 0.82 (CV = 30%)	2.42 ± 0.78 (CV = 32%)	
	Barley	1.94 ± 0.43 (CV = 23%)	2.18 ± 0.44 (CV = 20%)	
	Fodder oat	5.4 ± 1.1 (CV = 20%)	5.1 ± 1.3 (CV = 25%)	
	Forage mixture	8.88 ± 1.65 (CV = 18%)	5.4 ± 1.02 (CV = 19%)	Crop diversification
3	Water use efficiency wheat (kg grain/mm water)	7.88 ± 0.52 (CV = 6.6%)	7.09 ± 0.72 (CV = 10.9%)	Conservation Agriculture (no tillage + residue retention)
4	Soil erosion (kg/ha/yr)	22	364	Conservation Agriculture
5	Soil organic carbon (g/kg soil) at top 10 cm soil layer	18 ± 0.1	12 ± 0.1	Conservation Agriculture
6	Net return (US$/ha) wheat)	168	158	Conservation Agriculture
7	Benefit cost ratio for wheat	2.16	1.92	Conservation Agriculture
B Algeria				
1	Crop yield (t/ha)			Conservation Agriculture + improved variety
	Wheat	1.99 ± 0.5(CV = 25.1%)	1.65 ± 0.38 (CV = 23%)	
	Barley	2.72 ± 0.49 (CV = 18%)	2.6 ± 0.69 (CV = 26%)	
	Vetch	1.23 ± 0.19 (CV = 16%)	1.07 ± 0.24 (CV-23%)	
2	Water use efficiency kg grain/mm water (wheat)	6.98 ± 1.51 (CV-21.6)	6.02 ± 0.99 (CV-16.4%)	Better water management

(continued)

Table 1 (continued)

A. Tunisia				
S. No	Crop and practices	With CLCA innovations	Farmers' practice	Type of innovation
3	Soil organic matter (%)	1.95 ± 0.15	1.25 ± 0.25	Conservation Agriculture
4	Input cost (US$/ha)			Conservation Agriculture + better fertilizer and weed management
	Wheat	409.09	424.89	
	Barley	309.91	387.76	
5	Benefit cost ratio for wheat under rainfed -2018–2019 cropping season	Wheat/lentil 1.4	Wheat/barely 0.4	Conservation Agriculture

The values are average ± standard deviation (SD), values in bracket are coefficients of variance (CV)

production into farming system in ways that reduce water consumption, protect the soil, diversify crops, increase yields and farm profits, and eventually make their farms more resilient to the effects of climate change. But questions linger over factors that have hindered upscaling despite the sound technical, agronomic, and environmentally friendly merits of CA.

Upscaling CLCA technologies and practices will certainly entail changing the behavior, strategies and agricultural practices of small-scale producers engaged in crop-livestock systems. Critically, these farmers need to become better informed about the impacts of climate change so that they are motivated to adopt more climate-smart strategies within an integrated crop-livestock system and in the conservation agriculture (Sekaran et al. 2021). Extension and advisory services have traditionally served as a bridge between research and farming and have supported farmers through the delivery of knowledge about new technologies. Yet the successful upscaling of CLCA requires strategies that go well beyond changing farm-level agronomic practices. Accelerated adoption should progress through the development of delivery systems and participatory farmer-led extension systems should inform the development of contextually relevant CLCA technologies and practices (Kassam et al. 2014). Indeed, this process requires the identification and promotion of the most appropriate and smart practices through more effective and productive research-development interfaces, technologies, and/or models (new, improved, and adapted) within favorable enabling environments. It also needs to comprise constructive institutional arrangements, policies, and financial investments at different intervention levels (local, regional, and national).

Extension and advisory services, therefore, need to be backed by comprehensive expertise and skills to foster interaction and encourage the flow of knowledge among a broader range of stakeholders and actors than at present. The stakeholders and experts in question include both those involved in policy formation and those engaged in the actual practice and uptake of farming CLCA concepts. The central questions we are trying to answer here are: which extension and advisory services are most effective? What actions need to be taken to upscale of innovations? And what type of tools will ensure wider adoption and optimize impact? In this context, a management framework identifies three elements that are critical for innovation: functions, actions, and tools all of which have been used as reference points in the development of CLCA technologies. Innovation in this context refers to the process by which new knowledge is generated, adapted, disseminated, and adopted by a large number of stakeholders.

6.1 Strategies for Scaling Innovations in Integrated Crop-Livestock System

Scaling up crop-livestock integration under CA is a complex task due both to the diversity of solutions proposed for such integration and to the diversity of farmers' needs and capacities to invest and engage. For this reason, the scaling up of enhanced integration practices needs to be considered through a wide range of delivery systems that are context, and sometimes technology specific. This range of delivery systems and approaches includes:

- A strong and comprehensive partnership for scaling, where pilot project leaders promoting integration packages need to engage with a diversity of actors which are key for leveraging the scaling up of practices.
- Facilitation of public–private partnerships, especially for seed production, machinery design, and other types of investments needed to enhance the access of farmers to appropriate farming inputs which favor crop-livestock integration.
- Capacity development of extensionists, local lead farmers, proximate influencers, and regional and national administrative staff.
- Training and empowerment of farmers groups to enhance their capacity for accessing technologies and information, as well as their social and managerial capacities for self-sustained scaling up of relevant information among their communities and adherents.
- Engagement of civil society and non-governmental organizations (NGOs) in thematic areas which are relevant for them and in line with the principles of crop-livestock integration.
- Engagement of private actors for enhancing the delivery of farming inputs appropriate and relevant to CLI.
- Coordination across ongoing development and research projects (including national programs and investments projects) and other relevant actors and beneficiaries for more efficient information exchange and investment patterns.

Combining these actors together will lead to an enhanced partnership approach for scaling up. This is based on cooperation and coordination across four main types of actors, also called the 'four-wheels approach' of partnership for scaling

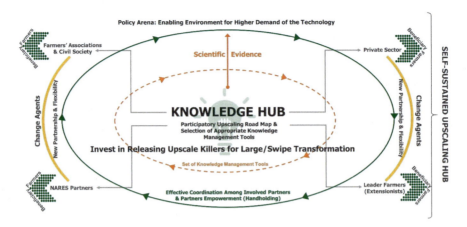

Fig. 7 Partnership for upscaling crop-livestock integration. *Note* NARES — national agricultural research and extension systems. *Source* Frija and Idoudi (2020)

up by Frija and Idoudi (2020) (Fig. 7). The approach mainly focuses on generating higher demand for the technology by building on: (i) three main stakeholders arenas including the research for development (R4D) projects responsible for effective design of upscaling activities; (ii) four key change agents which are necessary for stimulating transformative changes (the four-wheels); and (iii) the policy arena responsible for stimulating enabling environments. There should be a clear understanding for dividing partners—also called 'change agents'—into four categories including: (i) farmers' groups and associations of a different type; (ii) civil society (including NGOs) and the private sector; (iii) national public development partners; and (iv) lead farmers and extensionists who are key for local spread of the technologies.

The upscaling approach of CLI should also consider tools for strong interaction within and between these arenas and partners. These tools include knowledge management, coordination mechanisms, handholding of partners, and monitoring and evaluation. Africa teams are now exploring the effective implementation of this framework to frame ongoing upscaling roadmap activities in five innovation hubs. The upscaling approach for CLI should also build on the existing levels of readiness in CLI technology and build scaling activities accordingly. The challenge is making sure that real changes and transformation are generated through a set of identified upscaling activities implemented by 'upscaling leaders' in a given context. This, in turn, depends on the effectiveness of activities implementation and the strength of upscaling partners. For that reason, it is very important to empower all types of possible upscaling partners so that they can deliver on the scaling up of technologies which are most appropriate for them.

6.2 Integrated Activities to Promote the Crop-Livestock Innovations Under Conservation Agriculture

In this section, we discuss the importance of the various functions and tools that underpin the process of innovation management (Berthet et al. 2018), by which new knowledge from the CLCA initiative is widely disseminated and upscaled. Our assessment revealed the breadth and range of the functions and tools that were mobilized for the upscaling of CLCA. They are summarized in the following elements:

- Testing and adapting new CLCA practices in a participatory manner.
- Implementing capacity development of all stakeholders (such as small-scale farmers, women, SMSAs, GDAs, etc.).
- Establishing up scaling hubs for CLCA innovations and practices.
- Establishing and empowering public–private partnership development and coordination (i.e., for providing mechanical no-till seeders).
- Enhancing and identifying incentives and inputs delivery to create a feed seed system for livestock.
- Facilitating policy engagement and advocacy, steering, and influencing policy are important parts of the process of upscaling knowledge during the implementation of CLCA and through national initiatives.
- Identifying policy incentives devoted to facilitating the implementation of crop-livestock integration beyond the farm level.
- Fostering reflective self-learning among farmers involved with CLCA packages.

The upscaling of CLCA practices in the two North-African countries we have looked at (Tunisia and Algeria) involves a broad array of activities:

- Convening and setting up platforms for stakeholder interaction and forming networks of strategic partners involved in CA-farming systems.
- Facilitating dialogue and the exchange of knowledge among operational partners through knowledge hubs and platforms for information exchange.
- Sharing knowledge and experiences among all actors in order to identify and improve policies, actions, and practices.
- Disseminating information on new knowledge/practices/products through various media channels, ICT, and person-to-person outreach through knowledge hubs and small-scale producers' groups.
- Undertaking adaptive and action research through participatory approaches.
- Organizing joint events for implementing specific and targeted activities and setting up and enforcing (when exists) user groups such as SMSAs, GDAs and farmers groups feed seed cooperatives and input providers.
- Training farmers, knowledge-intermediaries, and service-providers from the public and private sectors, cooperatives, farmers groups, input providers, etc.
- Identifying and provisioning whenever possible incentives, inputs, and guidelines for encouraging adoption and partnerships.

- Advocating for policy recognition, greater public private partnership, targeting R4D investments, and the harmonization of tools and guidelines to accelerate the process of upscaling.

6.3 Tools to Promote Technologies of the Crop-Livestock System Under CA Initiative

We used a wide range of tools (formats, mechanisms, concepts, etc.) to manage CLCA innovations and to promote their upscaling. These include: tools for co-ordination at the project level (annual meetings, project team meetings, steering committees, advisory meetings, etc.); tools for encouraging interaction, planning and knowledge-sharing; tools for enhancing science delivery (surveys, frameworks, guidelines, etc.); tools for public private partnerships (agreements, country and regional networks, etc.) annual review and planning meetings, quarterly updates, and innovation learning hubs (scaling hubs); tools for learning and evaluation, such as 'monitoring, evaluation, and learning' indicators; tools for collective learning and action by farmers (SMSAs, GDA's, women groups, upscaling maps) and other farmer groups formed for the diffusion of knowledge; and tools to incentivize production and performance, such as subventions and the provision of free or subsidized inputs (e.g. feed seed resources for livestock).

6.4 Up-Scaling Actors and Their Respective Roles

Since the implementation of the CLCA project in Tunisia, a broad spectrum of actors has been engaged in upscaling and several research and development organizations (for example: Institut National des Grandes Cultures (INGC), Institut National de Recherche Agronomique de Tunisie (INRAT), Office de l'Élevage et des Pâturages (OEP), Institution de la Recherche et de l'Enseignement Supérieur Agricoles (IRESA), Institut National de la Recherche en Génie Rural, Eaux et Forêts (INRGREF), etc.) have been involved in the promotion of CLCA systems. Several other actors also supported the generation, adaptation, promotion and use of CA in North Africa, in general, and in Tunisia, in particular. Several donors, policy-maker and the private sector also played roles in promoting and upscaling CLCA, which was beneficial to the CA 'innovation trajectory'. A qualitative analysis of the results indicates that the adoption of CA was positively influenced by the development of a good responsive relationship with farmers. This was achieved through building confidence, formulating extension strategies, establishing strong development partnerships, strengthening farmers' education and trust, and carrying out capacity development of CLCA actors in addition to building on existing networks.

6.5 Use of ICT for Scaling the Crop-Livestock Innovations

New opportunities are emerging relating to the use of ICT for the dissemination of information about technologies (Dhehibi et al. 2021). The use of ICT for upscaling, extension, and dissemination will depend on partners' capacities to access and use these tools. However, it has been proved that the use of Short Message Service (SMS)-based messages and radio messages are very useful in some contexts, especially for enhancing women's access to information. SMS messages can be used to send health, pest, and disease alerts which can help farmers make better management decisions. Other tools, such as call centers and Open Data Kit, can further help policymakers and local managers better collect data and feedback from farmer's field and adjust their upscaling strategies accordingly. Some of these initiatives are being used in pilot R4D projects (Rekik et al. 2021).

7 Conclusions and Remarks

Climate change, land degradation, a decrease in soil fertility, and the increasing cost of inputs (feed resources, fertilizers, seeds, etc.) associated with changing demand (for animal products, legumes, etc.) may create new opportunities for change by encouraging practices, including those promoted through CLCA, that provide farmers with greater resilience to all types of external shocks such as drought. Integrated crop-livestock systems under CA principles have the potential to address several agro-ecological and socio-economic objectives mainly for rainfed agriculture: stabilization yields, reducing climate vulnerability, promoting more diverse on-farm crops (suggesting that opportunities exist to expand crop-livestock outputs without employing additional inputs or improved production technologies.), and reducing vulnerability to market fluctuations (for fuel, labor, seeds, feed resources, etc.).

The systems integrating crops and livestock under CA are recognized as resilient, smart, and sustainable models for production in fragile and conflict-affected situations, where critical inputs such as fertilizers and seeds become unavailable as local or international imports dry up or prices rocket. Key integrating factors for the sustainable intensification of crop-livestock systems include alternative grazing/feeding systems, crop diversification, integration of tree-crops and livestock, CA (which balance the tradeoff between leaving residues as feed for livestock and leaving them as mulch for the soil), scale-appropriate mechanization, and herd health management. Combining all or a few of these components helps improve overall farm incomes, crop productivity, soil quality, input-use efficiency, and the provision of healthy protein in the human diet and fodder for livestock consumption. Integrating livestock into cropland also provides the potential advantages of a sustainable intensification strategy.

For wider adaptability, innovations at farm level need to consider the following: (i) developing upscaling approaches and innovation hubs, as collective agricultural systems for crop-livestock integration; (ii) understanding and clarifying the different forms (complementarity, synergy, substitutions, etc.) of crop-livestock integration specially under a CA-based approach, assessing benefits (economic, social, etc.) and limitations, and identifying mechanisms to support these forms beyond the farming context; (iii) establishing and empowering the public–private partnership development and coordination (e.g. for mechanical no till seeders, mobile seed cleaning and treatment machines, and mobile feed grinders); and (iv) investing public funds in supporting forage-based crop-livestock farming and research (including forage seed systems) rather than cereal monocropping or industrial-scale meat and milk production.

Acknowledgements This study was supported by the CLCA project, Phase II (Use of Conservation Agriculture in crop–livestock systems in the drylands for enhanced water use efficiency, soil fertility and productivity in NENA and LAC countries) funded by the International Fund for Agricultural Development (IFAD) (ICARDA's agreement No. 2000001630).

Author's Contribution Study concept and design: all the authors. Drafting of the manuscript: Mina Devkota, Aymen Frija, Mourad Rekik, Zied Idoudi, Boubaker Dhehibi, Udo Rudiger. Critical revision of the manuscript: Mina Devkota, Boubaker Dhehibi, Udo Rudiger, Mourad Rekik, Zied Idoudi, Hatem Cheikh M'hamed.

References

Abidi S, Benyousse S, Ben Salem H(2020) Foraging behaviour, digestion and growth performance of sheep grazing on dried vetch pasture cropped under conservation agriculture. J Anim Physiol Anim Nutr (Berl)

Abidi S, Benyoussef S, Ben Salem H, Nasri S, Frija A (2019) Vetch summer grazing (VSG) under conservative agriculture (CA): promising alternative to cereal residue grazing for better Barbarin lamb's response. In: 6th international conference on sustainable agriculture and environment. Proceeding book, pp 477–483

Alary V, Moulin C-H, Lasseur J, Aboul-Naga A, Sraïri MT (2019) The dynamic of crop-livestock systems in the Mediterranean and future prospective at local level: a comparative analysis for South and North mediterranean systems. Livest Sci 224:40–49

Ates S, Cicek H, Bell LW, Norman HC, Mayberry DE, Kassam S, Hannaway DB, Louhaichi M (2018) Sustainable development of smallholder crop-livestock farming in developing countries. In: IOP conference series: earth and environmental science. IOP Publishing, p 12076

Ayeb N, Seddik M, Hammadi M, Barmat A, Atigui M, Harrabi H, Khorchani T(2013) Effects of feed resources in arid lands on growth performance of local goat kids in southern Tunisia. In: Ben Salem H, López-Francos A (eds) Feeding and management strategies to improve livestock productivity, welfare and product quality under cl

Bahri H, Annabi M, M'Hamed HC, Frija A (2019) Assessing the long-term impact of conservation agriculture on wheat-based systems in Tunisia using APSIM simulations under a climate change context. Sci Total Environ 692:1223–1233

Bell LW, Moore AD, Thomas DT (2021) Diversified crop-livestock farms are risk-efficient in the face of price and production variability. Agric Syst 189:103050

Ben-Hammouda M, M'Hedhbi K, Kammassi M, Gouili H (2007) Direct drilling: an agro-environmental approach to prevent land degradation and sustain production. In: International workshop on conservation agriculture for sustainable land management to improve the livelihood of people in dry areas. Damascus, Syria, pp 4–37

Berthet ET, Hickey GM, Klerkx L (2018) Opening design and innovation processes in agriculture: insights from design and management sciences and future directions

Cheikh M'hamed H, Annabi M, Ben Youssef S, Bahri H (2016) 'L'agriculture de conservation est un système de production permettant d'améliorer l'efficience de l'utilisation de l'eau et de la fertilité du sol'. Annales de l'INRAT 89:68-71

Cheikh M'hamed H, Bahri H, Annabi M (2018) Conservation agriculture in Tunisia: historical, current status and future perspectives for rapid adoption by smallholder farmers' Second African congress on conservation agriculture. Johannesburg, South Africa, pp 57–60

Christiansen S, Ryan J, Singh M, Ates S, Bahhady F, Mohamed K, Youssef O, Loss S (2015) Potential legume alternatives to fallow and wheat monoculture for Mediterranean environments. Crop Pasture Sci 66:113–121

Corsi S, Muminjanov H (2019) Conservation Agriculture: Training guide for extension agents and farmers in Eastern Europe and Central Asia. Food and Agriculture Organization United Nations

Costa S, Souza ED, Anghinoni I, Carvalho PCF, Martins AP, Kunrath TR, Cecagno D, Balerini F (2014) Impact of an integrated no-till crop–livestock system on phosphorus distribution, availability and stock. Agric Ecosyst Environ 190:43–51

Demir N, Yavuz F (2010) An analysis on factors effective in benefiting from forage crops support. Sci Res Essays 5:2022–2026

Devkota M, Patil SB, Kumar S, Kehel Z, Wery J (2021) Performance of elite genotypes of barley, chickpea, lentil, and wheat under conservation agriculture in Mediterranean rainfed conditions. Exp Agric 57:126–143

Devkota M, Yigezu YA (2020) Explaining yield and gross margin gaps for sustainable intensification of the wheat-based systems in a Mediterranean climate. Agric Syst 185:102946

Dhehibi B, Rudiger U, Sondos D, Khemaies Z (2021) Les TIC au service du changement dans l'agriculture tunisienne: exploiter le pouvoir de l'innovation dans le domaine agricole pour un impact à grande échelle, et tirer les enseignements de l'expérience du projet ICT2Scale en Tunisie. (31/3/2021). R4D Brief

Diffenbaugh NS, Giorgi F (2012) Climate change hotspots in the CMIP5 global climate model ensemble. Clim Change 114:813–822

Dixon J, Gulliver A, Gibbon D, Hall M (2001. Farming systems and poverty. Improving farmers' livelihood in a challenging World. http://www.fao.org/3/ac349e/ac349e.pdf

Dixon JA, Li X, Msangi S, Amede T, Bossio DA, Ceballos H, Ospina B, Howeler RH, Reddy BVS, Abaidoo RC (2010) Feed, food and fuel: Competition and potential impacts on small-scale crop-livestock-energy farming systems

Doré T, Makowski D, Malézieux E, Munier-Jolain N, Tchamitchian M, Tittonell P (2011) Facing up to the paradigm of ecological intensification in agronomy: revisiting methods, concepts and knowledge. Eur J Agron 34:197–210

Ezeaku IE, Mbah BN, Baiyeri KP, Okechukwu EC (2015) Integrated crop-livestock farming system for sustainable agricultural production in Nigeria. African J Agric Res 10:4268–4274

FAOSTAT (2021) United Nations Food and Agricultural Organisation [WWW Document]

Fernández-Ugalde O, Virto I, Bescansa P, Imaz MJ, Enrique A, Karlen DL (2009) No-tillage improvement of soil physical quality in calcareous, degradation-prone, semiarid soils. Soil Tillage Res 106:29–35

Frija A, Idoudi Z (2020) Self-sustained "Scaling Hubs" for agricultural technologies: definition of concepts, protocols, and implementation. In: Lebanon: international center for Agricultural Research in the Dry Areas (ICARDA). https://hdl.handle.net/20.500.11766/12248

FSI (2015) Food Security Statistics. Food and agriculture organization of the United Nations (FAO). Available at: http://www.fao.org/economic/ess/ess-fs/en

Guesmi H, Salem H, Moujahed N (2019) Integration crop-livestock under conservation agriculture system. New Sci 65:4061–4065

Harb OM, Abd El-Hay GH, Hager MA, Abou El-Enin MM (2015) Studies on conservation agriculture in Egypt. Ann Agric Sci 60:105–112

Herrero M, Thornton PK, Notenbaert AM, Wood S, Msangi S, Freeman HA, Bossio D, Dixon J, Peters M, van de Steeg J (2010) Smart investments in sustainable food production: revisiting mixed crop-livestock systems. Science (80-) 327:822–825

ICARDA (2017) Milestones for development of conservation agriculture discussed in Tunisia. ICARDA blogs. https://www.icarda.org/media/news/milestones-development-conservation-agriculture-discussed-tunisia

Idoudi Z (2020) Forage for biomass and sustainable crop-livestock systems: a private public partnership for scaling forage seeds in Tunisia. In: International Center for Agricultural Research in the Dry Areas (ICARDA) (Executive Producer). https://hdl.handle.net/20.500.11766

Kassam A, Friedrich T, Derpsch R (2019) Global spread of conservation agriculture. Int J Environ Stud 76:29–51

Kassam A, Friedrich T, Shaxson F, Bartz H, Mello I, Kienzle J, Pretty J (2014) The spread of conservation agriculture: Policy and institutional support for adoption and uptake. F Actions Sci Reports J F Actions 7

Lloveras J, Santiveri P, Vendrell A, Torrent D, Ballesta A (2004) Varieties of vetch (Vicia sativa L.) for forage and grain production in Mediterranean areas. Cah Options Méditerranéennes 62:103–106

Magnan N (2015) Property rights enforcement and no-till adoption in crop-livestock systems. Agric Syst 134:76–83

Matson, P.A., Parton, W.J., Power, A.G., Swift, M.J. (1997). Agricultural intensification and ecosystem properties. Science (80–) 277:504–509

Moraine M, Duru M, Nicholas P, Leterme P, Therond O (2014) Farming system design for innovative crop-livestock integration in Europe. Animal 8:1204–1217

Moujahed N, Abidi S, Youssef S, Darej C, Chakroun M, Salem H (2015) Effect of stocking rate on biomass variation and lamb performances for barley stubble in Tunisian semi arid region and under conservation agriculture conditions. African J Agric Res 10:4584–4590

Moussadek R, Mrabet R, Dahan R, Zouahri A, El Mourid M, Van RE (2014) Tillage system affects soil organic carbon storage and quality in Central Morocco. Appl Environ Soil Sci

Moussadek R, Mrabet R, Zante P, Lamachere JM, Pepin Y, Le Bissonnais Y, Ye L, Verdoodt A, Van Ranst E (2011) Influence du semis direct et des résidus de culture sur l'érosion hydrique d'un Vertisol Méditerranéen. Can J Soil Sci 91:627–635

Mrabet R (2011) Effects of residue management and cropping systems on wheat yield stability in a semiarid Mediterranean clay soil. Am J Plant Sci 2:202

Mrabet R, Moussadek R, Fadlaoui A, Van Ranst E (2012) Conservation agriculture in dry areas of Morocco. F Crop Res 132:84–94

Mrabet R, Rachid M, Devkota M, Lal R (2021) No-Tillage farming in maghreb region: enhancing agricultural productivity and sequestrating carbon in soils. In: Soil organic carbon and feeding the future: ccrop yield and nutritional quality

Murendo C, Gwara S, Mazvimavi K, Arensen JS (2019) Linking crop and livestock diversification to household nutrition: Evidence from Guruve and Mt Darwin districts, Zimbabwe. World Dev Perspect 14:100104

Rekik M, Frija A, Idoudi Z, López Ridaura S, Louahdi N, Dhehibi B, Najjar D, Rudiger U, Bonaiuti E, Becker LDZ, Cheikh M'hamed H, Devkota M, Rischkowsky B (2021) Use of Conservation Agriculture in Crop-Livestock Systems (CLCA) in the Drylands for Enhanced Water Use Efficiency, Soil Fertility and Productivity in NEN and LAC Countries – Progress Highlights: Year (3)—April 2020 to March 2021. Lebanon: International.

Rekik M, Santiago López R, Cheikh M'hamed H, Djender Z, Dhehibi B, Aymen F, Devkota M, Rudiger U, Bonaiuti E, Najjar D, Idoudi Z (2019) Use of Conservation Agriculture in Crop-Livestock Systems (CLCA) in the Drylands for Enhanced Water Use Efficiency, Soil Fertility and Productivity in NEN and LAC Countries–Project Progress Report: Year I-April 2018 to March 2019. https://hdl.handle.net

Rudiger U, Idoudi Z, Frija A, Rekik M, Elayed M, Cheikh M'hamed H, Zaim A (2020) Locally adapted machinery solutions for sustainable intensification of crop-livestock systems in Tunisia. https://hdl.handle.net/20.500.11766/12453

Ryan J, Singh M, Pala M (2008) Long-term cereal-based rotation trials in the Mediterranean region: implications for cropping sustainability. Adv Agron 97:273–319

Ryschawy J, Choisis N, Choisis JP, Joannon A, Gibon A (2012) Mixed crop-livestock systems: an economic and environmental-friendly way of farming? Animal 6:1722–1730

Salem HB, Smith T (2008) Feeding strategies to increase small ruminant production in dry environments. Small Rumin Res 77:174–194

Sekaran U, Lai L, Ussiri DAN Kumar S, Clay S (2021. Role of integrated crop-livestock systems in improving agriculture production and addressing food security–a review. J Agric Food Res 100190

Silva LCR, Lambers H (2020) Soil-plant-atmosphere interactions: structure, function, and predictive scaling for climate change mitigation. Plant Soil 1–23

Souissi A, Bahri H, Cheikh M'hamed H, Chakroun M, Benyoussef S, Frija A, Annabi M (2020) Effect of tillage, previous crop, and N fertilization on agronomic and economic performances of durum wheat (Triticum durum Desf.) under Rainfed Semi-Arid Environment. Agronomy 10:1161

Tibbo M, van de Steeg J (2013) Climate change adaptation and mitigation options for the livestock sector in the Near East and North Africa. In: Sivakumar MVK et al (eds) Climate change and food security in West Asia and North Africa. Springer, Berlin, pp 269–280

Yigezu YA, El-Shater T, Boughlala M, Bishaw Z, Niane AA, Maalouf F, Degu WT, Wery J, Boutfiras M, Aw Hassan A (2019) Legume-based rotations have clear economic advantages over cereal monocropping in dry areas. Agron Sustain Dev 39:58

Chapter 14
Realizing Food Security Through Agricultural Development in Sudan

Sharafeldin B. Alaagib, Imad Eldin A. Yousif, Khaled N. Alrwis,
Mirza Barjees Baig, and Michael R. Reed

Abstract Agriculture and livestock are the main sources of livelihood in the country and agriculture plays an important role in Sudan's economy. The main objective of this chapter is to investigate the role of agricultural development in realizing food security in the context of Sudan. Descriptive and analytical statistics were used to obtain the main results, which included the annual growth rate of GDP increased from 8.3% in 1990 to 10.5% in 2007, and then decreased to 1.9% in 2011 with a rate of decrease of 63.5% compared to 2010 due to the loss of 75% in oil production as a result of the secession of the south of Sudan. The total production of cereals has increased from 2.1 million ton in 1990 to 5.6 million tons in 2010, then decreased in 2013 by 2.3 million tons. The production of sorghum, wheat, and millet has declined due to drought and economic instability. Record cereal production was recorded in 2016 due to good rains. Livestock in Sudan has increased from 59.8 million heads in 1990 to 141.9 million heads in 2010, but decreased in 2011 to 104.3 million heads. The percentage of workers employed in agriculture decreased from 61.8% in 1991 to 49.2% in 2010, and then increased in 2017 to 53.3%. This trend may be due to more people moving into the flourishing service and industrial sectors during this period. Such a trend is commonly considered as a natural economic

---

S. B. Alaagib (✉) · I. E. A. Yousif · K. N. Alrwis
Department of Agricultural Economics, College of Food and Agriculture Sciences, King Saud University, Riyadh, Saudi Arabia
e-mail: salaagib@ksu.edu.sa

I. E. A. Yousif
e-mail: imyousif@ksu.edu.sa

K. N. Alrwis
e-mail: knahar@ksu.edu.sa

M. B. Baig
Prince Sultan Institute for Environmental, Water and Desert Research, King Saud University, Riyadh, Saudi Arabia
e-mail: mbbaig@ksu.edu.sa

M. R. Reed
Agricultural Economics, University of Kentucky, Lexington, KY, USA
e-mail: Michael.Reed@uky.edu

© The Author(s), under exclusive license to Springer Nature Switzerland AG 2022
M. Behnassi et al. (eds.), *Food Security and Climate-Smart Food Systems*,
https://doi.org/10.1007/978-3-030-92738-7_14

development process. The analysis concludes by developing some public policy-oriented recommendations.

Keywords Food security · Agricultural development · Poverty · Rural development · Sudan

1 Introduction

Sudan suffers from political instability, food insecurity, and a high incidence of poverty, with 46.5% of the population living below the poverty line (Ahmed 2015). The economic crisis, compounded by seasonal hardship and heightened conflict, has led to a deterioration in food security and nutrition. Continued political instability, increased food prices, hyperinflation, and declining economic growth are driving food insecurity in the country. Figure 1 shows that the annual growth rate of GDP increased from 8.3% in 1990 to 10.5% in 2007, and then decreased to 1.9% in 2011 with a rate of decrease of 63.5% compared to 2010 due to loss of 75% of the oil production as a result of the secession of the south of Sudan.

The latest Integrated Food Security Phase Classification (IPC) estimates that 3.9 million people in Sudan are classified as 'food insecure' and in 'crisis' and 'emergency' phases. In addition, the country hosts 1.2 million refugees and 2 million Sudanese are internally displaced (USAID 2019). Sudan is currently classified as a "low human development" country with the ranking of 168th among 189 countries

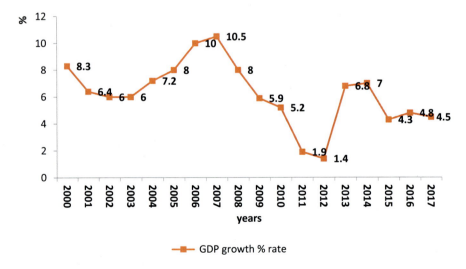

Fig. 1 GDP growth rate during the period 2000–2017

Table 1 Total and rural population in Sudan during 2014–2018 (in millions)

Year	Total population	Rural population	% of total population
2014	37.977	25.183	66.3
2015	38.902	25.717	66.1
2016	39.847	26.251	65.9
2017	40.813	26.785	65.6
2018	41.801	27.320	65.4

Source World Bankdata (2019)

on Human Development Index compared to 141 in 2006 (UNDP 2019). The population of Sudan reached 41.8 million, of which two-thirds live in rural areas, and its annual growth rate is 2.4% (Table 1).

Sudan occupies the north-eastern part of the African continent between 4° and 22° North of the equator and longitudes 22° and 38° East. It is bordered by two Arab countries and five African countries. With an area of 1.869 million square kilometers, after the cession of the south of Sudan, it is the second largest country in Africa after Algeria; the third largest Arab country after Algeria and Saudi Arabia; and the 16th largest in the World (Central Bureau of Statistical 2017). Sudan is divided by rivers, valleys, and numerous seasonal and permanent water tributaries, the most famous of which is the Nile River. Sudan's water resources consist of rivers, rain, surface, and groundwater. The country is in the tropics and its climate varies from the desert climate in the north to the rich savannah in the south. Sudan's land use includes: 73.5 million hectares of arable land, 71 million hectares of forests, 52 million hectares of pasture, and 12 million hectares of uncultivated area (AOAD 2017).

2 Agricultural Sector in Sudan

Agriculture and livestock are the main sources of livelihood in the country and agriculture plays an important role in Sudan's economy. However, the contribution of agriculture to the GDP declined from 49.9% in 1999 to 35.9% in 2008, and then to 24% in 2018 (Central Bureau of Statistical 2018). The agricultural sector is composed of crops and animal production (livestock and fisheries); and it is divided into two sub-sectors: irrigated and rain-fed agriculture. The rain-fed sub-sector is also divided into traditional and semi-mechanized sectors. Sudan has an animal wealth of 108.2 million heads, including 30.9 million heads of cows, 40.7 million heads of sheep, 31.7 million heads of goats and 4.9 million heads of camels. Sudan annually produces on average 1.5 million tons of red meat, 4.5 million tons of milk, 36 thousand tons of fish, 60 thousand tons of eggs, and 65 thousand tons of poultry meat (Ministry of Animals and Fisheries 2019).

Sudan's crop portfolio is quite diversified, including cereals (such as sorghum, millet, wheat, rice and maize), oilseeds (mainly sesame, groundnuts and sunflower), industrial crops (cotton and sugarcane), fodder crops (alfalfa, fodder sorghum and Rhodes grass), pulses (broad beans and pigeon peas) and horticultural crops (okra, onions, tomatoes, citrus, mango, etc.). Sudan is the third largest African producer of sugarcane after Egypt and South Africa, yet the country imports sugar, especially from India and Thailand. Its total planted area for sugar is around 82,000 hectares.

The agricultural sector supplies food and raw material to the industrial sector and plays an important role in achieving food security and reducing poverty in Sudan through food production and employment (70% of job opportunities), especially in the rural areas (Central Bureau of Statistical 2018). The food gap in Sudan has deteriorated in the last decade as the value of food imports increased from 843.8 million Sudanese pounds in 2000 to 13 billion pounds in 2017 and the inflation rate reached 62.68% in 2018 (Central Bureau of Statistical 2018). Sudan suffers from the food gap as a result of wars, large numbers of displaced people and refugees, and unbalanced development. Sudan ranked 99 out of 113 countries in its global food security index in 2018 (Economist Intelligence Unit 2019).

Agricultural production in Sudan faces many challenges such as the lack of finance, low productivity, little quality control, poor extension services, marketing problem, and heavy taxes. Therefore, improvements in agricultural production are vital to increase food security and reduce poverty. Sudan has vast and diverse agricultural resources that allow the agricultural sector to be the leading sector for the country's economic development. Furthermore, the global rise in food prices can be used to increase agricultural investment in Sudan and help the country benefit from modern technology.

Crop production in Sudan is practiced under three main patterns:

The irrigated sub-sector

It is estimated that one million hectares are in the irrigated sub-sector. It includes major irrigation schemes such are Gezira, Rahad, New-Halfa and Suki, using river flow from the Nile River and its tributaries. Large scale irrigation schemes (Gash and Tokar) also use water from seasonal floods.

This sub-sector contributes about 21% of the total value of agricultural production; 100% of wheat and 25% of sorghum are produced by this sub-sector in the country. Although its contribution to sorghum production is low relative to the rain-fed sub-sector, its production is more stable in meeting the consumption requirements.

The semi-mechanized rain-fed sector

This sector covers about 6 million hectares. It was encouraged during the mid of 1940s in the Gadarif area and it is in Blue Nile, Sinnar, Kosti, and Dallang areas. The two main crops produced by this sector are sorghum and sesame. The crop yield depends on rainfall availability, formal credit, and the level of prices in the previous year. However, this sector suffers from many limiting factors including low yields, high production costs, limited formal credit and poor infrastructure (Karrar and Elhag 2006).

Traditional rain-fed agriculture

It is estimated that its cultivated area is about 23 million feddans, and it depends on manual equipment, local seeds, the pattern of mobile cultivation, and the non-use of fertilizers, which led to less production and productivity. In spite of this, it plays a big role in providing food in rural areas and in the production of crops such as sorghum, millet, and sesame. It also contributes to agricultural exports by exporting sesame, gum Arabic (all Sudanese production), peanuts, hibiscus, and melon seeds. Production fluctuates from one season to another according to the amount of rain and its distribution. Most of the livestock in Sudan are intertwined with this type of agriculture, where the unharvested area is used as feed for livestock and other animals.

3 Forest Sector

Forests have a growing role in protecting agricultural land, especially in traditional marginal cultivation areas in the western states, and they also play an important role in protecting agricultural areas in southern White Nile, Gedaref, and Blue Nile. Forests are a natural resource for the most important forest products, which are gum Arabic, wood, charcoal, and some small fruits. In addition to that, they are the shelter of wildlife in Sudan, and the forest products and rare wildlife in forests have good economic return.

The Table 2 and Fig. 2 reflect the decline of the forest area in Sudan, especially after the secession of the south of Sudan in 2011, which covers most of its land forests. Where the area of forests decreased from 70 million hectares in former Sudan to 20 million hectares from 2010 to 2015, and the area of forests decreased to fifth in this period, due to the forested nature of most of southern Sudan land.

Table 2 Forest Area in Sudan, 1990–2015

Year	Forest area (000 hec.)
1990	76,381
2000	70,491
2005	70,220
2010	69,949
2015	19,210

Source: Ministry of Agriculture and Forestry (2019)

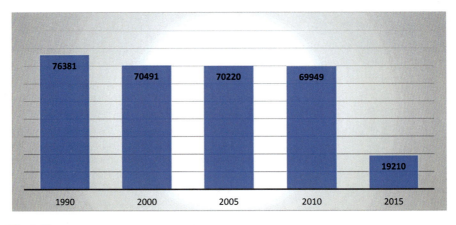

Fig. 2 Forest Area in Sudan, 1990–2015

4 The Economic Importance of Agricultural Sector for Food Security

Figure 3 shows that the total production of cereals has increased from 2.1 million tons in 1990 to 5.6 million tons in 2010, then decreased in 2013 to 2.3 million tons (a 26.5% decrease). Production of sorghum, wheat, and millet fell by 50%, 26%, and 67%, respectively, due to drought and economic instability. Record cereal production of 8.5 million tons was recorded in 2016 due to good rains.

Figure 4 shows the number of livestock in Sudan. The total number increased from 59.8 million heads in 1990 to 141.9 million heads in 2010, but decreased in

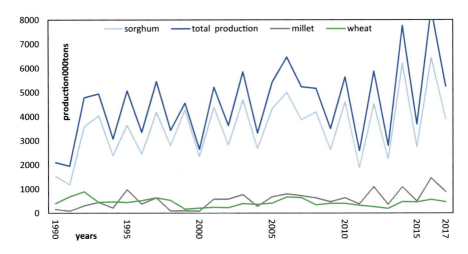

Fig. 3 Cereal production in Sudan (000 tons), 1990–2017

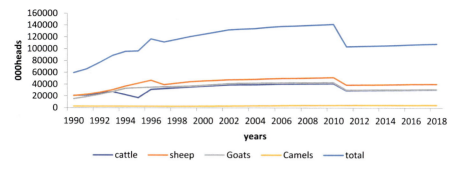

Fig. 4 Animal numbers during 1990–2018 (1000 heads)

2011 alone to 104.3 million heads (a 26% reduction in only one year). Between 2010 and 2011, the number of cattle, sheep and goats decreased by 29.1%, 24.5% and 29.5%, respectively, due to the secession of the south of Sudan and drought. There was a slight recovery in the numbers in 2017 and 2018.

Table 3 shows that animal meat production increased from 455 thousand tons in 1990 to 1634 thousand tons in 2010, but then decreased to 1427 thousand tons in 2011 (a fall of 12.7% in one year). Milk production increased from 2405 thousand tons in 1990 to 5840 thousand tons in 2010, but then decreased to 4273 thousand tons in 2011 (a drop of 26.9%), this is due to the secession of the south of Sudan in July 2011, where the rate of decline represents the percentage of the southern contribution in the livestock sector, before separation, which is estimated at about 30%. There was a significant, sustained increase in poultry production (chicken meat and eggs) over the 1990–2018 periods. Production increased from 40 thousand tons in 1990 to 145 thousand tons in 2018.

Table 4 and Fig. 5 show that the percentage of workers employed in agriculture decreased from 61.8% in 1991 to 49.2% in 2010, and then increased to 53.3% in 2017. This trend may be due to more people moving into the flourishing service and industrial sectors during this period. Such a trend is commonly considered as a natural economic development process.

5 Food Security Situation in Sudan

Achieving food security is a national objective and requires the attention and care of concerned state and non-state actors. Table 5 indicates the present situation of food security and production/consumption of the most important strategic food commodities in Sudan. The records show that:

- local production of wheat decreased from 895 thousand tons in 1992 to 172 thousand tons in 1999 with an average of 454.1 thousand tons and coefficient of variation (CV) of 41.2%. Domestic wheat production declined at an annual rate

Table 3 Animal products in Sudan, 1990–2019 (1,000 tons)

Year	Meat	Milk	Fish	Chicken meat	ggs
1990	455	2402	30	18	22
1991	610	4060	32	19	28
1992	868	4403	36	25	30
1993	137	5198	40	27	33
1994	1287	5841	45	29	35
1995	1281	5813	46	25.5	30
1996	1275	5785	47	22	25
1997	1331	6001	50	18	20
1998	1422	6230	52	15	18
1999	1473	6650	53	15	20
2000	1552	6879	56	15	21
2001	1569	7095	58	16	22
2002	1628	7298	60	18	22
2003	1663	7387	58	20	25
2004	1672	7405	63	22	28
2005	1694	7534	65	24	30
2006	1711	7253	57	18	20
2007	1725	7298	65	27	31
2008	1808	7360	70	27	32
2009	1841	7406	70	28	32
2010	1634	5839.5	66.5	34	35
2011	1427	4273	63	40	38
2012	1456	4318	87	45	40
2013	1467	4359	23	55	45
2014	1476	4391	28	62	52
2015	1484	4451	33	60	55
2016	1502	4507	36	65	60
2017	1517	4553	38	68	63
2018	1589	4591	41	70	65
2019	1543	4623	–	75	70

Source: Ministry of Livestock and Fisheries—Information Center (2019)

of 0.75% during the study period. Domestic consumption of wheat ranged from a minimum of 500 thousand tons in 1993 to a high of 2,651 thousand tons in 2014, with an average of 1,341 thousand tons and CV of 46.4%. Domestic consumption of wheat increased at an annual growth rate of 2.6% during the same period mentioned above.

Table 4 Sudan employment in agriculture, 1991–2017

Year	Employment in agriculture (%)
1991	61.8
1992	61.2
1993	61.1
1994	60.4
1995	61.0
1996	60.6
1997	60.5
1998	60.2
1999	59.9
2000	58.7
2001	58.6
2002	57.9
2003	56.8
2004	55.8
2005	54.8
2006	53.4
2007	51.4
2008	50.8
2009	50.0
2010	49.2
2011	50.3
2012	53.8
2013	53.9
2014	53.4
2015	52.8
2016	52.5
2017	53.3
Average	56.1

Source Sudan Annual Reports (2018)

- local production of sorghum ranged between a minimum of 1,180 thousand tons in 1991 to a high of 6,441 thousand tons in 2016, with an average of 3,544 thousand tons and CV of 37.1%. Domestic production of sorghum increased at an annual rate of 2.13% during the study period. Domestic consumption of sorghum ranged from a minimum of 1,420 thousand tons in 1991 to a high of 6,387 thousand tons in 2016, with an average of 3,580 thousand tons and CV of 36.8%. Domestic consumption of sorghum increased at an annual growth rate of 2.30% during the same period mentioned above.

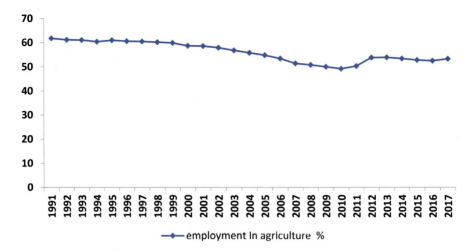

Fig. 5 Sudan employment in agriculture, 1991–2017

Table 5 Production and domestic consumption of important food commodities in Sudan, 1990–2016 (1,000 tons)

	Minimum	Maximum	Average	Standard deviation	Coefficient of variation (%)
Wheat production	172	895	454.1	186.9	41.2
Wheat consumption	499.8	2651	1341	622.6	46.4
Sorghum production	1180	6441	3544	1321	37.1
Sorghum consumption	1420	6387	3580	1319	36.8
Millet production	85	1457	536.7	344.9	64.26
Millet consumption	85	1457.08	528	353.09	66.88
Sugarcane production	428	842.6	655.4	100	15.3
Sugarcane consumption	484.4	2555.1	1086.2	521.8	48
Red meat production	455	1860	1432	337	24
Red meat consumption	422	1400	926	321	35

- local production of millet ranged from a minimum of 85 thousand tons in 1991 to a high of 1,457 thousand tons in 2016, with an average of 536.7 thousand tons and C.V of 64.3%. Local millet production increased at an annual rate of 5.69% during the study period. Domestic consumption of millet ranged from a minimum of 85 thousand tons in 1991 to a maximum of 1,457.08 thousand tons in 2016, with an average of 528 thousand tons and C.V of 66.9%. Domestic consumption of millet has increased at an annual growth rate of 6.5% during the same period mentioned above.
- local production of red meat ranged between a minimum of 455 thousand tons in 1990 to a high of 1,860 thousand tons in 2010, with an average of 1,432 thousand tons and C.V of 24%. Local production of red meat increased at an annual rate of 2.5% during the study period. Domestic consumption of red meat ranged from a minimum of 422 thousand tons in 1990 to a maximum of 1,400 thousand tons in 2010, with an average of 926 thousand tons and C.V of 35%. Domestic consumption of red meat increased at an annual growth rate of 3.4% during this study period.

Table 6 shows the results of a simple linear regression of production and consumption of important agricultural commodities over the study period. In each case, the

Table 6 Trend in production and consumption of important agricultural commodities in Sudan, 1990–2016

Statement	Growth rate (%)	F	R^2	Model
Wheat production	−0.74	0.47	0.018	Ln y = 6.136 - 0.0074 T (35.11)** (-0.68)^{ns}
Wheat consumption	4.6	37.05	0.60	Ln y = 6.451 + 0.046 T (53.27)** (6.09)**
Sorghum production	2.13	5.11	0.169	Ln y = 7.801 + 0.021 T (51.75)** 2.260)**
Sorghum consumption	2.30	6.82	0.214	Ln y = 7.790 + 0.022 T (55.23)** (2.61)**
Millet production	5.69	11.39	0.313	Ln y = 5.234 + 0.057 T (19.394)** (3.375)**
Millet consumption	6.46	15.238	0.379	Ln y = 5.087 + 0.065 T (19.172)** (3.904)**
Sugarcane production	1.24	7.345	0.259	Ln y = 6.324 + 0.0124 T (100.919)** (2.710)**
Sugar cane consumption	3.10	5.909	0.2196	Ln y = 6.520 + 0.031 T (37.413)** (2.431)**
Red meat production	2.48	18.290	0.413	Ln y = 6.868 + 0.024 T (71.26)** (4.277)**
Red meat consumption	3.44	26.695	0.507	Ln y = 6.263 + 0.034 T (56.70)** (5.167)**

Source Calculated and compiled from Tables 4 and 5

trend coefficient was significantly different from zero except for wheat production, where the trend was not significant. All significant trends were positive, varying from 2.13% for sorghum production to 5.69% for millet production.

6 Factors Affecting Food Supply Stability

There are some several important factors affecting food supply stability in Sudan, such as:

- The climate diversity in Sudan from the desert north to the equatorial south, which creates an agricultural environment conducive to agricultural activity in different and diverse conditions. This enables multiple production environments to specialize in the production of various essential food crops, on which the population depends for food.
- The climate diversity is reflected in the creation of diverse environments to attract various types of agricultural investments in many food crops, ensuring the stability of food supplies. Sudan has also established a strategic stock system to address food supply imbalances between regions and the fluctuation of agricultural commodity prices in the country. On the agricultural policy side, to ensure continuous food supply, Sudan has adopted a $ 367 million agricultural renaissance program for food security, poverty reduction, and rural development projects under the direct patronage of World Food Program (WFP). This program concentrates on: developing and protecting natural resources; increasing the incomes of workers and those associated with the agricultural sector; ensuring the stability of production in agricultural areas; and boosting agricultural export development through production directed to foreign markets while considering quality and specifications required.

7 Conclusion and Recommendations

Sudan suffers from a food gap as a result of wars, large numbers of displaced people and refugees, and unbalanced development. The main objective of this chapter was to evaluate the current food security situation in Sudan by studying the main agricultural food crops through using descriptive statistics and economic analysis. The total production of cereals has increased from 2.1 million tons in 1990 to 5.6 million tons in 2010, then decreased in 2013 to 2.3 million tons (a 26.5% decrease). As production of sorghum, wheat and millet fell by 50%, 26%, and 67%, respectively, due to drought and economic instability. Record cereal production of 8.5 million tons has been reached in 2016 due to good rains. Animal meat production increased from 455 thousand tons in 1990 to 1634 thousand tons in 2010, but then decreased to 1427 thousand tons in 2011 (a fall of 12.7% in one year). Milk production increased from 2405 thousand tons in 1990 to 5840 thousand tons in 2010, but then decreased to 4273 thousand tons in 2011 (a drop of 26.9%). This is due to the secession of the south of Sudan in July 2011, where the rate of decline represents the percentage of the southern contribution in the livestock sector. The trend coefficient was significantly different from zero except for wheat production, where the trend was not significant. All significant trends were positive, varying from 2.13% for sorghum production to 5.69 for millet production.

Therefore, the main recommendations are: to establish a strategic stock system to address regional food supply imbalances in the country's regions and the fluctuation of agricultural commodity prices in the country. On the agricultural policy side, government should ensure the continuous food supply, poverty reduction, and rural development. Also, measures should be put in place by government to ensure stability of producers in agricultural areas and promote agricultural export development through production directed to foreign markets considering quality and specifications required.

Acknowledgements The authors extend their sincere appreciation to the Deanship of Scientific Research at King Saud University for supporting the work through the College of Food and Agriculture Sciences Research Center.

References

Abbadi Karrar AB, Ahmed AE (2006) Brief overview of sudan. Economy and future prospects for agricultural development. Khartoum food aid forum from 6–8 June 2006, Sudan
Ahmed N (2015) Fighting poverty in sudan, central agency for public mobilization and statistics. Policy Brief on poverty in Sudan
Arab Organization for Agricultural Development (AOAD) (2017a) Statistical year book, Khartoum, Sudan
Arab organization for Agricultural development (AOAD) (2017b) Statistical year book (37), Khartoum, Sudan
Central Bank of Sudan (2018) Different annual Reports (2000–2017).). Sudan http:www.chos.gov.sd
Central Bureau of Statistical (2016) Statistical Year Book (2016), Sudan
Central Bureau of Statistical (2017) Statistical Year Book (2016). Sudan
Central Bureau of Statistical (2018) Statistical Year Book (2017). Sudan
Economist intelligence unit (2019) Global food security index. https://foodsecurityindex.eiu.com
Ministry of livestock and Fisheries (2019) Information Centre. Sudan
Ministry of Agriculture and forestry (2019) National Forestry commission, Sudan
Sudan annual Report (2018) central bank of Sudan, Sudan
UNDP (2019) Human Development report 2017, Inequalities in Human Development in the 21st Century
USAID (2019) Sudan: Food Assistance Fact Sheet, US Agency for international development, USA. http://www.usaid.gov/
World Bank data (2019) databank. worldbank.org/home.aspx.

Chapter 15
Agriculture in Fragile Moroccan Ecosystems in the Context of Climate Change: Vulnerability and Adaptation

Larbi Aziz

Abstract Climate disruption is among the factors degrading agro-systems in fragile areas of Morocco. Such a situation and its effects are felt by farmers; this is particularly the case for agro-systems in oasis and mountain areas that are considered more vulnerable to such effects. In these contexts, adaptation for small producers is not a choice but an imperative. To respond to such emerging effects while maintaining the agricultural production for subsistence and income stabilization, farmers often deploy their reactive capacities. To apprehend these dynamics, empirical studies were carried out in the Anti-Atlas Mountain and the oases of Ouarzazate, Morocco. The vulnerability of populations and production systems were analysed and the adaptive practices deployed by farmers to reduce the sensitivity of local systems to natural risks and increase their climate resilience were observed. The results show that farmers have detected variations in the local climate such as: irregularity of precipitation resulted in a decrease of water quantity and poor distribution over the year; increase in temperature; and reduction in the cold period. These variations had negative impacts on their production systems. To deal with these effects, farmers have adopted different strategies according to their socioeconomic contexts, soil, and climatic conditions, including: water management and use; diversification of agricultural activities; increasing the share of non-agricultural activities in income generation; and seasonal migration.

Keywords Climate Change · Oases · Mountain · Vulnerability · Adaptation Strategies

1 Introduction

Agriculture is the main pillar of Moroccan economy. The strong contribution of the agricultural sector to the economy is illustrated by the correlation between national GDP and agricultural GDP (Akesbi 2006). Indeed, the share of agriculture in the

L. Aziz (✉)
Department of Rural Development Engineering, National School of Agriculture of Meknes, Meknes, Morocco
e-mail: laziz@enameknes.ac.ma

national GDP varies from 16 to 20%, depending on annual rainfall, and the contribution of the sector to employment covers around 80% of jobs in rural areas and around 40% at the national level (Ministry of Environment 2016). However, agriculture remains a vulnerable activity that is strongly tied to climatic conditions as more than 80% of agricultural land is located in dryland areas (Harbouze et al. 2019). Moreover, the country is severely affected by climate change[1] since precipitation has recorded a downward trend with time and geographic variabilities (Driouech 2009). Since the 1960s, the average annual temperature in Morocco has increased by 0.16 °C per decade, while a significant decrease in the number of cold days has been observed (maximum temperature of less than 15 °C). Spring rains have also decreased by more than 40% and the maximum period without rain has increased by 15 days (Khattabi 2014). According to the World Bank (2011), Morocco is affected by a moderate level of drought every three years, a medium level every five years, a severe level every 15 years, and an extreme level every 30 years. This has turned agriculture susceptible to climate change (Ministry of the Environment 2016). Such changes are likely to affect all components of food security, including food production, access, use, and price stability (IPCC 2014). The situation is likely to become more serious for small farming in fragile agro-ecosystems such as mountains and oasis. Therefore, the question is how do farmers in these vulnerable areas cope with these changing dynamics? What adaptation strategies are they implementing to strengthen the resilience of their production systems and meet the food needs of their households?

To illustrate the vulnerability and resilience of small mountainous and oasis farming systems to climate change, we conducted an empirical study in two stages: first in Skoura, at the Skoura oasis (province of Ouarzazate); and then in Taliouine in the anti-Atlas region (province of Taroudant). The areas are administrated by the Ouarzazate Regional Agricultural Development Office (ORMVAO). The study chose two different territories where the impacts of climate change are different in terms of manifestation and magnitude. Besides, such territories differ in terms of exposure to the impacts of climate change (depending for instance on physical, social, economic, cultural, and political factors) and response capacity (capacity of understanding, anticipation, reparation, etc.) (Bertrand and Richard 2012).

Operationally, the key production system of the two study areas was taken as an input with particular emphasis on two plant species—namely the date palm for Skoura and the saffron for Taliouine—which are water dependent and constitute the backbone of respective agricultural economy. Date palm plays an important role in oasis territories, not only because of its economic importance, but also because of its ability to adapt to the harsh climatic conditions of the Saharan and sub-Saharan regions (Sedra 2015). While saffron is the most profitable crop in the Taliouine area since it is the most expensive spice in the world (Gresta et al. 2008; Aït Oubahou and El Otmani 2002). It is considered as a new origin-linked resource for the socio-economic development of the area (Landel et al. 2014).

---

[1] Climate change is considered here to be a set of variations in climate characteristics in a given location over time due to processes that are intrinsic to the Earth or that result from human activities (UNDP 2012).

This study aims at identifying the knowledge mobilized by farmers in Taliouine and Skoura to understand the resilience mechanisms and ensuring food security of local populations. It hypothesised that the building of a new knowledge results from the interweaving of imported 'technical' and traditional knowledge inherited from previous generations. Moreover, these case studies show that adaptation strategies of local populations also translate their responses, in a more global way, to the development challenges of their territories. This is because there is a direct relationship between development, which addresses current non-climate related issues, and adaptation activities, which seek to reduce the climate vulnerability of social-ecological systems (Locatelli 2010).

2 Methodology

2.1 Profile of Study Areas

In this section, the two study sites are presented below (Fig. 1).

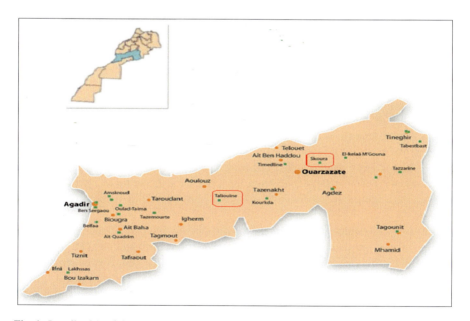

Fig. 1 Localisation of the two study sites

2.1.1 Skoura

The Municipality of Skoura belongs to the Caïdat of Skoura (circle of Ouarzazate) and consists of 56 *douars*.[2] The total area of the municipality is 105,900 ha, including 3,000 ha of useful agricultural land, 3,000 ha of rangelands, and 1,500 ha of forest, while wasteland covers 96,205 ha (ORMVAO undated). The average temperature in Skoura is 18.0 °C and the precipitation average is 171 mm per year. The driest month is July with an average of 2 mm of rain, and the November rainfall is the most important of the year with an average of 28 mm (ORMVAO undated).

According to the 2014 general population and housing census (High Commission for Planning 2018), the total population of Skoura is 24,055, including 4,332 in urban and 18,723 in rural areas. Drinking water reaches to almost 60% of the town's population, and 99% of this population has access to electricity.

The main crops are cereals, fodder crops and/or legumes (beans, peas), date palm trees, and alfalfa. In addition, there are fruit trees in the valleys and plains (Ibnelbachyr et al. 2014).

2.1.2 Taliouine

Taliouine (Taroudant province, Souss-Massa region)) is located in the Siroua massif at the junction of the High Atlas and the Anti-Atlas at a latitude of 30° 26 'N, a longitude of 8° 25' W and an altitude of 1200–1650 m.

The area is characterized by a semi-arid climate, with low rainfall (200 mm/year on average), low winter temperatures (-5 °C for the coldest month) causing night frosts (14–25 days of frost / year) between January and March, high summer temperatures (34–39 °C for the hottest month) and a relatively dry air throughout the year. It covers a total area of 327,500 ha and has a total population of 63,784 inhabitants living in 9100 homes and 297 douars (ORMVAO undated). The useful agricultural area (UAA) is 9,645 ha of which 5,985 ha are rain-fed and 3,030 ha are irrigated and it is characterized by a great fragmentation since 52% of the farms have an average size not exceeding one ha divided into small plots of 500 to 1300 m^2 (ORMVAO undated).

Subsistence farming is the main activity of the inhabitants of the region (cereals, saffron, market gardening and fodder crops). Saffron is the backbone of this agrarian system and is the main source of income for local populations. It is cultivated over an area of around 565 ha and its plots are located at altitudes ranging from 1500 m to over 2000 m. Water resources are limited and irrigation is practiced around water sources, rivers and private wells. Along with this plant production system, farmers are engaged in extensive breeding based on the Siroua goat breed and a local sheep breed.

---

[2] A *douar* is a set of households united by real or fictitious kinship ties, corresponding to a territorial unit with or without community modes of exploitation.

Table 1 The surveyed population in the two sites

	Skoura	Taliouine
Responsibles	5	4
Farmers	35	60

2.2 Surveyed Populations and Investigation Tools

As mentioned above, we conducted an empirical study in two stages: first in Skoura, and then in Taliouine. The surveyed population consists of two categories or decision levels: the first represents the staff responsible for the design and implementation of agricultural projects at both sites (5 people for Skoura and 4 people for Taliouine); and the second involved 35 farmers from the oasis of Skoura and 60 farmers from Taliouine (Table 1).

It should be noted that in the Taliouine area, two rural communes (RC), *Sidi Hssain* and *Agadir Melloul*, were studied. The first RC is located in the agro-ecological zone of low altitude (1500–1600 m) while the second RC belongs to the agro-ecological zone of medium altitude (1700–1800 m). Farmers in the two zones (oasis and mountain) were randomly surveyed and without seeking statistical representativeness.

For data collection, surveys were conducted for farmers and semi-structured interviews for institutional managers. The farmers' survey covered: the identification of production systems and their improvements; the distribution of tasks at farm level; the perception of respondents with regard to changes in the local climate; and the impact of these changes on their production systems (animal and plant), lifestyles, and managing strategies. The interviews of institutional managers considered: the identification of actions carried out for the benefit of farmers and particularly those related to climate change; the respondents' perception of the evolution of the local climate; the identification of effects of this development on the agriculture of the area and on the local population; and coping strategies undertaken by local actors. Descriptive analysis to analyse the collected data was used.

3 Climate Vulnerability of Two Fragile Ecosystems in Morocco

3.1 Vulnerability of Mountain Ecosystems to Climate Change

The mountains occupy a large part of the territory of Morocco. They extend over 187,741 km^2, or nearly 26% of the national territory; they are home to 7,548,000 inhabitants, or nearly 30% of the population with a density[3] of 40 inhabitants/km^2

---

[3] The national average density is estimated at 37 inhabitants/km^2.

(IFAD 2014). They shelter 62% of the Moroccan forest and are important centers of biodiversity. However, socially, these areas have a very high poverty rate, a reduced level of human development and are a major source of emigration to cities and abroad. Their direct economic contribution is very limited: 5% of the GDP and 10% to consumption, for 24% of the total population of the Kingdom and 25% of the national area and the incomes per inhabitant are twice lower than the national average (Harbouze et al. 2019).

Mountains constitute a fragile and vulnerable space due to their own natural constraints, often amplified by strong human influence. Humans and animals pressure is all the more harmful to the conservation of soils and forests. Thus, we are witnessing acute erosion problems in these areas due to population growth, overgrazing and clearing leading to the degradation of plant cover and soil and subsequently to the desertification advancement. In addition, these areas are highly exposed to natural disasters and benefit from a level of equipment below the national average. They constitute the national water tower and therefore they are at the heart of the issue of climate change. In fact, at the national level, the climate projections disseminated by different institutions (such as FAO 2017; World Bank 2011) indicate an increased frequency of extreme events and a gradual increase in aridity due to the decrease in rainfall (from −10 to −20%) and increase in temperature (from +2 to +3 °C) (Schilling et al. 2012).

In terms of agriculture, mountain areas have 450,000 farms, representing 30% of Morocco's total. They mainly concentrate a small family agriculture and production systems oriented towards cereals, arboriculture and agro-pastoral farming. These agro-pastoral ecosystems are facing strong pressures (demographic growth, growth in the number of livestock, extension of crops to the detriment of rangelands).In addition, the negative effects of climate change on yields, especially from 2030, are assessed by the International Center for Agricultural Research in the Dry Areas (ICARDA) to be 15 to 20% reduction (Dahan et al. 2012).

3.2 Vulnerability of Oases to Climate Change

The total area of Moroccan oases is 107,324 km^2, or 15% of the national area (710,850 km^2), of which 2% is cultivated and 98% are desert areas (Kabiri 2014). These oases are considered as an ecological rampart against the advance of desertification. They are an important refuge for biodiversity because they contain a variety of plant resources (dates, roses, saffron, vegetables, etc.), some of which are endemic. Nevertheless, this heritage is in perpetual degradation due to the combination of factors linked to drought, desertification, soil salinization, loss of biodiversity, low productivity, and inefficient use of palm trees (Mouline 2010).

On the bio-climatic level, the Moroccan oasis space is arid to Saharan with very irregular rainfall from one year to another. The rains are often intense and concentrated over time in the form of rainstorms, causing violent floods. The annual average rainfall is only 132 mm and the number of rainy days is just 20 days (UNDP 2011a).

However, in some valleys, there are microclimates characterized by a reduction of aridity due to the presence of water and vegetation and the protection of these valleys by high topography. Nevertheless, these oases remain vulnerable and fragile ecosystems that can be altered by the effects of exogenous factors, such as climate change, which cause warming and desertification (Ben Salah 2014). In Ouarzazate, for example, over the period 1961–2008, the climate warming was 0.3 °C per decade and the highest water deficits recorded reached -84% in 2000 (Driouech 2010). The 2000s were also characterized by frequent droughts despite the occasional occurrence of non-deficit years (compared to the average). In addition, the seasonal and annual warmings projected for 2030–2050 are between 1 °C and 2.2 °C, and the cumulative winter rainfall could decrease over the entire oasis zone (UNDP 2011b). This decrease could reach, over this period, 35% in Ouarzazate, compared to the period 1971–2000 (Driouech 2010). The effects of these changes would be very significant on both local populations and economy, as farming is their main activity.

4 Results and Discussions

4.1 Production System in the Two Study Areas

The IPCC (2014) noted that rural areas are expected to experience major impacts of climate change on water availability and supply, food security, infrastructure, and agricultural incomes, including shifts in production areas of food and non-food crops around the world. This is in correlation with the results of this study (see below). However, to understand how the local impacts of climate change have affected farmers and their production systems in the two study areas and how they coped with these impacts, it would be useful to present such production systems first.

4.1.1 Production System in Skoura

Like all other oases, the economic activity in Skoura is dominated by agriculture and livestock (77% of the surveyed population). This is essentially irrigated agriculture carried out on small areas. In fact, 89% of surveyed farmers have areas of less than 3 ha and 11% of them have areas greater than 3 ha. To make the most of their plots, thus securing their food and diversifying their income, most farmers cultivate their land in different 'layers'. The first layer represents the date palm[4] that constitutes the main production in the cropping system. The second layer or stratum consists of fruit trees (mainly almond, olive, and apple) planted under the date palm trees. The last stratum is made up of underlying crops (cereals, legumes, vegetables, and fodder crops, especially alfalfa). This is the case for 39% of respondents, while 29% of

---

[4] Which creates a microclimate that protects the other underlying crops against the severity of the outdoor climate and against sudden changes.

them no longer cultivate cereals as the third stratum, and 32% do not practice either cereals or vegetables due to water shortage. However, all surveyed farmers practiced the date palm whose production is largely intended for market. Date palm production provides the major part of producers' income, while the production of other crops is intended for self-consumption and households' food security. Cultivated lands are generally located in the valleys, where water can be mobilized by the development of irrigation canals.

All agricultural equipment at the farms in the study area remains traditional because of the small size of cultivable areas and the high cost of 'modern' agricultural equipment (tractor, seed drill, combine harvester, etc.) for farmers. Only a minority of these farms (6% of respondents) use such equipment. They are generally farmers with relatively large areas (3–5 ha) and whose production is primarily sold in local or external markets.

Livestock is an essential component in the oasis production system. Used as a labour power (traction of traditional agricultural equipment) and for commercial purposes (sale of animals or their by-products), this activity remains a source of income for farmers. It is also an intensive oasis-type breeding (in permanent housing) combining sheep (the Dman breed), goats (the Deraa goats), and cattle (a local breed). Milk production is completely self-consumed by all surveyed farmers due to the absence of a dairy cooperative and a milk collection centre that can enhance local production.

4.1.2 Production System in Taliouine

The majority of respondents (84%) are engaged in agriculture and livestock and the marketing of saffron is their main source of income. For land tenure, private property is predominant and represents 83% of the total area, while collective land occupies 17%. At low and medium altitudes, farms are small; 40% of farmers have small areas not exceeding 5 ha, while those of medium size (between 5 and 10 ha) and large size (more than 10 ha) represent 23% and 37% of respondents, respectively. Consequently, the sole allocated to saffron remains limited: less than one ha for more than three-quarters of respondents and one to two ha for less than a quarter of respondents. To make better use of water and soil resources, farmers manage their plots in terraces built perpendicular to the slope. Thus, they prevent water runoff and soil erosion. To cultivate their crops, 75% of respondents still use traditional tools and equipment, while only a quarter of respondents have modern equipment (tractors and ploughs, combine harvesters, saffron dryers, etc.). As saffron requires more water, most of the farmers interviewed have a well on their farms. Regarding crop yields, they vary between 5 and 6 kg/ha depending on the season (the more favorable are the climatic conditions, the better are the yields) and provide significant incomes for locals. Thus, saffron plays the role of an origin-linked resource that is a major source of employment and income of local populations.

Livestock is an essential component of the local production system and is considered as source of family cash flow. It is characterized by the abundance of the Siroua sheep breed, which makes the area famous for both the quality meat and wool.

4.2 Impacts of Climate Changes on Local Populations and Economy

While climate change (CC) has, above all, been understood as a global phenomenon, the sources of carbon emissions as well as the impacts of CC are often finely localized (Bertrand and Richard 2012). In fact, the surveyed farmers in both oasis and mountain ecosystems noted the same manifestations of the changing local climate but with a differentiation of their expression from one ecosystem to another.

4.2.1 The Situation in Skoura

Like their ancestors, farmers have always observed environmental changes, including changes in the local climate, which are being noticed more clearly by surveyed farmers. In fact, 43% of those farmers felt the existence of disturbances in the oasis climate such an increase in temperature and a decrease in precipitation over recent years. They estimated that the rainy period has been reduced, delayed than usual, and ended early before the crops attain maturity. Sometimes the rains are totally absent, and sometimes too abundant. Today, they cannot predict anything about rainfall. As a result, farmers are increasingly uncertain about both rains and harvests.

The same observations, relating to climate change and its impacts on agricultural production, were also raised by surveyed institutional managers. In fact, an official from the Centre of the Agricultural Council of Skoura pointed out that the increase in temperature in recent years has caused the reduction of the areas planted with rosacea trees such as apple, apricot, and peach. These crops were quite widespread between the 1970s and the 1980s when the number of chilling hours needed for fruit production was sufficient. Currently, with the rise in temperatures, the rosacea trees no longer receive the necessary chilling hours and their production is affected. This could worsen in the future since climate projections for 2030 predict a decrease in the number of cold days in winter from 2 to 4 days and an increase in the number of days of summer heat waves from 2 to 10 days for Ouarzazate (Driouech et al. 2011). On the other hand, farmers and agents of the Centre for Agricultural Development declared that they noticed, in recent years and in places, an early emergence of the spathes[5] of the date palm.

For farmers practicing livestock, the changes observed concern mainly the drop in forage production. They also noticed that the soils were losing more and more of their vegetation cover. This reduction in fodder availability has consequently affected

---

[5] In the date palm, the inflorescence has thousands of flowers enclosed in a bract called a spathe.

milk production: "there is less milk since there is less fodder" affirm respondents. On the other hand, a short rainy season often induces massive sales of cattle or sheep due to the increase in the fodder deficit.

4.2.2 The Situation in Taliouine

Climate change in the region is a reality that is obvious and visible to the naked eye for almost all of the farmers surveyed. The members of the Saffron Economic Interest Group (GIE) interviewed for their part confirmed these findings. In fact, the climate and its variations is a daily concern that it becomes easy for farmers to see the significant changes. This opinion is found among older farmers. For them, climate change is discernible from several manifestations, the most remarkable of which are: an increase in temperatures, a decrease in precipitation and the amount of snow in the mountains. They have noted a clear decrease in rainfall in recent years, although they point out that there have always been periods of more or less dry years in the region. They noted an irregularity in rainfall reflected by the late arrival or the early end of the rains with bad distribution over the year. The prevailing feeling is that the changes are such that farmers live in the greatest uncertainties and cannot predict rainfall. Precipitation was, in the past, and by their personal experience, more important and more assured. They also noted an increase in temperatures as early as April and May. For them, the temperature regime was fairly consistent from year to year.

This reflects the vulnerability of these mountain populations and their production system to the increasing climatic variability observed. This raised some concerns among the farmers interviewed about their agricultural production. In fact, for three quarters of them, the new climatic conditions in the area have led to a drop in the annual production of saffron. For them, this is due to the drying out of the saffron bulbs following the increase in evapotranspiration caused by the rise in temperatures recorded in recent years. This caused a lag in the reproduction phase of the crop, which impacted the productivity and the quality of the saffron filaments. On the other hand, the high temperatures in September (recorded in recent years) as well as the recurrence of frost and cold in winter affected the flowering of the crop. Consequently, the quantity and the quality of the production are affected.

According to the producers surveyed, the climate change observed locally has had negative effects on their natural environment and on their production system. For them, the lack of rainfall has led to a drop in agricultural yields, losses in livestock, a drying up of rivers. Which translated into a decline in farmers' income and an increasing risk to households' food security.

4.3 Farmers' Practices to Strengthen Climate Resilience

The effects of climate change observed by farmers in both regions have given rise to some concerns about the sustainability of their agriculture. To cope with such a situation, they have developed adaptation[6] practices based on their adaptive capacity. Practices deployed by surveyed farmers focused on adjustments that affected, in general, both their cultivation system, resource management, and ways of life. However, as the impacts of climate change are heterogeneous in their expressions as well as in their local implications, the responses of the territories also appear variable (Bertrand and Richard 2012). This is why the adjustments in question have been translated on the ground in a differentiated manner in the form of practices varying from one area to another.

4.3.1 Rethinking the Cropping System to Ensure Food Security and Secure Producers' Income

a. *In Skoura, a revisited cultivation system*

As mentioned above (cf. production system) for the case of Skoura, the third stratum is intended for different crops. Farmers grow cereals, almost equally, between barley and wheat. Since barley, according to surveyed farmers, adapts to local climatic conditions due to its relative salinity tolerance and its low demand for water, they devote a larger area for its cultivation. Some farmers have resorted to the use of drought-tolerant varieties which provide better yields. However, they pointed out that, in addition to recent droughts, cereals have low profitability as their production cost become higher due to the increasing cost of labour and the decline of traditional forms of mutual aid between farmers. This is why some of them (20%) have abandoned cereal farming and converted to more profitable crops by installing drip irrigation systems at their farms under the framework of ORMVAO's subsidies. Thus, 6% of surveyed farmers introduced fodder crops and 14% of them introduced vegetable crops. These farmers have large areas compared to what exists in the region (with an average of 3.5 ha) and have some level of education which has enabled them to attend training sessions organized by institutional actors. This shows that these farmers are open to technological innovation and take advantage of the available institutional support to adapt to changing environmental and climatic conditions.

However, this reconversion concerned only a part of respondents; other farmers who were more affected by the impacts of current changes have opted for opposite strategies and abandoned these crops. Thus, 28% of surveyed farmers abandoned both vegetables and cereals, and 7% abandoned fodder crops. In short, since water resources have become scarce due to the droughts experienced by the region during

---

[6] 'Adaptation' means here the process of adjustment to the current or expected climate, as well as its consequences, so as to reduce or avoid the harmful effects and to exploit the beneficial effects (IPCC 2014).

the last three decades, and in addition to the growing shortage of labour (see below), cropping systems in Skoura have become much less intensive. This is why in many farms only the date palm trees are grown. As the latter constitutes the main profitable crop at the local level, farmers have sought to make it as profitable as possible by optimising the use of water resources at their disposal. Thus, many have taken advantage of public subsidies to equip their parcels of land with drip irrigation equipment. On the other hand, as the surveyed persons in charge underlined, the ORMVAO opted for a restructuring of the old palm groves within the framework of the Green Morocco Plan (GMP)[7] through the distribution for the benefit of phoeniculturists of certain new varieties (Najda, in particular), considered as well adapted to the soils and climatic conditions of the region.

b. *In Taliouine, opening up of local knowledge to 'exogenous' knowledge*

Faced with the impacts of climate change on local farming system, farmers are mobilized to manage such risks. Thus, animal husbandry and mixed farming are among the adaptation practices used by farmers in the region to improve their income and meet their food needs. On the other hand, farmers are starting to introduce new plant species to cope with the new climatic conditions. An official in the agricultural subdivision of Taliouine indicated that farmers have introduced new plantations of olive and almond trees in the lower valleys and walnut in the high valleys. Some of them have established a few modern orchards of olive, almonds or even apple trees.

While saffron is the backbone of this system and of the local economy as a whole, farmers are more concerned about this crop as a main source of income. Indeed, some farmers (young members of a professional organization) have made changes in the technical management of the crop to strengthen its resilience, including what follows: they changed the sowing date from July to September to prevent the bulbs from drying out; they have increased the frequency of irrigations when there is no rainfall; and they opted for staggered harvesting to avoid problems related to the unavailability of labour during the harvest period.

The adoption of these practices has been supported and accompanied by many development actors working in the area (ORMVAO, FAO, the National Institute of Agronomic Research (INRA), the NGO Migration and Development, local professional organizations, etc.). This shows that local producers do not limit themselves only to their local know-how but remain open to "exogenous" technical knowledge carried by development organizations.

In addition, producers seek to strengthen their resilience by limiting costs and expenses. For example, during periods of saffron harvest which requires labor, some farmers avoid recruiting external workers and resort to family members or mutual assistance between families because the production of saffron requires a great deal of manual work, particularly for flowers' harvesting and pruning (Gresta et al. 2008; Wyeth and Malik 2008). This affirms the return of the local population to traditional

---

[7] GMP is the national agricultural strategy, which has been adopted by Morocco for the period 2008–2020, with the objective to make the agricultural development as the main engine of the country's economic growth.

forms of mutual aid, thus strengthening their social cohesion, which is one of the four factors influencing adaptation capacities identified by Magnan (2009).

4.3.2 Management of Irrigation Water

a. *Rehabilitation of local knowledge in Skoura for mobilization of new water resources*

In this arid environment, water is always an issue. Thus, when asked about the main limiting factors for local agriculture, 93% of surveyed farmers stressed that water remains the most determining one. According to them, this resource is becoming scarce. They are then faced with the problems of lowering the water table and drying up of the *khettaras*,[8] which supply water for most farms. In fact, Skoura is characterized, like other Moroccan oases, by the use of *khettaras* as the main mean for accessing water. Unlike canals, which tend to be invaded by vegetation and are very sensitive to bad weather and prone to siltation by sand, *khettaras* provide safe water with minimal evaporation. However, the viability of this traditional irrigation system is threatened by the reduction in groundwater recharge due to limited rainfall and floods. In fact, the great droughts of the 1980s and 2000s, which dried up most of the *khettaras* of Skoura, affected both the traditional smooth agricultural operations of the oasis and farmers' will to collectively invest in the mobilization of new water resources. (reinforcement of *khettara* galleries or dams on wadis) (Mahdane et al. 2011). Today, a resurgence of interest in these *khettaras* is occurring and farmers are opting for their restoration and maintenance to better manage available water resources. The ORMVAO contributes to this action by helping local companies to take care of water tables and their recharging capacity. For those with wells, and in order to access groundwater, they have to dig deeper compared to the past while using electric pumping newly introduced in the area.

b. *Taliouine: The "drip" for watering bulbs*

In the context of uncertainty around water resources, the search for viable options in agriculture is a daily concern for local farmers. They particularly opted for the digging of new wells or the development of existing ones and installation of the drip system. This is the way to reduce dependence on irregularities in rainfall. Thus, two thirds of the interviewees proceeded to convert the surface irrigation system to localized irrigation in order to save water, labor and time. To make these investments, these farmers benefited from a subsidy granted by the State under the Agricultural Development Fund (FDA). Thinking that the drip irrigation system is not suitable for cultivating saffron, the other third of the respondents kept the gravity irrigation system and opted for the digging of new wells or the restoration of wells. existing

---

[8] A *khettara* is an ancient system of groundwater mobilization, which is brought to the surface by a gallery using gravity. It is a system designed by the population to manage water and deal with climatic hazards.

systems to capture more water. To minimize production costs, some of them have opted to reduce the irrigated area of their farms and rely on water sources.

4.3.3 Income Diversification, a Common Strategy in the Two Agro-Systems

In addition to their strategy based on polyculture, making the best use of the soil and water resources at their disposal, farmers in Skoura and Taliouine are opting for more income diversification. This is the case in Skoura where 64% of surveyed farmers said that the main source of additional income is contribution by the family. For them, agricultural income remains uncertain since local agriculture suffers from both climatic, biophysical, and socio-economic constraints that hinder its development. Thus, they (especially young people) engage, at the same time, in other income-generating activities. They perform some small businesses (sellers of used clothes, traders) in the village centre and other jobs (carpenters, mechanics, shoemakers, etc.). Some of them migrate seasonally to the nearest towns (Ouarzazate and Marrakech for the people of Skoura, and Agadir, Taroudant, Tata and Ouarzazate for those of Taliouine) to carry out works that hardly require professional qualification (construction, industry or trade).

Some migrants from these zones manage to transfer remittances to their families, allowing them to support themselves. In this context, migration becomes an adaptation strategy to reduce risks and compensate for losses due to poor harvests. However, this migration to foreign countries and exodus to urban centres have a negative impact on local agricultural development as they cause a loss of labour and skills. Through these practices, local populations tend to enhance economic diversification and improve their living conditions, which constitute, according to Magnan (2009), two of the four factors influencing adaptation capacities (social cohesion, economic diversification, political-institutional structuring, and living conditions).

5 Conclusions

The oasis and mountain ecosystems remain fragile and vulnerable to climate change. The findings of this study support this statement for the case of Skoura and Taliouine. Farmers in both areas, working on small, marginalized, and degraded land, have felt the impacts of environmental and climatic changes on their farming activities and are developing relevant adaptation practices to foster their resilience. These practices are the result of building a 'new' collective knowledge based on the integration of imported 'technical' knowledge with inherited local knowledge. In both study areas, such practices are particularly concerned with the adoption of polyculture (to ensure food security for local households), the diversification of income sources (trade, agricultural wage labour, and migration), and water management. Regarding this last point, producers have introduced drip irrigation for saffron in Taliouine, and for the date palm in Skoura. In addition, in Skoura' oasis, producers use ancestral know-how in the mobilization and management of water by rehabilitating the *khetaras*. The interest shown for the saffron and date palm trees confirms their importance in the local economy of each of the two zones. In fact, if other crops grown in the two zones have the main function of ensuring households' food security and livelihoods, date palm and saffron remain the most profitable species. They allow producers, through the generated income, to meet their needs and finance the agricultural campaign. However, the adaptation practices put in place by producers are only ad hoc responses to local climate variability. In fact, it is a question of moving from spontaneous adaptation to planned adaptation (Brodhag and Breuil 2009). Indeed, it is essential to develop the anticipation capacity, especially since the forecasts for 2030 would be serious for the two agro-systems, which helps act in a preventive way instead of reacting to the impacts already experienced in a curative manner.

It then becomes necessary to support agriculture in the two zones, especially saffron and date palm, which are the most profitable crops capable of creating wealth at the local level. These crops should be more productive and more resistant to various risks, including climate change. This is why many actions must be carried out, with the support of other local and regional institutions, to achieve the expected development, in particular:

- The application of good practices in the technical management of crops (especially saffron and date palm). The focus should be on those practices developed through research and which are resilient to climate change and on promoting the successful experiences of certain local producers; and
- Improving harvesting, post-harvesting, and production packing practices. Support to producers and their organizations in implementing labelling (PDO for saffron and PGI for dates) should be provided.

These challenges are worth addressing so that agriculture, in Skoura and Taliouine, continues to contribute to the integrated development of these two areas under present and future climatic conditions.

References

Aït Oubahou A, El Otmani M (2002) The cultivation of saffron. Technical sheet. Technology Transfer in Agriculture, Monthly bulletin of information and liaison of PNTTA MADREF/DERD

Akesbi N (2006) Evolution and prospects of Moroccan agriculture. In: 50 years of human development and 2025 prospects, pp 85–198

Ben-Salah M (2014) Recycling of oasis by-products: achievements and prospects. In: MENA-DELP Project, Knowledge Sharing and Coordination on Desert Ecosystems and Livelihoods. Sahara and Sahel Observatory. Tunis, p 85

Bertrand F, Richard E (2012) Climate change adaptation initiatives, between maintaining development logic and strengthening cooperation between territories. Territory in movement Revue de géographie et aménagement [Online], 14–15 I 2012, posted on July 01, 2014, consulted on June 22, 2020. http://journals.openedition.org/tem/1799, https://doi.org/10.4000/tem.1799

Brodhag C, Breuil F (2009)Climate glossary. Institute of Energy and the Environment of the Francophonie, p 70

CLIMATE-DATA.ORG. http://fr.climate-data.org/location/37264/

Dahan R, Boughlala M, Mrabet R, Laamari A, Balaghi R, Lajouad L (2012) A review of available knowledge on land degradation in Morocco. Ed. International Center for Agriculture Research in the Dry Areas (ICARDA). ISBN: 92–9127–265–5

Driouech F (2009) Climate variability and change in Morocco: Observations and projections. Department of National Meteorology. National conference: climate change in Morocco: challenges and opportunities

Driouech F (2010)Distribution of winter precipitation over Morocco in the context of climate change: downscaling and uncertainties. Doctoral thesis in "Sciences of the Universe, Environment and Space". University of Toulouse. France, p 164

Driouech F, Kasmi A, El-Hadidi A, Bari W (2011) Assessment of future climate changes in the Moroccan oasis areas. In: Project Adaptation to climate change in Morocco for resilient oases

FAO (2017) Towards sustainable agriculture and food in Morocco within the framework of the 2030 agenda for sustainable development. In: Rapid diagnosis of the sustainability of agriculture in Morocco, Rome, p 100

Gresta F, Lombardo GM, Siracusa L, Ruberto G (2008) Saffron, an alternative crop for sustainable agricultural systems. A review. Agron Sustain Dev 28:95–112

Harbouze R, Pellissier J-P, Rolland J-P, Khechimi W (2019) Synthesis report on agriculture in Morocco. In: ENPARD Mediterranean Initiative Support Project, pp 104

High Commission for Planning (2018) Data from the 2014 general population and housing census, National Level. Rabat

Ibnelbachyr M, Chentouf M, Benider M, Elkhettaby A (2014) Adaptation of FAO-CIHEAM indicators to the intensive goat farming system in South-East Morocco (Ouarzazate). In: Chentouf M, López-Francos A, Bengoumi M, Gabiña D (eds) Technology creation and transfer in small ruminants: roles of research, development services and farmer associations. Zaragoza: CIHEAM/NRAM/FAO, pp 481–488 (Options Méditerranéennes : Série A. Séminaires Méditerranéens; n. 1 08)

IFAD (2014) Mountain Rural Development Program (PDRZM), Near East, North Africa and Europe Division

IPCC (2014) Climate change 2014: synthesis report. In: Core Writing Team, Pachauri RK, Meyer LA (eds) Contribution of working groups I, II and III to the fifth assessment report of the intergovernmental panel on climate change. IPCC. Geneva. Switzerland, p 151

Kabiri L (2014) Comparative study of the management, conservation and enhancement methods of the natural resources of oases and desert areas. In: MENA-DELP project, knowledge sharing and coordination on desert ecosystems and livelihoods

Khattabi A (2014) Climate vulnerabilities and development strategies. In: Summary and strategic recommendations for taking climate risk into account in sectoral policies and strategies. Royal Institute of Strategic Studies

Landel PA, Gagnol L, Oiry-Varraca M (2014) Territorial resources and tourist destinations: couples in the making? Journal of Alpine Research [En ligne], 102–1 | 2014, mis en ligne le 23 mars 2014. http://rga.revues.org/2326, https://doi.org/10.4000/rga.2326

Locatelli B (2010) Local, global: integrate mitigation and adaptation. Perspective, 3, CIRAD

Magnan (2009) Proposal for a research framework to understand the capacity to adapt to climate change. VertigO Electron J Environ Sci 9(3). [Online: http://vertigo.revues.rog/9189]

Mahdane M, Lanau S, Ruf T, Valony MJ (2011) The management of drainage galleries (khettaras) in the oasis of Skoura, Morocco. In: Dahou T, Elloumi M, Molle F, Gassab M, Romany B (eds) Powers, societies and nature in the South of the Mediterranean. Editions Karthala, pp 209–234

Ministry of the Environment (2016) 3rd national communication on cclimate change. Rabat Morocco, p 296

Mouline MT (2010) What overall strategy for the Moroccan oasis system? royal institute for sstrategic studies. Rabat Morocco

ORMVAO (undated) Draa's goat

Sedra MH (2015) Date palm status and perspective in Mauritania. In: Date palm genetic resources and utilization: Volume 1: Africa and the Americas. Springer Netherlands, pp 328–368. https://doi.org/10.1007/978-94-017-9694-1_9

Schilling P, Freier KP, Hertige E, Scheffran J (2012) Climate change, vulnerability and adaptation in North Africa with focus on Morocco. In: Research Group Climate Change and Security (CLISEC). Institute of Geography and Klima Campus, University of Hamburg

UNDP (2011a) Adaptation to climate change for resilient oases. Study evaluation of future climate change in the Moroccan oasis areas

UNDP (2011b) Prospective assessment of climate risks and vulnerabilities for 2030 and 2050. Project to assess the vulnerability and impacts of climate change in the oases of Morocco and structuring of territorial adaptation strategies

UNDP (2012) Mainstreaming Climate Change into National Development and Country Programming Processes of the United Nations, Handbook to Support United Nations Country Teams in Mainstreaming Climate Change Risks and Opportunities, UNDP, New York, USA.

World Bank (2011). Land markets for economic growth in Morocco. Volume 1: Inheritance and land structures in Morocco. p. 40.

Wyeth, Malik (2008) A strategy for promoting Afghan saffron exports, Report, Afghanistan, ICARDA and Washington State University. http://www.icarda.org/RALFweb/FinalReports/G_Marketing_Afghan_Saffron_Strategy_RALF02-02.pdf

Chapter 16
Digitalization and Agricultural Development: Evidence from Morocco

Hayat Lionboui, Abdelghani Boudhar, Youssef Lebrini, Abdelaziz Htitiou, Fouad Elame, Rachid Hadria, and Tarik Benabdelouahab

Abstract The agricultural sector plays a major role in the socio-economic development of several countries around the world, including African countries. Advances induced by digital innovation have positively impacted this sector and could have a considerable impact by reducing the risk of food insecurity. In this chapter, applied research experiences conducted in Morocco have been reported to highlight some of the opportunities that digital transformation can offer in agriculture. First, agricultural land management was approached through remote monitoring of agricultural land and mapping of cropland using satellite imageries and machine learning. Then, the question of what digital can offer for the management of agricultural water resources was discussed through two examples: the first deals with the use of remote sensing and modeling to support irrigation water management; and the second is related to another aspect of multi-year bioeconomic modeling for predicting changes in water management indicators. Finally, experiences on the importance of digital transformation in risk management were discussed in order to analyze the risks in a more informed way and to relate aspects that could not be connected before in the past, in an efficient and relevant manner. Indeed, an experiment on the evaluation of spatial variability of wheat yield has been reported and an example of the spatial analysis of risk related to losses in production value using satellite data was discussed. With

---

H. Lionboui (✉) · Y. Lebrini · A. Htitiou · F. Elame · R. Hadria · T. Benabdelouahab
National Institute of Agronomic Research (INRA), Rabat, Morocco
e-mail: hayat.lionboui@inra.ma

F. Elame
e-mail: fouad.elame@inra.ma

R. Hadria
e-mail: rachid.hadria@inra.ma

T. Benabdelouahab
e-mail: Tarik.benabdelouahab@inra.ma

A. Boudhar · Y. Lebrini · A. Htitiou
Faculty of Sciences and Techniques, Sultan Moulay Slimane University, Beni Mellal, Morocco

A. Boudhar
Center for Remote Sensing Applications (CRSA), Mohammed VI Polytechnic University, Ben Guerir, Morocco

© The Authors(s), under exclusive license to Springer Nature Switzerland AG 2022
M. Behnassi et al. (eds.), *Food Security and Climate-Smart Food Systems*,
https://doi.org/10.1007/978-3-030-92738-7_16

steps added or substituted compared to the research presented in this chapter, digital innovation allows precise monitoring of agricultural land and offers socio-economic conditions that are more advantageous for farmers. Overall, all of the experiences cited on the use of digital innovation in the agriculture sector can be extended to other contexts, particularly in Africa where sustainable agricultural development remains the ultimate goal.

Keywords Agricultural land · Agricultural risks · Agriculture · Development · Digital transformation · Food security · Morocco · Risk management · Water resources

1 Digital Transformation in Agriculture

Digital has become an essential part of the sustainable development strategy of many countries. It would facilitate the participation of stakeholders in an eco-responsible strategy that meets Sustainable Development Goals (SDGs) (ElMassah and Mohieldin 2020). In agriculture, in order to make informed and sustainable decisions, Unmanned Aerial Vehicles or Drones (UAVs), for example, can allow farmers to better map fields and thus get to know precisely which places need water or nitrogen (Talaviya et al. 2020). Likewise, software uses data on the climate, crop control, seasonal forecasts and local specifics to achieve better yields in a sustainable manner (Peng et al. 2018).

In general, 'digitization of agriculture' refers to the use of information and communication technologies (ICT) in addition to new digital technologies (El Bilali and Allahyari 2018). These include, for example, videos, radio, television, Internet, remote detection, digital broadcasting, the use of smartphones (mobile agriculture, also called m-agri), artificial intelligence, etc. All these technologies make it possible to not only access, store, transfer and manipulate information, but also to analyze it and give it a meaning, in order to transform agriculture into a more modern, profitable, and sustainable sector. Nevertheless, the massive introduction of digitization has significantly impacted the organization of work and jobs (Dorn 2017). In some cases, it is responsible for job destruction and rising unemployment, while in others it essentially creates new activities and stimulates the economy (Brasseur and Biaz 2018). The success of digitization in other sectors has encouraged its adoption in agriculture around the world and also reflects the new expectations of farmers. However, the introduction of technology into the system that farmers form with their environment risks overturning it and making technology a problem rather than a solution.

In Africa, where agriculture is a strategic activity for many countries (Fig. 1), the digital transformation of the agricultural sector, despite everything, is an opportunity to promote its socio-economic development and keep its commitments to achieve SDGs (Hinson et al. 2019).

16 Digitalization and Agricultural Development: Evidence from Morocco 323

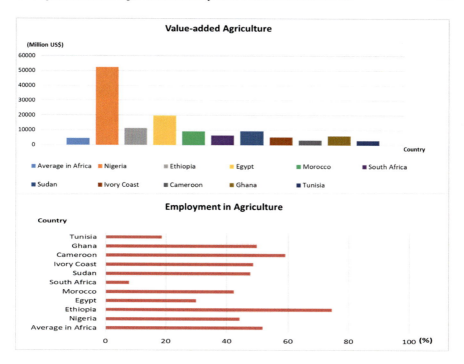

Fig. 1 Socio-economic importance of agriculture in Africa. Average 1991–2020 calculated based on the World Bank data. *Source* World Bank (2021a)

Located in the north western part of Africa, Morocco has been endowed with the Green Morocco Plan (GMP) during the period 2008–2020 as an ambitious agricultural strategy, which provided a clear and precise roadmap, giving a new impetus to the country's agricultural sector (Lionboui et al. 2020b). In such a policy framework, the contribution of digital technology to the governance of agricultural sector, like other sectors, has shown its effectiveness in the monitoring and evaluation of the various actions. Thus, digital technologies had made it possible to report actions carried out in the field and to have centralized dashboards with a view to steering and monitoring the agricultural policy (GMP) impact indicators. In fact, the Moroccan Ministry of Agriculture has set up several programs during the implementation of GMP, including the national register of farmers, the system of agricultural aid and subsidies (SABA), the agricultural GIS, the market price monitoring system (Asaâr), the international market monitoring system (EACCE), the Crop Growth Monitoring System-Maroc (CGMS-Maroc), and the irrigation warning system.

The GMP, which expired in 2020, has generally produced tangible results (Treguer and Pachon 2019). Based on the globalization of trade and its fluidity, this Plan has favored certain export sectors (early vegetables, citrus fruits, olive oil) allowing the country to acquire foreign currency. However, it has not been able to ensure a noticeable increase in the production of basic foodstuffs like cereals, pulses, milk, vegetable oils, etc. (Sraïri 2021). During the ongoing Covid-19 pandemic, evidence shows that countries dependent on imports of basic foodstuffs are the most affected (Alvaro et al. 2020). Hence the importance of reviewing the country's agricultural choices with regard to its food security. This context coincided with the launch of a new strategy for the next ten years, called 'Green Generation 2020–2030', which provides solutions for the revitalization of the agricultural sector in order to improve food security and make rural areas more attractive. This agricultural strategy aims at facilitating the emergence of a new generation of agricultural middle class through the consolidation of the gains made in the agricultural sector and the creation of new activities generating jobs and income, particularly for young people in rural areas. The digital transformation of the agricultural sector is also one of the main objectives of this new strategy. This strategy includes the digitization of the agricultural support system granted through the Agricultural Development Fund (FDA). This digitization has accelerated within the framework of the health provisions related to Covid-19 pandemic with the establishment of an electronic platform dedicated to the submission of grant application files.

In the area of scientific research, there is not enough published research on digitization policies (Trendov et al. 2019). Nevertheless, a lot of work from a technical point of view exists in the international literature on the different forms of digitization in agriculture (Klerkx et al. 2019). The digitization of agriculture is, therefore, a very topical issue and offers many solutions – such as weather forecasting, irrigation planning, plant disease control, market organization, information on agricultural extension services…—to the questions we are making today about climate change, agriculture, and sustainable food security. This is where this chapter intends to make a contribution, by providing an overview of some examples of research carried out in Morocco, and which may constitute a step towards the development of agriculture in Africa. An exploratory review of the literature shows that three thematic groups on digitization in agriculture can be identified: agricultural land management; agricultural risk management; and agricultural water management. These themes, which offer a large margin for a future multidisciplinary science on digital agriculture, will be developed in this chapter in order to show the results already achieved and the challenges still open for the development of the agricultural sector.

2 Digital Agricultural Land Management

Sustainable management of agricultural land is fundamental to food security around the world, especially in the context of climate change and increasingly unpredictable weather conditions (Misra 2014). In the African continent, sustainable land management is at the heart of the development challenge (Emerton and Snyder 2018). The introduction of digital innovation in the management of agricultural land can help increase agricultural production and strengthen food security in Africa, including Morocco. Indeed, land degradation in this country impedes agricultural growth and increases the poverty and vulnerability of rural populations (Kouba et al. 2018). For this, the digital transformation of agricultural sector could constitute a major lever for improving the management of agricultural land and minimizing the impact of this degradation on the economic growth of agriculture.

Sustainable land management is a holistic approach that involves several disciplines and has the potential for sustainable transformation in the short and long terms. The goal here is, in particular, to identify, analyze, and discuss promising experiences adapted to the specificity of the African continent while including digital to promote the management of agricultural land.

2.1 Remote Monitoring of Agricultural Land

Monitoring agricultural land is a very complex task for African countries, especially in arid and semi-arid regions where water scarcity and droughts are common. However, even if field data collection is a time-consuming and expensive task, large-scale information on farming systems is essential for decision-makers to better manage agricultural land. Digital innovation has shown its ability to offer solutions that can facilitate better decision-making regarding livelihoods and food production and security. Two studies conducted in the Oum Er Rbia River basin in Morocco by Lebrini et al. (2020) and Lebrini et al. (2019), have confirmed this digital innovation capacity. In this research, remote monitoring of agricultural land was carried out using satellite data and machine-learning methods. On the one hand, the authors assessed the performance of machine learning methods in mapping agricultural systems at large spatial scales using phenological metrics derived from long-term spatial remote sensing data acquired by MODIS sensor. On the other hand, they studied the spatial and temporal changes between the four classes of agricultural systems in the studied region (Fig. 2).

To do this, the TIMESAT software (Jönsson and Eklundh 2004) was used for both filtering noisy NDVI time series and extracting the main phenological metrics for each

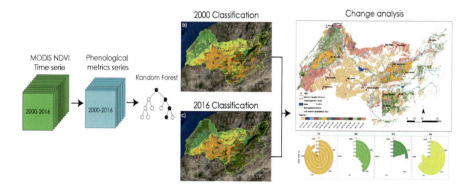

Fig. 2 Agricultural land monitoring steps using remote sensing. *Source* Developed by the authors

season. Afterwards, the produced metrics were used through three machine learning classifiers based on the CARET package of R (Kuhn 2008) for land classification. The accuracy of each classifier results was ensured using ground reference data collected through a fieldwork during the 2015–2016 season and from Google Earth images for the rest of series.

The four identified classes of land cover were: irrigated perennial crops, irrigated annual crops, rainfed areas, and fallow land. The authors found that the Random Forest classifier produced satisfactory results on mapping and monitoring the mentioned agricultural systems. Finally, a land cover change analysis was performed to quantify the spatial and temporal dynamics between the selected land cover classes in the studied area.

The results of this study are encouraging since they can inform the dynamics and the effect of regional policy on changes in agricultural systems at the spatial scale over a longer or shorter period. This can represent for decision-makers a precise and inexpensive tool for the management of agricultural land in the context of environmental and climatic changes.

2.2 Cropland Classification Using a Machine Learning Approach

The classification of cropland and crop types is very important for monitoring land use and land cover. Nonetheless, land use is a changing reality. Indeed, plants are subject to natural laws of lifespan, in addition to a large category of crops also obeying cultural principles based on crops rotation for better production and maintenance of crop soil fertility (Baldivieso-Freitas et al. 2018). Usually, information on crop rotation types and their areas is done in the conventional way through extensive census and ground surveys, which makes the process both expensive and time-consuming. Remote sensing can offer cost-effective solutions that can facilitate the collection and standardization of data. In the same logic, we present, in this part, a research experience carried out in Morocco by Htitiou et al. (2019, 2020) that can be very useful for monitoring agricultural activity using a machine learning approach. This research mainly aims to compare and identify the potential of Sentinel-2 (S2) and Landsat-8 (L8) multi-temporal data to identify different crop types over a heterogeneous agricultural area. Indeed, the spectral reflectance of a field varies depending on the phenological stage of each type of crop as well as other factors such as their phytosanitary state. The integration of these types of data increases the information available to distinguish between the spectral signatures of each crop class for a more efficient classification.

For this, various phenological information were derived from NDVI time series in order to explore key variables to identify both cropland and crop types. Indeed, 10 feature sets were developed and evaluated to discriminate different types of crops via the RF classifier based on derived Sentinel-2 phenological metrics and smoothed

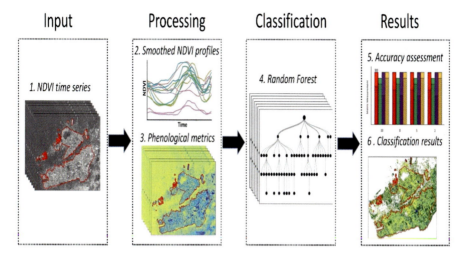

Fig. 3 Schematic presentation of agricultural land classification steps

vegetation indices (Fig. 3). The results showed that the combined use of the adjusted IV and the phenological characteristics gave the best performance. Additionally, the result of using the optimal features was the most accurate among the 10 feature sets, with an overall accuracy of 88% and a kappa of 0.84. In addition to providing a synoptic view, this research constitutes a considerable contribution in the identification of crop types, with a view to ensuring the efficient management of the agricultural sector during the growing season based on updated information.

In order to improve the temporal datasets frequency for crop type classification in high heterogeneous context, Htitiou et al. (2019, 2020) assessed the quality of various fusion models to take advantage of the combined use between S2 and L8 images. Compared to the standard methods of spatiotemporal data fusion (STARFM and FSDAF), mapping results from the Very Deep Super-Resolution (VDSR) method performs well.

These approaches can be successfully applied to other edapho-climatic contexts in Africa. In addition, improvements and adjustments could be made to improve the accuracy of the results depending on the context of each area studied, in particular with the launch of the European Sentinel-2B sensor and the upcoming Landsat-9 (proposed for launch in middle 2021), which will substantially increase moderate-resolution satellite observations available for agricultural monitoring. These new advances will improve the quality of cultivated land maps in heterogeneous agricultural areas and facilitate the decision-making process.

3 Using Digital in Water Resource Management

Water resources have become increasingly scarce in a context where the gap between water supply and demand continues to widen sharply in Morocco and other North-African countries (Schilling et al. 2020). Furthermore, non-rational use of irrigation water resources constitutes a major constraint to development in agricultural areas (Lionboui et al. 2016a). A better valuation and adoption of more water-efficient practices are becoming increasingly necessary, especially in arid and semi-arid regions. Encouraged by the orientation of the implemented agricultural policy (GMP) during the study period in terms of water saving, the research presented in this section offers methodological frameworks for managing and organizing the control and allocation of water resources, thus ensuring their protection and conservation.

3.1 Support Irrigation Water Management Using Spatial Remote Sensing and Modeling

The scientific advances made possible by the use of satellites in the field of monitoring the moisture content of crops are not a new phenomenon. Indeed, when there is not enough water and the crop is under water stress, it no longer manages to lose heat quickly enough through perspiration, which increases its temperature. Remote sensing can detect this using heat-sensitive electromagnetic spectra. In addition, it provides very frequent data, which allows the evolution of processes on a spatiotemporal scale to be closely followed.

In the experiment carried out by Benabdelouahab et al. (2019b) in the Tadla irrigation perimeter (center of Morocco), remotely-sensed reflectance was used to estimate the water content of soil and vegetation for various crops and to monitor water irrigation per unit of surface, relying on the high temporal and spatial resolution of satellite images. In this study, the capacity of two spectral indices (NDWIRog and MSI), derived from SPOT-5 images and backscatter values derived from SAR images, was evaluated. Then, these indices were compared with the corresponding in situ measurements of soil moisture and vegetation water content in 30 plots sown with wheat, with which they showed a strong correlation. The results of this research stipulate that the NDWIRog index could be used as an operational index for monitoring irrigation during the main stages of crop growth at field and regional level in the Tadla irrigation scheme. In another step, remote sensing was combined with the field crop growth model (AquaCrop v4.0) developed by the Food and Agriculture Organization (FAO), to provide an operational tool for irrigation and crop management. Under the conditions of the Tadla irrigation scheme, AquaCrop was adjusted and tested to simulate the impact of irrigation on durum wheat yields. Analysis of irrigation scenarios has shown that the model can optimize the frequency and timing of irrigation water supplies. This could promote the efficiency of water use and guarantee optimal growing conditions during the different stages of the crop.

The approach presented is an important step towards the establishment of an effective irrigation management system to assist decision makers and managers in planning irrigation supplies. However, this approach must be verified before judging its suitability for application in other regions and for other crops. For this, new perspectives for improving crop modeling, by integrating weather forecasting into the decision process and by adopting spatial modeling using data layers on a grid format derived from satellite data, are important.

3.2 Predict Changes in Water Management Indicators Using Multi-Year Agro-Economic Modeling

Modeling is a representation of the information necessary for a decision-making process such as water resource management. Information related to a natural resource such as water is variable over time and space and depends on several factors. For this, the modeling of such resource can present many solutions by being part of an exhaustive vision of management and monitoring of irrigation water. Such a vision may integrate the agronomic, economic, and hydrological dimensions in the analysis of the management of this resource in the regions under study.

The example presented here focused on the case of Tadla sub-basin in central Morocco, where a marked decrease in the supply of irrigation water is noticed (Lionboui et al. 2018). The aim of this work is to analyze water resource sustainability in the current context of agricultural policies by estimating the likely changes in water management parameters and in the sensitivity of the agricultural sector to possible external shocks. Given the multidimensional nature of water management, this research presents integrated agronomic, economic, and hydrological modeling for the Tadla sub-basin, which classifies agricultural areas according to different sources of irrigation water (Fig. 4).

To feed the database of this modeling framework and to give a clear idea about the agro-economic system functioning underlying the current water management in the study area, Technico-economic efficiency and water valuation were characterized through the study of the main agricultural crops in the Tadla region. Hence, the technico-economic efficiency was analyzed for the Tadla farms using the Data Envelopment Analysis method (DEA) (Lionboui et al. 2016b). The characterization of the agro-economic system was followed by a modeling of this system and then by an analysis of water management in the study area. The proposed model is disaggregated by territorial unit, municipalities, water irrigation sources, and by crops.

This tool allows decision-makers to predict the inter-annual variations of selected socio-economic indicators of water management according to different scenarios. To test the sensitivity of water valuation to external shocks, the impact of an increase in the equilibrium rate of irrigation water applied to surface water at the level of the irrigated perimeter of Tadla was simulated. Then, another type of exogenous shocks

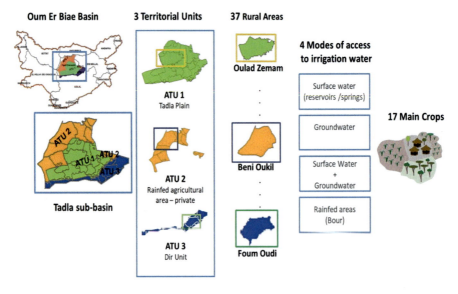

Fig. 4 The disaggregation levels of the proposed water management modeling framework

relating to climate change was simulated in order to examine their potential impacts on the long-term agricultural water management in the region (Lionboui et al. 2018).

Among the main results of this research is irrigation water shadow price which reflects water scarcity. It is around 0.15 US \$/m^3 in the Tadla sub-basin, which means that irrigation water is underpaid, since its equilibrium price, paid in this region, is only USD0.04 m^3 (Lionboui et al. 2016a). The developed model can constitute a tool for analysis and decision-making support in terms of political choices concerning the allocation of water resources and allows an assessment of water valuation at the scale of the Tadla basin. Finally, this research provides a viable modeling framework for sustainable water resource management, applicable to other agro-climatic contexts, in order to help decision-makers improve irrigation water management and mitigate the exogenous shocks' effects on the sustainability of the system.

4 Digital Transformation and Risk Management

Agriculture must face climatic, health, economic, and environmental risks which can have consequences on the viability of agricultural activity. In order to increase the resilience of agricultural systems and minimize the impacts, particularly socio-economic, of these risks, digital technology can support the agricultural sector towards better risk management by providing reliable and quality strategic indicators, which are decisive in decision-making. Thus, digital transformation is now shaping up to be a real necessity to identify risk areas in order to improve the competitiveness of farms in a context of growing risks.

4.1 Assessment of Spatial Variability of Wheat Yield Using Remote Sensing

Crop yields variability has been extensively studied in the world literature in relation to several factors such as soil and climate. The analysis of yields, generally measured on small elementary areas of a few m^2, conventionally assumes the statistical independence of observations (Carter et al. 1983; Snedecor and Cochran 1980). However, given the strong interactions between yields and other factors, the validity of this basic assumption seems very problematic. Indeed, risk in agriculture is now one of the most important elements to be taken into consideration in the overall management of a farm.

This example offers an opportunity to meet the needs of current agricultural policies (GMP and Green Generation) in which Morocco has put in place instruments for monitoring the productivity of agricultural land, in particular for cereals, which constitute a fundamental element of the country's food security. Digital innovation can provide important tools for monitoring the variability of yields in order to optimize the profitability of farms in a risky environment. Thus, a study on the spatial variability of yields was carried out by Benabdelouahab et al. (2019a) using spatial remote sensing as a data source to understand the variability of wheat yields in Morocco. Based on sixteen-year phenological parameters (2000–2016), a yield estimation model was developed. Phenological metrics were derived from Moderate Resolution Imaging Spectroradiometer (MODIS)/NDVI data. In addition to developing the yield estimation model, a spatio-temporal analysis of variability and analysis of wheat yield trends were carried out for the main cereal areas in Morocco (Fig. 5).

The historical data used in this study contain 400 records of wheat grain yield estimated at the level of the plots sampled by the Department of Strategy and Statistics in Morocco for the period 2000/2016. Random and stratified sampling was adopted in the selection of plots in order to guarantee the spatial representativeness of rainfed and irrigated agricultural areas. The model developed showed a good correlation with the field data (R^2 and RMSE were 0.62 ($p < 0.01$), 4.01 qs/ha, respectively). The results of the spatial variability analysis showed an increase in the instability of wheat grain yields in the central and southern regions, especially in the rainfed areas which represent more than 51% of the cultivated area. Nevertheless, irrigated areas have shown a positive trend in wheat production mainly linked to irrigation and technological improvement that the country has been experiencing for years (Fig. 6).

The approach taken in this research overcomes missing data on weather, soil, and irrigation by providing quality data at a spatial scale, which reflects production conditions including physical or human factors.

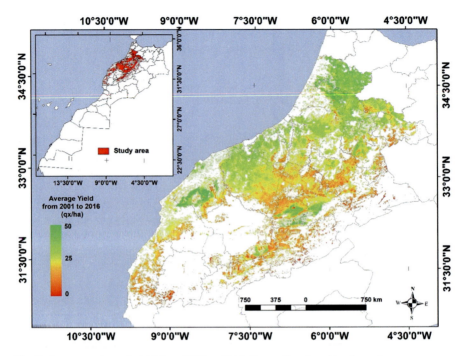

Fig. 5 Average estimated yield from 2001 to 2016. *Source* Developed by the authors

The proposed model can be applied in different agro-climatic contexts, especially in Africa, and could replace empirical models requiring climate data. In addition, it constitutes an inexpensive decision-making aid tool that allows managers and decision-makers to analyse the impact of an agricultural policy, to monitor the dynamics of agronomic potential at the spatial scale, and optimize the choices of agricultural land use.

4.2 Spatial Analysis of the Risk Related to Losses of Wheat Production Value Using Satellite Data

For many Africans, the ability to access sufficient, nutritious, and safe food capable of meeting their dietary needs has been diminished by a succession of natural disasters and epidemics. Currently, among the crucial questions that arise is: in addition to climatic conditions, how to support food security in Africa in the era of Covid-19 pandemic, especially with uncooperative trade policies on global food markets?

In Morocco, wheat production is particularly vulnerable to climate change dynamics and fluctuates from year to year. Considered a dry year, the year 2020 was marked by a major drop in cereal production of 57% compared to an average

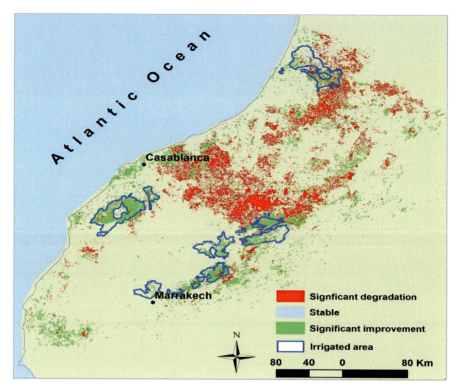

Fig. 6 Trend of wheat production in the main agricultural areas (2001 to 2016). *Source* Developed by the authors

year in last ten years (17.7 million Qx of soft wheat; 7.9 million Qx of durum wheat and 6.4 million Qx of barley) (MAPMDREF 2020), particularly wheat, which poses major challenges for the country given the food security challenge, especially in the context of Covid-19 pandemic. Hence the need to come up with innovative approaches, particularly digital ones, which could help design measures and increase food security at the country level. through an innovative approach that uses digital innovation, the experiment conducted by Lionboui et al. (2020a) is a contribution which aims to analyse the risks of losses in wheat production value with a view to helping decision-makers to develop a risk management strategy and achieve an optimal level of socio-economic security.

The study looked at thirty-four provinces at the Kingdom level representing about 70% of the total area sown to wheat. In order to achieve the objective assigned to this research, the adopted methodology used data from remote sensing in addition to field data. On the one hand, a new index (PV_I) has been developed to estimate the intensity of the risk related to losses in production value on a spatial scale. On the other hand, the probability of occurrence of the risk of losses in wheat production value was calculated for sixteen crop years (2001–2016) (Fig. 7).

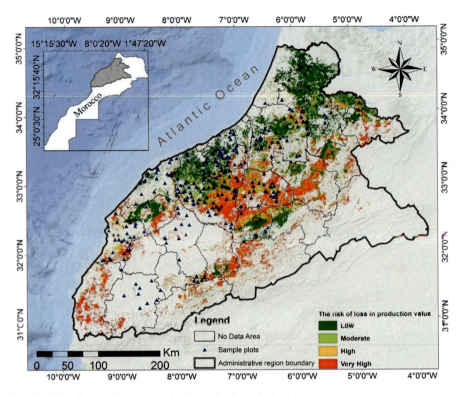

Fig. 7 Risk of losses in wheat production value in agricultural areas

To perform calculations, wheat yield was estimated from its phenological parameters as detailed in the previous section, in order to overcome the variability of the farm income risk caused by the growth phase shifts over time and in space and by technological differences between production units (Benabdelouahab et al., 2019a). In order to make significant the market prices of wheat over time for making comparisons in the studied period, constant prices were calculated. Then, the value of the annual wheat production was calculated spatially by multiplying the estimated annual yield per pixel and the calculated constant annual price of wheat in the corresponding province.

Overall, the results allow us to analyse variability and estimate the severity of differences in terms of the value of wheat production in the provinces studied. The results showed that the value of wheat production varied from one agricultural region to another over the years studied. Its average value is US$533,51/ha calculated for the period 2001–2016, with great variability, i.e. a standard deviation of US$180,14/ha. In addition, four categories of agricultural areas according to their level of risk of losses in wheat production value have been identified:

- Areas that present a risk of losses in wheat production value with low frequency and severity (45.34% of the total study area);

- Areas with a risk of losses in wheat production value with low frequency and high severity (2.7% of the total study area);
- Risk areas with high frequency and low severity (31.83% of the total study area); and
- Areas with a high risk of frequency and severity (20.13% of the total area of the study). This last category is the most affected area and requires urgent intervention by agricultural policies to ensure socio-economic viability and fight against rural exodus.

By using digital innovation, this research offers a global and quantified view of the risks related to losses of production value and makes it possible to detect spatio-temporal differences. In order to prevent risks, the approach adopted could be applied in other context especially in Africa and constitute a basis to help policy makers and producers identify and reduce unacceptable losses in production value and ensure agricultural activity sustainability.

5 Conclusion

Due to several factors, people in Africa live on the brink of food insecurity. To face such a challenge, decision-makers should prioritize actions aimed at reducing the risks to food security. Fostering digital innovations in agriculture could present a real solution and would be much less costly than frequent disaster relief and action, both for human lives and for the development of socio-economic conditions in the agricultural sector.

In this chapter, examples of scientific research in Morocco were presented, where digital technology continues to develop and guide decision-making in the agricultural sector. For years, Morocco have taken preventive measures to protect its food supply by increasing the productivity of crops and reducing their sensitivity to adverse climatic conditions. Indeed, adaptation to climate change is essential to preserve the achievements that Morocco has accomplished in recent decades thanks to the considerable efforts of its Ministry of Agriculture and its partners in the economic and social development of the agricultural sector. However, due to several environmental, financial, human, and technical factors, this adaptation remains difficult.

Digital innovation can offer even more advantages for analysing, with precision, the current situation of the agricultural sector, making forecasts by considering climate change and agricultural policies and proposing the necessary measures to ensure the development of the agricultural sector and favourable socio-economic conditions for rural population. Nonetheless, farmers are still less involved, hence the importance of adopting participatory approaches. In addition, the problem of illiteracy in rural areas, 60.1% for women and 34.9% for men (HCP 2018), makes the situation more delicate since the acquisition of digital data requires the use of social networks by farmers, which seems to be expanding rapidly, but whose development remains conditioned by their access to these technologies.

Whether in terms of management of agricultural land, water resources or agricultural risk, the experiences carried out in Morocco can constitute a basis for the emergence of other digital initiatives and could be applied in different contexts, particularly in Africa.

In order to benefit from this research, it is important to broaden the application of the proposed approaches to other crops and regions. In addition, the launch of new sensors could contribute to the strengthening and relevance of the proposed tools by ensuring regularly updated agricultural land monitoring. These new advances could improve the effectiveness of the approaches presented and potentially benefit the decision-making process. In this quest for performance, digital transformation is now seen as a real necessity in order to develop agriculture. However, the sector must acquire the human skills and technical capacities necessary for data management in order to provide a shared and more agile vision of reality to making more informed decisions.

References

Alvaro E, Rocha N, Ruta M (2020) COVID-19 and Food Protectionism: The Impact of the Pandemic and Export Restrictions on World Food Markets. Policy Research Working Paper No. 9253. World Bank, Washington, DC. © World Bank. https://openknowledge.worldbank.org/handle/10986/33800. License: CC BY 3.0 IGO

Baldivieso-Freitas P, Blanco-Moreno JM, Armengot L, Chamorro L, Romanyà J, Sans FX (2018) Crop yield, weed infestation and soil fertility responses to contrasted ploughing intensity and manure additions in a Mediterranean organic crop rotation. Soil Tillage Rese 180:10–20

Benabdelouahab T, Lebrini Y, Boudhar A, Hadria R, Htitiou A, Lionboui H (2019a) Monitoring spatial variability and trends of wheat grain yield over the main cereal regions in Morocco: a remote-based tool for planning and adjusting policies. Geocarto International, pp 1–20

Benabdelouahab T, Lionboui H, Hadria R, Balaghi R, Boudhar A, Tychon B (2019b) Support irrigation water management of cereals using optical remote sensing and modeling in a semi-arid region. In: El-Ayachi M, Elmansouri L (eds) Geospatial technologies for effective land governance. IGI Global, Hershey, PA, USA, pp 124–145

Brasseur M, Biaz F (2018) L'impact de la digitalisation des organisations sur le rapport au travail: entre aliénation et émancipation. Question(s) de management 21:143–155

Carter T, Burton J, Cappy J, Israel D, Boerma H (1983) Coefficients of variation, error variances and resource allocation in soybean growth analysis experiments. Agron J 75:691–696

Dorn D (2017) La montée en puissance des machines: comment l'ordinateur a changé le travail. Revue Française Des Affaires Sociales 1:35–63

El Bilali H, Allahyari MS (2018) Transition towards sustainability in agriculture and food systems: Role of information and communication technologies. Inf Process Agric 5:456–464

ElMassah S, Mohieldin M (2020) Digital transformation and localizing the Sustainable Development Goals (SDGs). Ecol Econ 169:106490

Emerton L, Snyder KA (2018) Rethinking sustainable land management planning: understanding the social and economic drivers of farmer decision-making in Africa. Land Use Policy 79:684–694

HCP (2018) Les indicateurs sociaux du Maroc. In: High commission for planning, directorate of statistics. Rabat, Morocco.

Hinson R, Lensink R, Mueller A (2019) Transforming agribusiness in developing countries: SDGs and the role of FinTech. Curr Opin Environ Sustain 41.1–9

Htitiou A, Boudhar A, Lebrini Y, Hadria R, Lionboui H, Benabdelouahab T (2020) A comparative analysis of different phenological information retrieved from Sentinel-2 time series images to improve crop classification: a machine learning approach. Geocarto Int 1–24

Htitiou A, Boudhar A, Lebrini Y, Hadria R, Lionboui H, Elmansouri L, Tychon B, Benabdelouahab T (2019) The performance of random forest classification based on phenological metrics derived from sentinel-2 and landsat 8 to map crop Cover in an irrigated semi-arid region. Remote Sens Earth Syst Sci 2:208–224

Jönsson P, Eklundh L (2004) TIMESAT—a program for analyzing time-series of satellite sensor data. Comput Geosci 30:833–845

Klerkx L, Jakku E, Labarthe P (2019) A review of social science on digital agriculture, smart farming and agriculture 4.0: New contributions and a future research agenda. NJAS—Wageningen J Life Sci 90–91:100315

Kouba Y, Gartzia M, El Aich A, Alados CL (2018) Deserts do not advance, they are created: land degradation and desertification in semiarid environments in the Middle Atlas, Morocco. J Arid Environ 158:1–8

Kuhn M (2008) Building predictive models in r using the caret package. J Stat Softw 28:142704

Lebrini Y, Boudhar A, Hadria R, Lionboui H, Elmansouri L, Arrach R, Ceccato P, Benabdelouahab T (2019) Identifying Agricultural Systems Using SVM Classification Approach Based on Phenological Metrics in a Semi-arid Region of Morocco. Earth Syst Environ 3:277–288

Lebrini Y, Boudhar A, Htitiou A, Hadria R, Lionboui H, Bounoua L, Benabdelouahab T (2020) Remote monitoring of agricultural systems using NDVI time series and machine learning methods: a tool for an adaptive agricultural policy. Arab J Geosci 13:796

Lionboui H, Benabdelouahab T, Elame F, Hasib A, Boulli A (2016a) Multi-year agro-economic modelling for predicting changes in irrigation water management indicators in the Tadla sub-basin. Int J Agric Manag Dev 5:96–105

Lionboui H, Benabdelouahab T, Hasib A, Boulli A (2016b) Analysis of farms performance using different sources of irrigation water: a case study in a semi-arid area. Int J Agric Manag Dev 6:145–154

Lionboui H, Benabdelouahab T, Hasib A, Elame F, Boulli A (2018) Dynamic Agro-Economic Modeling for Sustainable Water Resources Management in Arid and Semi-arid Areas. In: Hussain CM (ed) Handbook of environmental materials management. Springer International Publishing, Cham, pp 1–26

Lionboui H, Benabdelouahab T, Htitiou A, Lebrini Y, Abdelghani B, Hadria R, Elame F (2020a) Spatial assessment of losses in wheat production value: a need for an innovative approach to guide risk management policies. Remote Sens Appl Soc Environ 18:100300

Lionboui H, Elame F, Boudhar A, Hadria R, Elboukari B, Benabdelouahab T (2020b) A modelling approach to assess technology effect on wheat farms performance in semi-arid areas. Int J Prod Quality Manag 30:561–577

MAPMDREF (2020) Une production céréalière définitive de 32 millions de quintaux pour la campagne 2019–2020. Ministry of Agriculture, Maritime Fisheries, Rural Development and Water and Forests, Morocco

Misra AK (2014) Climate change and challenges of water and food security. Int J Sustain Built Environ 3:153–165

Peng B, Guan K, Pan M, Li Y (2018) Benefits of seasonal climate prediction and satellite data for forecasting U.S. maize yield. Geophys Res Lett 45:9662–9671

Schilling J, Hertig E, Tramblay Y, Scheffran J (2020) Climate change vulnerability, water resources and social implications in North Africa. Reg Environ Change 20:15

Snedecor G, Cochran W (1980) Statistical methods. The Iowa State University Press 7:507

Sraïri MT (2021) Repenser le modèle de développement agricole du Maroc pour l'ère post Covid-19. Cahiers Agricultures 30:1–9

Talaviya T, Shah D, Patel N, Yagnik H, Shah M (2020) Implementation of artificial intelligence in agriculture for optimisation of irrigation and application of pesticides and herbicides. Artif Intell Agric 4:58–73

Treguer D, Pachon M (2019) Morocco digital and climate smart agriculture program, Program Information Document (PID). In: PIDC190843 (ed) The World Bank. Washington, USA

Trendov N, Varas S, Zeng M (2019) Technologies numériques dans le secteur agricole et dans les zones rurales. In: Food and agriculture organization of the United Nations (FAO). Rome, Italy

World Bank (2021a) Agriculture, forestry, and fishing, value added (current US$). The world Bank Group accessed. https://data.worldbank.org/indicator/NV.AGR.TOTL.CD

World Bank (2021b) Employment in agriculture (% of total employment) (modeled ILO estimate). The world Bank Group accessed. https://data.worldbank.org/indicator/SL.AGR.EMPL.ZS

Chapter 17
Natural and Regulatory Underlying Factors of Food Dependency in Algeria

Amel Bouzid, Messaoud Lazereg, Slimane Bedrani, Mohamed Behnassi, and Mirza Barjees Baig

Abstract Since its independence in 1962, Algeria adopted several public food policies with the objective to nourish its population. To do so, two options were proposed: either raise the local production or import from international markets. Nowadays, the country adopts a combination of these two options by supporting local producers while seeking new foreign suppliers. The main objective of such policies was to meet a growing food demand generated by a rising population, improved incomes, and changing food consumption patterns. This chapter aims to analyse the impact of those policies—mainly subsidies—on local market supply. The results show that Algeria's food consumption has increased significantly in the last half-century both quantitatively and qualitatively, resulting in an increased dependency on foreign suppliers. This dependency can be explained not only by population growth and rising households' incomes, but also by the agricultural policies in place and inefficient economic governance. Yields are low compared to what could technically be achieved because they are constrained by insufficient use of productive inputs, lack of (or outdated) equipment, irrational use of irrigation water, poor access to loans, land fragmentation, and the decrease in rainfall.

A. Bouzid · M. Lazereg (✉)
Research Centre in Applied Economics for Development, Algiers, Algeria

S. Bedrani
National Higher School of Agronomy, Algiers, Algeria

M. Behnassi
International Politics of Environment and Human Security, College of Law, Economics and Social Sciences, Ibn Zohr University of Agadir, Agadir, Morocco
e-mail: m.behnassi@uiz.ac.ma

Founding Director, Center for Environment, Human Security and Governance (CERES), Agadir, Morocco

M. B. Baig
Prince Sultan Institute for Environmental, Water and Desert Research, King Saud University, Riyadh, Kingdom of Saudi Arabia
e-mail: mbbaig@ksu.edu.sa

Keywords Algeria · Food dependency · Agricultural policies · Food policies · Food security

1 Introduction

According to the Food and Agriculture Organization (FAO 2006: 6): "Food security exists when all people, at all times, have physical and economic access to sufficient, safe and nutritious food that meets their dietary needs and food preferences for an active and healthy life". Self-sufficiency was a major goal of many developing countries in the 1960s and 1970s. According to Barker and Hayami (1976), self-sufficiency in food grains has been a publicized goal of government policy in many developing countries. Certainly, Algeria was not an exception in this regard. As the goal of *feeding the Algerian population* is a challenge that is likely to persist in the future, meeting such a challenge will face the dilemma of producing more food from less land and water resources (Porkka et al. 2013).

The food self-sufficiency[1] objective can be achieved best, in the long run, through reforms aimed at improving infrastructure and development institutions such as irrigation, research, and extension systems, which have the main effects of shifting production functions. However, because such reforms require heavy investments and long gestation periods, there was always a temptation from the Algerian government to adopt short-term policies, such as supporting product prices and subsidizing inputs, which are supposed to stimulate farm producers to increase food output along existing production functions (Barker and Hayami 1976).

The inefficiency of Algerian agriculture is perceptible today through the discordance between the rates of increase in domestic production with the rate of population growth. This inefficiency is explained by numerous factors, including the outdated equipment, the small size of farms, and the decrease in rainfall.

In the face of uncertainty about the future, this research aims to assess the impact of public policies on the food dependency[2] using indicators provided by the FAO. It also aims to provide evidence about the evolution of the food security indicators and how these public policies contribute to increase food dependency. Finally, the efficiency of the agricultural system via the inadequacy of input use is assessed.

2 Overview of Algeria's Food Security

The Economist Intelligence Unit (Global Food Security Index 2019) ranks Algeria among the five lowest countries in terms of food security in the Middle East and North Africa (MENA) region. The objective of food security is a part of all key agricultural

---

[1] Food self sufficiency is defined as the domestic production of sufficient per capita calories.
[2] Food dependency is used to represent the share of imports in the total consumption for the basic products: wheat and milk.

policies adopted by the country (Bessaoud 2013). Like almost all MENA countries, energy and food subsidies have been used for decades and constitute the main part of the social 'security safety net' in the region (Albers and Peeters 2011). As defined by the FAO, food security in Algeria is guaranteed by the compilation of four pillars: availability, accessibility, stability, and safety.

2.1 Availability

Through this pillar, the aim is to measure the sufficiency of the national food supply, the risk of supply disruptions, national capacity to disseminate food, and research efforts to expand agricultural output. According to FAO data, the food and nutritional situation has improved markedly over the past half-century both quantitatively and qualitatively. On the quantitative level, the average food intake per capita has doubled, but it is still unbalanced. The diet includes too much wheat and not enough proteins and fats. The insufficient nutritional balance of the available ration also leads to the spread of severely disabling non-communicable diseases such as diabetes or cardiovascular disease. In addition, food still accounts for more than 40% of the average household budget (41.8%), and exceeds half of the budget for the lower deciles 1 and 2 (53.7% and 51.5%, respectively), although an improvement has been observed since 2005. This improvement is linked to the dietary energy supply derived from cereals, roots, and tubers. The energetic supply from cereals has decreased from 59% (2000–2002) to 53% in the total dietary supply (2015–2017). In addition, the share of proteins in the Algerian diet has been improved during the past two decades. The three-year average of total proteins consumed passed from 77 g/capita/day in 2000–2002 to 92.7 g/capita/day in 2015–2017 (with 25% of this protein from animal origin) (Faostat 2020).

2.2 Accessibility

Through this pillar, the aim is to measure the ability of consumers to purchase food, their vulnerability to price shocks, and the presence of programs and policies to support them when such shocks occur. The accessibility is analysed by indicators: GDP and the prevalence of undernourishment and severe food insecurity (Fig. 1). It refers to the probability that a household can acquire the necessary quantities of food. Increased food access possibilities for all consumers were favoured by the establishment of the regulatory system for some sectors at the origin of products entering the average food ration (wheat, milk, potatoes) with the objectives of price stability and guaranteeing the protection of farmers' income and consumers' purchasing power (CREAD-PAM, 2018). Access to food was further facilitated for individuals and their families when they were not able to work through social assistance in the form of various allowances and food donations.

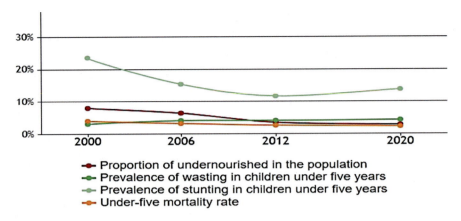

Fig. 1 Trend for indicator values—Algeria. *Source* Global Hunger Index (2020)

2.2.1 GDP Per Capita

As an oil-producing country, the GDP per capita is dependent on oil exports revenues. Many studies have shown that the efficiency of global economic growth will have a positive impact on food security (Manap and Smail, 2019; Świetlik, 2018). Food imports are generally covered by oil exports revenues, and this can constitute a major challenge for the future due the volatility of oil prices in international market. The GDP per capita has grown steadily since the early 2000s, which has allowed a comfortable margin of maneuver to ensure food supplies from the international market even during the 2007–2009 food crisis (Fig. 2).

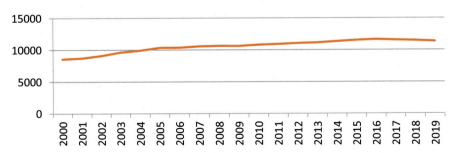

Fig. 2 Evolution of the Gross Domestic product per capita. *Source* Elaborated based on FAO's data (2020)

17 Natural and Regulatory Underlying Factors … 343

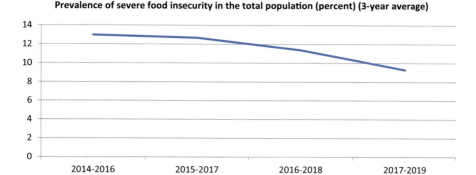

Fig. 3 Prevalence of severe food insecurity. *Source* Elaborated based on FAO's data (2020)

2.2.2 Prevalence of Undernourishment

Among those children under 5 years-old, 32% suffer from anemia, 15.9% are stunted, and 7% have vitamin A deficiency. Yet, 17.5% of adults are overweight (FAO, 2012). Nowadays, the situation has been improved with less than 3% of prevalence of undernourishment and 16% of adults are overweight (Faostat 2020).

2.2.3 Prevalence of Severe Food Insecurity

Quantifying poverty in Algeria is not easy. It can be very difficult to control people's resource levels in an environment where jobs are concentrated in the informal sector, which is made up largely of self-employment and where accounting and records are non-existent. In such circumstances, identifying the poor can be an expensive, imperfect, complex, and controversial exercise. In 1995, the World Bank estimated that 5.9% of the population lives below the international poverty line of US$1.90 per person per day (in 2011 purchasing power parity), but this estimate dropped to 0.5% in 2011. The World Bank also estimated in 2011 that 5.5% of the population was below the national poverty line, which is higher than the estimates for 2001, where the poor population was 0.5% in rural areas and 1.1% in urban areas (CREAD-PAM, 2018). Moreover, 62% of the municipalities (958) in Algeria were classified as poor in 2015, 31% (480) were classified as medium, and 7% (103) were considered rich, according to the Ministry of the Interior and Local Authorities.[3] The municipalities classified as poor are mainly agricultural or pastoral. The Fig. 3 clearly shows that there was a regression in the prevalence of severe insecurity from 13% to 9.3% (average of 3 years) of the total population and this is due to the various support programs for vulnerable populations.

---

[3] https://www.radioalgerie.dz/news/fr/article/20151216/61631.html.

2.3 Stability

This pillar is examined by the following indicators: cereal import dependency, irrigation equipment capacity, and the value of food imports in total merchandise exports (Castro et al. 2020; Capone et al. 2014; Du et al. 2015). Cereals are an important component of total calorie intake of the Algerian consumer and of the animal feed ration. To satisfy a growing demand for cereals, the government has relied on imports. The share of cereals imports is high because of low international prices and Algeria's varied climatic conditions. This share has always been high (around 70% of total consumption) with some variations depending on the annual cereal harvest. This increase of cereal imports is mainly due to population growth, urbanization, and changing consumption patterns (preference for more animal proteins). The current consumption habits are reflected in the high demand for animal proteins as a result of the revenue increase in developing and emerging countries (Abdelhedi and Zouari, 2018).

To improve per farm productivity, the government has subsidized the use of the irrigation equipment since 2000, and the surface area of arable land equipped for irrigation increased from 8% (2000–2002) to more than 18% (2015–2017) (Ministry of Agriculture 2017). This strategy had a double objective: the increase of on-farm productivity and the rationalization of water use.

2.4 Food Safety

There are dietary imbalances with excess salt, sugars, and insufficient fruits and vegetables. Food quality in Algeria remains insufficient due to an insufficient number of quality control laboratories (24 laboratories for quality control and fraud repression) and the regulatory rules to fight food frauds. Food safety control and monitoring is the responsibility of four ministries that operate at different levels: agriculture, trade, health, and industry. However, only the Ministry of Trade has a detailed scheme of intervention by two tools:

- *National surveillance systems—food-borne disease in humans*: data on food-borne illness in humans are collected from the Ministry of Health.
- *National monitoring systems—food-borne hazards in the food chain*: this is the mandatory reporting system for food-borne and water-borne diseases.

Few studies and research are done on the issue of food safety. In the written press, especially during the summer, there are reported cases of food poisoning which are generally due to many factors such as: unsanitary conditions of food factories (mainly in the informal sector); the increase in fraud and pressure tactics; administrative delays; lack of control officers and poor working conditions; disregard of scientific evidence and insufficient infrastructures in the area of R&D. All these factors increase the risks associated with food. Consumers need more information and assurance about food safety so that they can feel safe when purchasing food. The WHO Cooperation Strategy with Algeria (2016–2020) suggests that the emergence, despite an obvious underreporting, of food poisoning, testifies the perfectible nature of the management and respect of hygiene conditions within communities, schools and universities, as well as during one-off events, in particular family events.

3 From Food Self-sufficiency to Food Security in Algeria

As a concept, food self-sufficiency has moved higher on the policy agenda for many countries following the extreme food price volatility experienced during the 2007–08 food price crisis and its aftershocks (Clapp 2017). Algeria is highly dependent on oil export revenues to ensure its food supply. If the decline in these resources continue over the next few years—ceteris paribus—there is a risk that the country will not be able to be food secure (Bessaoud et al. 2019).

In addition, even though the situation of international markets remains currently favourable to importing countries (low prices, abundance of supply), this situation seems volatile due to many factors such as: the sensitivity of wheat markets to climate change; and China's demand for milk and wheat[4] may increase in the future, thus affecting the level of prices on world markets (FAO 2019). Indeed, the countries producing the basic products imported by Algeria are increasingly vulnerable to climatic risks that can affect the world food supply. Will stocks then be able to absorb the shocks caused by a drop in production or an increase in international demand and maintain price levels compatible with the country's purchasing capacity?

This research on food and nutritional security is carried out in Algeria at a time when important concerns at the national level weigh on the capacities of its agro-food system. There are concerns regarding the issue of linking the necessary food supplies to domestic resources, while the country's purchasing power is strongly challenged by declining oil revenues. In recent years, the improvement in availability, access, and stability has been mainly based on substantial public support for agricultural production and domestic supply, with imports used to meet the increase in food demand.

Food policy has been linked to food industry development and should, within the framework of industrialization, favor the maintenance of the workforce at low cost. This was made possible by subsidizing the agricultural sector and downstream by subsidizing food consumption. The government's actions had two objectives: *firstly*, it was imperative to correct food and nutritional failures through preferential access to food, through price subsidies and income supports; *secondly*, to provide a variety of food and a balanced ration capable of eradicating diseases attributable to food deficiencies and deficits (Lebech 2012).

---

[4] These two commodities are the more important in the Algerian trade balance.

According to FAO's global indicators (availability, prevalence of undernourishment, stability), the positive progress made in recent years by Algeria in the food security area is remarkable and recognized, but not sustainable in future. This complexity arises from the multiplicity of socio-economic factors which influence food security and their constantly changing nature. Factors such as population growth, climate change, the impact on human health, and food of an intensive and industrialized production model, are all future challenges for the food and nutritional security of vulnerable populations (urban, rural, poor, children, breastfeeding women, etc.).

In Algeria, the issue of food availability has been a priority for public authorities since independence. It arose dramatically during 1962–1965, in the countryside and cities (1569 kcal/inhabitant/day), but the food availability deficit could not be alleviated except by international food aid. In fact, per capita nutritional energy was estimated at 1.758 cal per day between 1965 and 1969, while the normalized ratio was 2.100 cal/day.

Subsequently, successive governments have taken care to guarantee, at all costs, a regular supply of basic foodstuffs to the domestic market. With predominantly rainfed agriculture with low and irregular yields, increasingly massive food imports have been a solution for supplying the market, although the option of improving domestic supply has never been abandoned.

Until the end of the 1980s, this option was reflected in official speeches with reference to the goal of food self-sufficiency, and it was only in the 2000's that the stated objective became *food security*. This objective was explicitly taken as a reference during the national consultation on agriculture organized in 1992. Agriculture and food are now considered as two elements contributing to national food sovereignty.[5] As part of the Renewal policy, production targets were set for 20 agricultural production chains, by year and wilaya (department), over the period 2009–2014. These objectives involved achieving a doubling of the growth rate of agricultural production compared to the previous five years (CREAD-PAM 2018).

Even if food producers act under a private statute, public authorities' actions will always be necessary to stimulate, direct, and support their activities in order to obtain a better match with consumer demand. This is why public policies in agriculture, fisheries, and agri-food sectors need to be examined. These policies can largely explain the behaviour of producers – farmers, stockbreeders, fishermen, processors – and consumers whose behaviour can be modified by the adoption of regulatory mechanisms.

---

[5] Defined by transnational social movements as: the right of peoples to healthy and culturally appropriate food produced through ecologically sound and sustainable methods, and their right to define their own food and agriculture systems (Schiavoni 2016).

4 Food Model Boosted by the Consumption of Milk and Wheat

The government's choices to keep food prices low over time have created an imbalance between local food supply and demand that has become structural. Demand has continued to grow, driven by significant population growth and food purchasing power boosted by these low prices, especially for basic products (wheat, milk, sugar, and vegetable oil) which are heavily subsidized (Daoudi and Bouzid 2020).

Over the last half-century (1967–2013), the average quantity of food intake per capita – expressed in calories—has improved significantly in some North-African countries: it has doubled for Algeria; increased by 60% for Morocco; and 40% for Tunisia (Table 1).

Table 1 Quantitative evolution and composition of average food intake per capita (in Kcal) in selected North-African countries

Countries		Annual Average 1963–1967		Annual Average 2009–2017		
		Total	Composition (%)	Total	Composition (%)	Multiplication factor
Algeria	Average ration	1577	100	3332	100	2
	Of which plant products	1428	91	2922	88	2
	Of which animal products	148	9	413	12	2,8
Morocco	Average ration	2126	100	3381	100	1,6
	Of which plant products	1976	93	3062	91	1,5
	Of which animal products	150	7	319	10	2
Tunisia	Average ration	2354	100	3443	100	1,5
	Of which plant products	2172	92	3040	88	1,4
	Of which animal products	182	8	402	12	2

Source Elaborated based on FAO's data (2020)

Table 2 The evolution of the protein and fat composition of the diet in selected North-African countries (gr/capita/day)

		1963–1967 Protein	1963–1967 Fat	2007–2017 Protein	2007–2017 Fat	Multiplication Factor Protein	Multiplication Factor Fat
Algeria	Total	**41,5**	**29,1**	**92,94**	**78,74**	**2,23**	**2,7**
	Vegetables	33,7	19,1	66,89	54,79	1,98	2,86
	Animals	7,8	10,1	26,05	23,95	3,3	2,37
Morocco	Total	**56,7**	**33,6**	**97,91**	**68,2**	**1,72**	**2,02**
	Vegetables	47,9	22,7	71,41	47,36	1,49	2,08
	Animals	8,8	11,0	26,5	20,84	3	1,9
Tunisia	Total	**60,6**	**59,8**	**98,4**	**95,8**	**1,62**	**1,6**
	Vegetables	49,1	47,5	71,08	70,9	1,44	1,49
	Animals	11,4	12,2	27,48	24,9	2,41	2,04
Spain	Total	**80,4**	**80,4**	**105,08**	**147,18**	**1,3**	**1,83**
	Vegetables	47,7	53,0	40,46	91,22	0,84	1,72
	Animals	32,7	27,4	64,62	55,98	1,97	2,04

Source Elaborated based on FAO's data (2020)

There is also a significant improvement in terms of quality. Indeed, even if the ration remains mainly based on plant products, the proportion of animal proteins has multiplied between the two periods by 3.3. As for fat consumption, it has doubled for Algeria, and multiplied by 3 for Morocco and 2.4 for Tunisia (Table 2). However, for both total fat and animal protein, the food intake remains well below the level found in Spain.[6]

Wheat is the main source of calories and protein in the Algerian diet, contributing 43% of the total calories consumed and 46% of proteins (Faostat 2020). Milk is the main source of animal protein with 16% of the average daily protein intake; much more than other animal products (red and white meats and eggs) which contribute only 10%.

This improvement in the average food intake cannot be attributed to an increase in local agricultural production. In fact, it is mainly based directly on food imports and indirectly on subsidies. In half a century, the share of imports in the composition of the ration has increased from 36 to 68%.

Quantitatively, Algeria has doubled its food intake (in calories) from 1577 kcal in 1963–67 to 3209 kcal during 2007–2017. Now it is within the standards set by the FAO and WHO. The same observation can be made for Morocco, Tunisia, and Spain, which started from a more advanced point than Algeria. Qualitatively, the Algerian

---

[6] According to experts from FAO and WHO, carbohydrates should meet most of the energy requirements, or 55 to 75% of the daily intake, and free sugars should remain below 10%. Protein should cover 10–15% of the calorie intake and the amount of salt absorbed should be less than 5 g per day (https://www.who.int/mediacentre/news/releases/2003/pr20/fr/).

diet is a Mediterranean diet that is mainly based on plants, which are more available in the market. The same observation is shared by other countries in the region.

The shares of animal energy in the food ration remains marginal (11% in the period 2009–2013) despite its multiplication by 2.5 in the studied period (from 148 to 396 kcal). This improvement is higher than recorded in Morocco and Tunisia, but remains far from Spain, which benefits from its membership in the European Union market.

Two products are specifically subsidized by the Algerian government: the bread baguette made from soft bread wheat flour; and pasteurized milk made from imported anhydrous milk powder (in one-liter packages). Considered as strategic products, the administration of their price and availability has always occupied a central place in food policies (Lazereg and Brabez 2019).

5 Causes of Increased Food Dependency

5.1 Agriculture in the Global Economy

In 2019, the growth rate of agriculture was appreciable and stood out from its underperformance of 2016 and 2017. Thus, after growth in agricultural value added of 3.5% in 2018, 2019 remains on a trend of improvement with, however, a slight slowdown with growth of 2.7% or 0.8 point of growth less. Cereal production remains high in 2019 with 56.3 million quintals, although down compared to 2018 which was 61 million quintals. Crop production, excluding cereals, also posted a significant growth rate of 8.9% in 2019 against 5.1% in 2018 and 0.4% in 2017. Conversely, animal production recorded decreases of 0.9% in 2019 and 1.2% in 2018, against a positive growth of 0.3% in 2017. The average volume growth rate was 6.4% between 2000 and 2015 and it experienced several variations due to interannual rainfall variation (Fig. 4). This growth resumed in 2018 after two years of decline (but still remains below the average of the last 15 years); 2019 and 2020 also saw a decline in agricultural growth due to the poor rainfall recorded.

When analysing food issues in Algeria, the geographical characteristic must be taken in consideration. Obviously, the aridity of the climate hampers agricultural productivity. In Algeria, as in the rest of the central Maghreb, where the agricultural land is relatively abundant but water is scarce, rain-fed agricultural yields remain low and uncertain (Djenane 2012). Indeed, water resources severely constrain agricultural development in Algeria. Its current scarcity explains why climate risk is a fundamental parameter of farmers' practices (intensification, mitigation, adaptation). Certainly, climate change will further aggravate the negative situation in agriculture. The natural basis of agricultural production is characterized by an appreciable natural potential that go hand in hand with a marked fragility of ecosystems, making the challenge of sustainability and stability very acute. The diversity of relief and climates translates into a wide variety of ecosystems and an undeniable richness of the plant

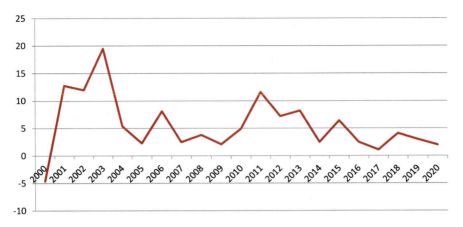

Fig. 4 Volume growth rate of agricultural production (2000–2015). Source Office for National Statistics (2017)

and animal genetic heritage, as well as its adaptation to natural conditions in certain extreme cases (from north to south, seven major types of ecosystems).

The assessment of climatic data during the last decades and all future scenarios predicts the increase of climate change negative impact. For the Mediterranean region, available research confirms that it will experience a sharp increase in aridity making it more vulnerable to water stress and desertification (Agrimonde 2009; Martin et al. 2015). Moreover, the development of the desertification sensitivity map by remote sensing has shown that more than 570.000 ha of land in the steppe zone are already totally desertified without the possibility of biological recovery, and that nearly 6 million ha are highly threatened by the effects of wind erosion. In addition, unsuitable cultivation practices subjected nearly 1.2 million ha to wind erosion annually (Abahussain et al. 2002).

Table 3 shows that there is a structural deficit between consumption and production. This structural deficit covers commodities that dominate Algeria's consumption pattern (wheat, pulses, milk, edible oils and fats, and sugar). This fact forces the government to increase imports of essential food products to meet national needs, despite the undeniable progress made in increasing domestic supply.

Several factors can explain the Algeria's increase in food dependency. In addition to population growth and rising households' incomes, the determining factors appear to be those of agricultural policies and economic governance. These latter factors translate into a poor performance of agricultural sector due to low productivity and, therefore, low yields.

The structure of food imports is diverse in terms of partners. In 2017, the European Union was Algeria's leading supplier of agricultural products with US$2.9 billion. Its percentage of Algeria's agricultural imports declined, however, from 52% in 2001 to 39% in 2010 and 31% in 2017. Argentina is the second leading supplier (mainly cereals and soybean meal) with a 6% market share in 2017 (Bessaoud et al. 2019). In addition, the country imported an average of 3.9 million tons of common wheat from

Table 3 The sources of supply of the national market (average 1990/2017)

	Food import/availability (%)		
	Average 1990–1999	Average 2000–2009	Average 2010–2017
Durum	62,3	82,3	82,9
Barley	28,7	25,0	36,9
But	90,8	123,5	104,3
Potatoes	13,7	8,2	3,2
Sugar	97,5	101,6	112,9
Pulses	72,1	78,7	71,5
Edible oils	79,8	111,5	107,5
Tomatoes	7,8	6,6	14,3
Dates	0,3	1,0	1,1
Milk	60,5	55,8	41,6

Source Elaborated based on FAO's data (2020)

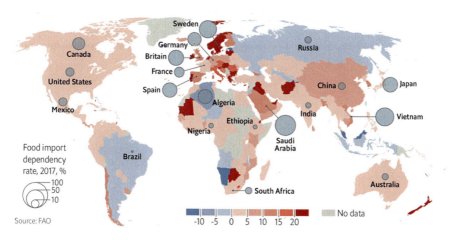

Fig. 5 Food Import Dependency rate 1997–2017, percentage points. *Source* The Economist (May 9th, 2020)[7]

France during 2005–2020, but imports reached a record 5.6 million tons in 2020. Algeria is the second leading importer of milk powder after China with 224 thousand tons. Daoudi and Bouzid (2020) estimate that the foreign markets contribute 55% of the calories consumed by Algerians, a fact that ranks Algeria in the top ten of food import-dependent countries (Fig. 5).

---

[7] https://www.economist.com/briefing/2020/05/09/the-worlds-food-system-has-so-far-weathered-the-challenge-of-covid-19.

5.2 Population Growth and Rising Incomes

The sharp rise in demand for food products in Algeria is primary due to population growth, which was very high (more than 3% annually) up to 1985, but started to quickly decline to 1.48% in 2000 (the rate has risen again and exceeded 2% in 2010 and 2.15% in 2015). According to Flici (2020) and Aissaoui (2020), this new 'baby boom' (the third in 50 years after that of 1968 and 1982) is due in part to a doubling of the marriage rate. Indeed, more than 369,000 marriages were recorded in 2015 against 177,000 in 2000, an average of 9 marriages per 1,000 inhabitants against 5 per 1000, which explains this growth in the birth rate. It seems that the sharp decline in the population growth rate observed after 1985 is explained by the start of the long economic crisis and the fears it placed on future employment and living conditions. The postponement of the creation of new households temporarily blocked population growth. From the 2000s, the new found security and stability, the ambitious programs of social housing construction, the creation of jobs, and the gradual improvement of purchasing power caused the observed demographic recovery. But this increase in population is obviously a major contributor to the problem of food dependency.

Demand was also boosted by the relatively strong income growth. Per capita GDP (expressed in current US$ PPPs) increased by 120% in Algeria between 1990 and 2015. Rapid urbanization of the country has led the force-work to move from the agricultural sector to other sectors (services, building, administration, etc.). In fact, employment in agriculture has fallen from 22% (2000) to 9.9% (2019) due to employment opportunities in other sectors.

Agriculture is the most impotant sector in rural areas in terms of employment, but the gap between urban and rural economic development is linked to rural residents migrating to cities in search of better-paid work. Table 4 shows that there is a relationship between the share of rural population and the employment in agriculture (principal source of income). So, the migration to other economic sectors has increased because the lack of workers' interest in the agricultural profession, which is regarded as a hard job, makes them poorly adapted to a modern life and to their hobbies and keeps them socially discredited (Abelhedi and Zouar, 2018). Similarly, creeping urbanization, which captures the best agricultural soils, can also cause abandonment of agricultural lands (Padilla 2008).

5.3 Agricultural Policies and Economic Governance

In the 1980s, the Algerian government transitioned towards a market economy after the policy of collectivization and self-management of the large agricultural estates resulting from colonization in the 1960s and direct institutional intervention by the state in economic activity in the 1970s. Direct state management has now given way to the implementation of regulatory instruments which, overall, have not had significant effects on strategic staple food production.

Table 4 Rural areas indicators

	Agriculture value added (% of GDP)	Employment in agriculture (% of total employment)	Rural population (% of total population)
2000	8,40	22,21	40,08
2001	9,75	21,71	39,29
2002	9,23	21,73	38,50
2003	9,81	21,72	37,72
2004	9,44	20,06	36,94
2005	7,69	18,43	36,17
2006	7,54	17,03	35,41
2007	7,57	15,59	34,65
2008	6,59	14,31	33,90
2009	9,34	13,02	33,17
2010	8,47	11,86	32,46
2011	8,11	10,77	31,76
2012	8,77	10,69	31,09
2013	9,85	10,61	30,42
2014	10,29	10,50	29,78
2015	11,58	10,35	29,15
2016	12,22	10,25	28,54
2017	11,95	10,16	27,95
2018	11,98	10,02	27,37
2019	11,97	9,86	

Source FAOSTAT (2020); World Bank (2020)

Under the effect of the 1986 shock on oil markets, direct consumer subsidies were replaced by direct support of incomes of the poorest categories (increase in the national minimum wage, family allowances, retirement pensions, and specific allowances to people without income). This period was then characterized by an economic recession that induced a decline in demand for products from the agro-food industries as well as by a reduction in the import capacities of raw materials and, therefore, in the rate of use of productive capacities. Except the dairy industry, whose production increased during the reform period, and the cereal industry, whose production remained more or less stable, all the others showed a sharp drop in their output.

In 2000, the government launched the National Agricultural Development Program (NADP) which aimed to ensure the country's food security, promote incomes and employment in rural areas, and manage fragile natural resources in a sustainable way. The NADP is part of the economic recovery support program whose goal was to achieve an annual growth rate of 10% against 4% over the last years of the previous decade.

Nowadays, agricultural policies are to blame for the rise in the country's level of food dependency, especially those pertaining to the management of the lands related to the former colonial agricultural sector, the treatment of the private sector, and the support of agriculture. The former colonial agricultural sector (owned by the state after Independence) covered approximately 2.3 million hectares and 28% of the country's utilized agricultural land (UAA) at Independence. This colonial sector had most of the fertile land yet it has never been exploited optimally since nationalization. From 1963 to 1987, the centralized management (dictated by the socialist orientation of the state) was poor and did not meet expected goals. The 1987 reform created Collective Farms (CF) and Individual Farms (IF), while abolishing state supervision of farms. The reform maintained the land in the state's private domain and the joint ownership of agricultural land in the collective farms. Yet this policy held back investments in and development of the collective farms because agreement was required of all beneficiaries. Most CFs have gradually broken their land into individual farms with the members staying formally in the CF. In this situation, farmers could not apply for a bank loan unless all other members of the CF agreed.

Another impediment to agricultural investment has been the fragile and changing legal underpinnings of this sector. Agrarian reforms took place in 1971 with the partial nationalization of large private properties; in 1981 with the reform of the self-managed sector aimed at reducing the size of farms and giving them the autonomy of management; the creation of CFs and IFs in 1987 with the reaffirmation of non-intervention of the agricultural administration in the management of these new farms (Baci 1999). Finally, the agricultural orientation law (2010) transformed the right of perpetual use (99 years) into a renewable 40-year right of use (Tatar 2013; Amichi et al. 2015).

Agricultural policies concerning the private sector were not very favourable until the early 1980s, particularly for the wealthy groups who were victims of the agrarian revolution. As a result, few investments were made before the 1980s. It was only after that time that the private sector was granted some subsidies in the fields of mechanical farm equipment, poultry farming techniques, and livestock feed, etc.

Finally, the economy has lagged in industrialization, and it has been unable to meet the growing supply of labour moving from poor rural areas. As a result, the agricultural sector has always occupied a large part of the total population on increasingly smaller farms (because of fragmentation by inheritance). These farms rarely have the means—financial, human and material—to increase their productivity.

6 The Inefficiency of the Agricultural System

The inefficiency of the Algerian agricultural system has been outlined by several researchers, particularly in strategic sectors (Djermoun and Chehat 2012; Lazereg et al. 2020) for milk sector; Hitouche et al. (2019) and Derderi et al. (2015) for

Table 5 Algerian yields compared to proximate countries (Spanish yield Index = 100)

	Wheat	Dry beans	Carrots/Turnips	Fresh beans	Onions	Bovine meat	Sheep meat	Poultry meat
Algeria	54	61	35	34	48	72	150	56
Morocco	60	57	60	119	53	75	121	66
Tunisia	65	84	32	23	43	83	116	76
Spain	**100**	**100**	**100**	**100**	**100**	**100**	**100**	**100**

Source Elaborated based on FAO data (2017)

potato sector; and Djermoun (2009) for cereal sector). The problems with agricultural production can be explained by the following factors.

6.1 Low Production and Yields

FAO has determined five important input resources for food production: land and irrigation, labour, machinery, fertilizers, and pesticide. Agricultural production and yields have been growing slowly in Algeria. In almost half a century, the index of agricultural production went up from 128.6 (2009) to reach 154.3 (2016)[8] (Knoema, 2020), which was lower than the population increase.

Yields for the main crops remain low compared to what it is technically possible. They are much lower than the yields obtained in Spain, a country with a similar climate, except for the yield of sheep meat (Table 5).

The prevalence of micro-farms is a likely cause for this lag in productivity relative to Spain. For the wheat, for example, Algeria shows the lowest yield compared to Tunisia and Spain, with 19.22 q/ha and 31 q/ha respectively.

6.2 Low Consumption of Productivity Inputs

Low-level yields have their origin in the low level of higher-quality farm inputs. In Algeria, the use of fertilizers in agriculture is not known exactly, except for farmers involved in the cereal intensification program and for those growing potatoes. The use of fertilizers fell sharply between 1987 and 1994 because implicit and explicit subsidies disappeared (Bedrani et al. 2001). Fertilizer use fell from 232.000 tons in 1987 to 113.000 tons in 1994. It also had a sharp drop between 1995 and 1998 (from 83.000 tons to 45.000 tons). This was likely due to the absence of *ammo-nitrate* on the national market for security reasons. A resumption of fertilizer use began in 1999 due to the support for fertilizer use in cereals and potatoes as part of the

---

[8] https://knoema.com/atlas/Algeria/Crop-production-index.

Table 6 Evolution of irrigated areas (in hectares)

	Gravity irrigation	sprinkler irrigation	Drip irrigation (localized)	Total
Year 2000	275.000	70.000	5.000	350.000
Year 2005	524.503	153.006	147.697	825.206
Year 2010	573.667	211.594	196.475	981.736
Year 2016	621.000	388.000	251.000	1.260.000
Multiplication factor from 2000 to 2016	2,3	5,5	50,2	3,6

Source MADR (2017)

NPAD[9] program. According to the estimated FAO's data, the use of fertilizer per area of cropland in Algeria was 5.41 kg/ha in 2000 to 8.24 kg/ha in 2018, which is too low compared to Morocco (27.4 kg/ha) and Tunisia (14.38 kg/ha).

Algerian farmers have been hit hard by rising prices for industrial products used in agriculture. While it took 43.8 tons of durum wheat to buy a 62 HP tractor in 1974, it takes 1.4 times more in 2016, or 62.2 tons, which can explain the low level of productivity. Only 67% of farms are mechanized and the agriculture sector has only 8500 combine harvesters. In Algeria, the number of tractors was at level of 140 units per 100 sq. km in 2008, up from 139 units per 100 sq. km previous year; this is a change of 0.67%. In 2008, the number of tractors in Tunisia was 143 units per 100 sq. km. This number has increased from 86 units per 100 sq. km in 1989 to 143 units per 100 sq. km in 2008, growing at an average annual rate of 2.79%.

6.3 Inefficient Use of Irrigation Water

The majority of agricultural productions in Algeria (wheat, barley, pulses, fodder) are dry and dependent on rainfall, which is characterized by two rainy seasons— in winter and spring – and a pronounced summer drought—rainfall of more than 1000 mm from the high coastal reliefs of the north-east, and less than 100 mm in south of Sahara (Boughedaoui 2019).

Water remains an essential factor of agricultural production, but its importance is heightened due to Algeria's arid climate. The inefficient use of such a resource is the result of a poor water governance. While gravity irrigation – also called 'surface irrigation' – is the most wasteful way to irrigate, it is still highly practiced in the country. It is true that water-saving irrigation techniques have increased since the 2000s: spray irrigation areas have tripled between 2000 and 2016 and those with drip irrigation have increased by a factor of 50 from 5.000 ha to 251.000 ha (Table 6). Yet the area under gravity irrigation still accounts for 49% of total irrigated

---

[9] National Plan for Agricultural Development.

areas. Furthermore, water-saving techniques are not always well managed, especially sprinkler systems, which leads to overconsumption of water compared to plant needs.

With wastewater production of 1.2 billion m^3/year and a treatment capacity of 0.8 billion m^3, Algeria's water reuse is currently only 0.1 billion m^3/year (0.08%). This is due to three factors: the drastic conditions imposed by regulations on the use of treated wastewater which are not yet fulfilled by many wastewater treatment plants; the lack of irrigable land downstream of the wastewater treatment plants; and the prohibitive cost of bringing irrigable land upstream from the wastewater treatment plants.

6.4 Low Access of Farmers to Bank Loans

Access to credit is a prerequisite for the modernization of farming. This access, which is normally through a banking institution, has always been very limited for farmers. The 2001 General Census of Agriculture indicates that only 3.1% of farmers use bank credit. In the Maghreb countries, the situation is practically the same: less than 10% of farms in Tunisia have bank financing (Audinet Tunisia, 2007; Chebbi et al. 2019); and the Moroccan agricultural sector suffers the marginal contribution of banks to the financing of farm needs (Akesbi et al. 2008), and it was only after the launch of the Green Morocco Plan in 2008 that these credits were able to grow again and reached a wider range of categories of farmers – 1 million farmers have had access to credit, 3 times more than in 2008 (Harbouze et al.2019).

The campaign to increase investment credits in 2008 (triggered by the state), did not greatly improved the rate of bank lending for farmers. More recent studies confirm the persistence of this low lending (Daoudi et al. 2013; Chabour 2017). Daoudi and Wampfler (2010) specified that the lack of information on the operational details of farms, their inability to provide material guarantees, and the high transaction costs are the main reasons for this exclusion of the banking system. In the Maghreb countries, the reluctance of banks to lend in the agricultural sector is accentuated by the importance of climate risks which make production very uncertain (Mesli 2007).

6.5 Unsuitable Land Structures

The agricultural sector is characterized by highly fragmented land structures and many farms have an unsuitable legal status. This makes it difficult to create economically viable farms that are able to improve their productivity. In Algeria, farms of less than 10 ha made up 76% of the total in 2001. The 2001 General Agriculture Census indicated that 62% of owners – accounting for 51% of utilized agricultural land (UAL)—have an 'undetermined' legal status or no title to property. The UAA operators in the state private sector—with 49.3% of the total UAL—were not allowed to use their property as a guarantee to get loans from banks (lack of implementation texts).

The UAL per capita is today barely equal to 0.2 ha, which is small compared to other Mediterranean neighbours (Spain at 0.4 ha, Morocco at 0.35 ha, Tunisia at 0,47 ha). Algeria and Portugal are the countries with the smallest endowment of arable land per inhabitant (Faostat 2017).

The 2001 census of agriculture confirmed the dominance of small-scale farming in the land structures of the country. The majority of farms (70%) have an area of less than 10 ha, which is a handicap for the promotion of mechanization.

7 Conclusion

Population growth is and will continue to be a real challenge to Algeria's food and nutritional security. A United Nations study predicts that its population will grow by 52 percent to 63 million in 2050 before doubling to 90 million in 2100. Although Algeria has improved the nutrition of its population during the last half century, it remains very dependent on international markets for basic food products. It could improve its level of food self-sufficiency by pursuing more effective policies to sustainably increase the productivity of water, land, and labour. Decreasing the food import bill would allow the country to devote more foreign currency to the import of capital goods, semi-finished products, and services needed for its industrialization. This would allow the construction of a more integrated economy able to absorb low productivity labour still present in the agriculture sector.

Nowadays, due to the price volatility on international markets and the aggravated impact of climate change, Algerian agricultural policies must revitalize local agriculture by increasing subsidies to promote investment in farming (new technologies, water saving materials, etc.) and by integrating climate change factors into agricultural and rural development projects. In addition, the improvement of productivity will depend largely on investments in R&D and the innovations that may result for the management of climate change-induced risks.

Finally, the lack of public coordination for food security is an issue to be tackled. To date, there are still no coordinating mechanisms allowing effective operational management and information sharing between different ministries involved in food security: trade, agriculture, industry, health, national solidarity, and Interior. All these ministries conduct complementary or competing projects that need to be coordinated for more positive impacts.

References

Abahussaın AA,, Abdu AS, Al-Zubari WK, El-Deen NA, Abdul Raheem M (2002) Desertification in the Arab region: analysis of current status and trends. J Arid Environ 51: 521–545

Abdelhedi IT, Zouari SZ (2020) Agriculture and food security in North Africa: a theoretical and empirical approach. J Knowl Econ11:193–210. https://doi.org/10.1007/s13132-018-0528-y

Agrimonde (2009) Agricultures et alimentations du monde en 2050: scénarios et défis pour un développement durable, Éditions Quae, p 293

Aissaoui N (2020) Analysis of spatio-temporal variations and at-risk populations of road insecurity: case study of road traffic accidents in Algeria. J d'Economie, de Management, d'Environnement et de Droit (JEMED) ISSN 2605–6461. 3(2), mai

Akesbi N, Benatya D, El Aoufi N (2008) L'agriculture marocaine à l'épreuve de la libéralisation. Rabat : Economie Critique. http://www.ledmaroc.ma/pages/ouvrages/agriculture.pdf

Albers R, Peeters M (2011) Food and energy prices—Government subsidies and fiscal balances in South Mediterranean countries. EconomicPapers 437, EuropeanEconomy, European Commission, Bruessels

Amichi H, Mayaux PL, Bouarfa S (2015)Encourager la subversion: recomposition de l'État et décollectivisation des terres publiques dans le Bas-Chéliff, Algérie. Politique africaine 2015/1 N° 137, p 210

Audinet Tunisie (2007) www.investir-en-tinisie.net/news/article.php?id=1268

Baci L (1999) Les réformes agraires en Algérie. In: Jouve AM, Bouderbala N (eds) Politiques foncières et aménagement des structures agricoles dans les pays méditerranéens: à la mémoire de Pierre Coulomb. Montpellier: CIHEAM, p 285–291 Cahiers Options Méditerranéennes, 36

Barker R, and Hayami Y (1976) Price support versus input subsidy for food self-sufficiency in developing countries. Am J Agric Econ 58(4), Part_1:617–628. https://doi.org/10.2307/1238804

Bedrani S, Boukhari N, Djennane A (1997) Eléments d'analyse des politiques de prix, de subvention et de fiscalité sur l'agriculture en Algérie. Options Méditerranéennes, Série B 11:121–149

Bedrani S, Chehat F, Ababsa S (2001) L'agriculture algérienne en 2000. Une révolution tranquille: le PNDA. Revue Prospectives Agricoles, INRAA, Alger 1:7–60

Bessaoud O (2013) Aux origines paysannes et rurales des bouleversements politiques en Afrique du Nord : l'exception algérienne, Maghreb—Machrek, 1/2013. 215. 9–30

Bessaoud O, Pellissier JP, Rolland JP, AndKhechimi W (2019)Rapport de synthèse sur l'agriculture en Algérie. [Rapport de recherche] CIHEAM-IAMM, p 82. ffhal-02137632f.

Boughedaoui M (2019) Rapport thématique sur les changements climatiques. Coopération Technique Belge

Capone R, El Bilali H, Debs P, Cardone G, Driouech N (2014) Food economic accessibility and affordability in the mediterranean region: an exploratory assessment at micro and macro levels. J Food Secur 2(1):1–12. https://doi.org/10.12691/jfs-2-1-1

Castro CB, Ren Y, Loy J-P (2020) Scarce water resources and cereal import dependency: the role of integrated water resources management. Water 12:1750

Chabour D (2017 Le crédit bancaire agricole: cas de la daïra de Timezrit, Wilaya de Bejaia. Mémoire de master en sciences agronomiques. Ecole Nationale Supérieure Agronomique, Alger

Chebbi HE, Pellissier J-P, Khechimi W, Rolland J-P (2019) Rapport de synthèse sur l'agriculture en Tunisie. [Rapport de recherche] CIHEAM-IAMM, p 99. ffhal-02137636f

Clapp J (2017) Food self-sufficiency: Making sense of it, and when it makes sense. Food Policy 66:88–96

CREAD-PAM (2018)Revue stratégique de la sécurité alimentaire et nutritionnelle en Algérie. http://www.cread.dz/tmp/2018/07/Revue-SAN-Alg%C3%A9rie-Version-finale-1.pdf

Daoudi A, Bouzid A (2020) La sécurité alimentaire de l'Algérie à l'épreuve de la pandémie de la COVID-19. Les cahiers du CREAD. 36(3)

Daoudi A, Terranti S, Hammouda RF, Bédrani S (2013) Adaptation à la sécheresse en steppe algérienne: le cas des stratégies productives des agropasteurs de Hadj Mechri. CahAgric 22:303–310. https://doi.org/10.1684/agr.2013.0629

Daoudi A, Wampfler B (2010) Le financement informel dans l'agriculture algérienne: les principales pratiques et leurs déterminants. Cahiers Agric 19(4):243–324. https://doi.org/10.1684/agr.2010.0414

Derderi A, Daoudi A, Colin JP (2015) Les jeunes agriculteurs itinérants et le développement de la culture de la pomme de terre en Algérie. L'émergence D'une Économie Réticulaire. Cahiers Agric 24(6):387–395

Djenane A (2012) La dépendance alimentaire: un essai d'analyse. Confluences Méditerranée 2:117–131. https://doi.org/10.3917/come.081.0117

Djermoun A, Chehat F (2012) Le développement de la filière lait en Algérie: de l'autosuffisance à la dépendance. Livestock Res Rural Dev 2

Djermoun A (2009) La production céréalière en Algérie: les principales caractéristiques. Nat Technol (1):45

Du T, Kang S, Zhang J, Davies WJ (2015) Deficit irrigation and sustainable water-resource strategies in agriculture for China's food security. J Exp Bot 66(8):2253–2269. https://doi.org/10.1093/jxb/erv034

FAO (2005) Utilisation des engrais par culture en Algérie. Rome

FAO (2006) Plant nutrition for food security. A guide for integrated nutriment management, Rome

FAO (2019) Food Outlook—Biannual Report on Global Food Markets—November 2019. Rome. http://www.fao.org/3/CA6911EN/CA6911EN.pdf

FAO (2020) FAOSTAT statistical database. [Rome]. FAO

FAO (2012) La situation mondiale de l'alimentation et de l'agriculture. Rome. http://www.fao.org/3/i3028f/i3028f.pdf

Flici F (2020) Analyzing the trend of life expectancy evolution in Algeria from 1962 to 2018: the S-logistic Segmentation with Jumps. Pop Rev 59(1). https://doi.org/10.1353/prv.2020.0002

Harbouze R, Pellissier J-P, Rolland J-P, Khechimi W (2019) Rapport de synthèse sur l'agriculture au Maroc. [Rapport de recherche] CIHEAM-IAMM, p 104. ffhal-02137637f

Hitouche S, Pham HV, Brabez F (2019) Facteurs déterminant l'implication des opérateurs dans une politique de stockage incitative: Cas du dispositif de régulation Syrpalac en Algérie. New medit: Mediterranean J Econ Agric Environ Revue Méditerranéenne d'économie, Agric et Environ 18(1):65–78

Keulertz M, Woertz E (2015) States as actors in international agro-investments, international development policy revue internationale de politique de développement [Online], 6 | 2015. http://journals.openedition.org/poldev/2023; DOI: https://doi.org/10.4000/poldev.2023

Lazereg M, Brabez F (2019) Politique laitière et accès au marché formel des petits éleveurs dans la région de Sétif. Les Cahiers du Cread 4

Lazereg M, Bellil K, Djediane M, Zaidi Z (2020) La filière lait Algérienne face aux conséquences de la pandémie de la COVID-19. Les Cahiers Du Cread 36(3):227–250

Lebech R (2012) La Politique Alimentaire En Algérie : De L'autosuffisance à La Sécurité Alimentaire. Djadid El Iktisad 7(1):58–69

MADR (2017) Évolution des superficies irriguées de 2000–2018. Document interne, Alger

Manap NMA, Ismail NW (2019) Food Security and Economic Growth. Int J Modern Trends Soc Sci 2(8):108–118. https://doi.org/10.35631/IJMTSS.280011

Martin S, Eglin T, Bardinal M (2015) Analyse comparative de scénarios de lutte contre le changement climatique pour l'agriculture à l'horizon 2035, Rapport final, ADEME, p 38

Mesli ML (2007) L'agronome et la terre. Alger: éditions Alpha

Ministère de l'Agriculture et du Développement Rural (Alger, Algérie)(2003) Recensement général de l'agriculture 2001: rapport général des résultats définitifs. Alger (Algérie): Ministère de l'Agriculture et du Développement Rural, p 125

Office National des Statistiques (2017) Enquête sur les dépenses de consommation et le niveau de vie des ménages 2011. Dépenses des ménages en alimentation et boissons en 2011. Collections Statistiques N° 195, Série S: Statistiques Sociales

Padilla M (2008) Chapitre 8—Assurer la sécurité alimentaire des populations, dans: CIHEAM éd., MediTERRA 2008. Les futurs agricoles et alimentaires en Méditerranée. Paris, Presses de Sciences Po, Annuels, pp 231–249. https://www.cairn.info/mediterra-2008--9782724610642-page-231.htm

Porkka M, Kummu M, Siebert S, Varis O (2013) from food insufficiency towards trade dependency: a historical analysis of global food availability. PLoS ONE 8(12):e82714. https://doi.org/10.1371/journal.pone.0082714

Protein and amino acid requirements in human nutrition (2003) Report of a Joint WHO/FAO/ UNU Expert Consultation. Geneva, World Health Organization (in press). https://www.who.int/mediacentre/news/releases/2003/pr20/fr/

Schiavoni CM (2016) The contested terrain of food sovereignty construction: toward a historical, relational and interactive approach. J Peasant Stud. https://doi.org/10.1080/03066150.2016.1234455

Świetlik K (2018) Economic growth versus the issue of food security in selected regions and countries worldwide. Problems Agric Econ 3(356):127–149. https://doi.org/10.30858/zer/94481. Available at SSRN: https://ssrn.com/abstract=3256188

Tatar H(2013) Transformations foncières et évolution des paysages agraires en Algérie. Méditerranée 2013/1, 120–136.https://doi.org/10.4000/mediterranee.6660.

Chapter 18
Boosting Youth Participation in Farming Activities to Enhance Food Self-Sufficiency: A Case Study from Nigeria

Omowumi A. Olowa and Olatomide W. Olowa

Abstract The Food system in Nigeria is characterized by a demand–supply gap or a distorted balance between the market and the society, thus culminating in high food prices and endangering the much touted food self-sufficiency. While various investigations to unravel the possible causes have focused on other factors, assessment of this problem through the lens of youth involvement in farming production has been given little attention. This study focuses on factors that determined youth decision to participate in farming in Nigeria using Ogun State as a case study. Purposive and random sampling techniques were employed to obtain data from 300 youths spread across the four agricultural zones of the State. Collated data were analysed using descriptive, Logit and Poisson regression. The Logit model estimates revealed that years of youth in social organization, access to Information and Communications Technologies (ICTs), nature of land ownership, and youth access to State-owned agricultural programmes positively determine youth decision to participate in farming activities in the study area. On the contrary, male youth, years of formal education, and marital status of youth were negative determinants. The Poisson estimates showed that youth age, number of extension visits, and years in social organization as well as purpose of farming were positive determinants of the total hours spent by youths on farming activities per day in the study area. On the other hand, years of formal education, farm income of previous farming seasons, land ownership, and access to credit affect youth participation negatively. To increase youth participation in farming activities in the study area, it is recommended among others that the government should ease the access of youths to soft loans, tractor hiring services, land acquisition as a motivation to intensify their participation in farming activities.

Keywords Nigeria · Youth participation · Farming · Food self-sufficiency · Agricultural zones

---

O. A. Olowa (✉) · O. W. Olowa
Department of Agricultural Education, Federal College of Education (Technical), Akoka, Lagos, Nigeria

1 Introduction

Food self-sufficiency is generally taken to mean the extent to which a country can produce the quantity of food (calories) that equals or exceeds its food consumption (Gunnar 2018). This most basic definition can apply at the level of individuals, countries, or regions. Food self-sufficiency is under increasing stress in many countries. A convergence of issues—such as climate change, resource depletion, dysfunctional farm policies, loss of biodiversity, and aging of farmers—is now threatening the availability of healthy food for an ever-increasing population.

Nigeria is a very large economy with a population of about 200 million according to World Bank (2019). The country represents about 20% of the total population of sub-Saharan Africa and about 76% of this population lives in the rural areas and about 90% of the rural dwellers are engaged in agricultural production (World Bank 2016). Naturally, high rural population translates to increased farm and non-farm activities in a labour-intensive agriculture. Unexpectedly, the goal of food self-sufficiency in Nigeria remains an elusive target in spite of its population advantage. One of the problems behind the non-achievement of this goal is the condition of Nigerian farmers whose current average age is around 55 years and it is expected to rise to around 75 years by 2030 (Akpan 2010; Akpabio 2012). The age and low level of education of average Nigerian farmers correlate with their aversion of risks associated with the adoption of new innovations; hence the very low productive capacity. In the opinion of many, getting youths to engage in farming seems a possible solution to the problem (Sarah 2014).

The term 'youth' is sometimes considered by some schools of thought as a period of transition from the dependency of childhood to adulthood's independence. This period is often characterized by sexual maturity, peak of strength and emotion as well as growing social and economic independency from parents and guidance. In developing societies, the period is often prolonged due, among others, to various types of social, economic, and political uncertainties (Akpabio 2012). Generally, 'youth' as a social group is more often defined in terms of age. For this reason, the spectrum of youth has been variously defined as ranging from the ages of 10 or 11 year (as in some traditional societies in Africa) to as high as 35 years in some countries like South Africa and Tanzania. In an attempt to 'standardize' the concept of youth, international organizations such as the United Nations and the Common Wealth of Nations defined youth as encompassing those between the 15 to 24 age group (UNGA 2008). The African Youth Charter, with final ratification in 2019 by the African Union, considers that youth are people in the age range of 15 to 35 years of age (African Union 2019). In a similar vein, Nigeria's National Youth Development Policy encapsulates the youth as comprising all young persons of age 18 to 35 years. However, the tendency to extend the category of youth to 35 years and beyond in Nigeria seems to be a reflection of the emerging phenomenon of the prolonged period of youth dependency on the host. The foregoing as noted by Alhaji and Rusmawati (2019), is a metaphor for Africa's poverty.

This phenomenon is an indication of the inability of many young people in the country to be economically self-sustained, which is the result of the volatile economic situation in the country. Hence, for analytical purposes, and in corroboration with the definition of youth by the Nigeria's National Youth Development Policy, this study uses the age category 18 to 35 years as an acceptable definition of 'youth' in Nigeria.

The need to increase participation of youth population in farming activities is justified by many facts. *Firstly*, the current level of youth unemployment in Nigeria is alarming. According to the Nigerian Bureau of Statistics, the national unemployment rate is 23.9% with the youth accounting for more than 70%. Increased involvement of youth in farming activities will help reduce the problems of the ageing farm population and increasing youth unemployment. Youth unemployment incurred costs to the economy, society, and their families (Ajaegbu 2012; CBN 2014). *Secondly*, from researches, the current average age of a Nigerian farmer is around 55 years and it is expected to rise to around 75 years by 2030 (Akpan 2010; Akpabio 2012). The situation is worsened by the fact that by 2030, an estimated 50% more people will migrate to urban areas (Sarah 2014). It is doubtful if the present farming practiced by ageing farmers can produce enough food to feed the anticipating population of 230 million people in 2030. The ageing of farmers in Nigeria and globally has been identified as one of the major structural and policy challenges facing the future of food self-sufficiency (Sarah 2014). As the older generation retires, are the youth willing and ready to take over the food self-sufficiency challenges?

Within this perspective, this chapter attempts to empirically determine the expanded factors (pull, push, and economic-based factors) that model youth decision and actual participation in farming activities in Nigeria. More specifically, the chapter aims at: investigating the rate of youths' participation in farming activities; identifying the farming activities that youth participate in; estimating the factors that determine youths' decision to participate in farming activities; and isolating the factors that determine hours spent by youths in farming activities.

2 Methodology

2.1 The Study Area

The study was conducted in Ogun State in south-western Nigeria. The State has a total population of 7.1 million according to Ogun State website (ogunstate.gov 2016) and is located in the rainforest vegetation belt of Nigeria within longitude 2 45' and 3° 55' E and latitudes 7 01 N and 7° 8' N in the tropics. It is bounded in the west by Benin Republic, in the south by Lagos State and Atlantic Ocean, in the east by Ondo State, and in the north by Oyo State. It covers a land area of 16,432 km sq., less than 2% of the country's landmass (ogunstate.gov 2016). The rainy season starts around the middle of March and continues until late October whereas the dry season starts in November and lasts until February in most locations in the State. Rainfall

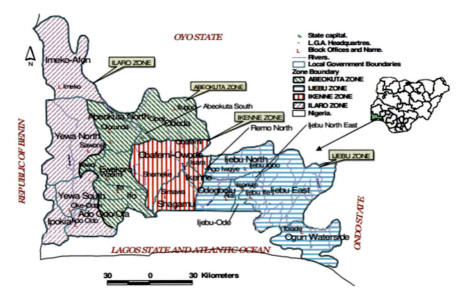

Fig. 1 Map of Ogun State ADP zones and blocks showing study location

ranges between 1600 and 900 mm annually. The State is warm throughout the year with a temperature between 28 °C and 35 °C. Humidity is between 85 and 95%. The main occupations of the people in the State are: agriculture, fishing, clothing, textiles, and civil service. The State was divided into four agricultural extension zones namely: Abeokuta, Ilaro, Ijebu-Ode, and Ikenne (OGADEP 2005) (Fig. 1). The four agricultural zones are well known for farming activities of different kinds.

2.2 Data Source and Sampling Procedures

Primary data were used and respondents were youths. Combinations of sampling methods were used to select respondents. First, the four agricultural zones—namely Abeokuta, Ilaro, Ijebu-Ode, and Ikenne—were purposively selected. In the second stage, three villages were purposively selected from each agricultural zone based on their popularity for farming activities, adding up to 12 villages in all. In the third stage, a total of 25 youths were randomly sampled in each of the 12 sampled villages. Hence, a total of 300 youths were randomly sampled and used for data analysis.

2.3 Data Analysis

2.3.1 Logit Model

A Logit model following Thakur and Jasral (2018) and modified was used to identify significant factors that influence youth decision to participate in farming activities in the rural areas of the State. Implicitly, the specified model is shown in Eq. 1. The Logit Model which captures youth decision to participate in farming activities is given below:

$$Dec = Ln\left(\frac{P_1}{1-P_1}\right) = Z_j = \beta_0 + \beta_1 x_1 + \beta_2 x_2 + \beta_3 x_3 + \beta_4 x_4 \\ + \beta_5 x_5 + \beta_6 x_6 + \beta_7 x_7 + \beta_8 x_8 + \beta_9 x_9 + \beta_{10} x_{10} + \mu \quad (1)$$

The marginal effect of the Logit model measures instantaneous effect that a change in a particular explanatory variable has on the predicted probability (i.e., the likelihood that a youth in the study area will choose to involve in farming activities or not); when the other covariates are kept fixed. They are obtained by computing the derivative of the conditional mean function with respect to explanatory variables.

$$\frac{\partial P_1}{\partial X_i} = \frac{E\{Y/X\}}{\partial X_i} = f(Z_1)\beta_1 = f(X\beta_1)\beta_1 \quad (2)$$

Variables used in Eq. (1) are defined as follows:

Dec   Youth decision to participate in farming (dummy; 1 for yes and 0 for no).
X_1 Age of youth farmer (years).
X_2 Gender of the farmer (1 = Male, 0 otherwise).
X_3 Formal educational (years).
X_4 Marital status of a youth farmer (1 for married and 0 otherwise).
X_5 Membership of social group (number of years).
X_6 Access to ICT (Number of times youth farmer browse in a week).
X_7 Nature of land ownership (dummy; 1 for owned land and 0 otherwise).
X_8 Number of non-farm occupations.
X_9 Perceived price of fertilizer (dummy; 1 for high and 0 for normal).
X_{10} Youth access to State-owned agricultural programme(s) (Number of programmes accessed).
U   stochastic error term.
Pi   Probability to participate in agricultural activity.
Ln   Natural logarithm function.

2.3.2 Poisson Model

To estimate the determinants of number of hours spent by youth in farming activities, Poisson model, following Santos Silva and Tenreyro, was adopted. The number of hour(s) spent by any youth was discrete and takes only non-negative integer values; therefore, the count-data model was specified. The model is explicitly expressed as:

$$\mu_i = E\{Y/X\} = \beta_0 + \beta X_1 + \cdots + \beta_n X_n \tag{3}$$

In Poisson model, the estimated coefficients correspond to semi-elasticity. Thus, coefficient estimates can be directly converted into marginal effects. For a continuous regressor Xi, the marginal effect is:

$$\frac{\partial \mu_1}{\partial X_i} = \frac{E\{Y/X\}}{\partial X_i} = \exp(X\beta)\beta = \beta_1 \mu_1 = \beta_1 e^{\beta_0 + \beta_1 x_1 + \beta_2 x_2 + \beta_n x_n} \tag{4}$$

Implicitly, the Poisson regression model is shown below:

$$Hrs = \alpha_0 + \alpha_1 x_1 + \alpha_2 x_2 + \alpha_3 x_3 + \alpha_4 x_4 + \alpha_5 x_5 \mid \alpha_6 x_6 + \alpha_7 x_7 \\ + \alpha_8 x_8 + \alpha_9 x_9 + \alpha_{10} x_{10} + \mu_i \tag{5}$$

where:

Hrs   Average number of hours spent in the farm in a day (discrete number).
X_1 Age of a youth farmer (years).
X_2 Farmer's years of formal education.
X_3 Last season farm income (N).
X_4 Number of times in contact with an extension agent in the last farming season.
X_5 Membership of social group (number of years).
X_6 Purpose of farming (1 for commercial and 0 for family used).
X_7 Nature of Land ownership (dummy; 1 for owned land and 0 otherwise).
X_8 Access to credit facilities (dummy 1 for access and 0 otherwise).
X_9 Marital status of farmer (1 for married and 0 otherwise).
X_{10} Average wage rate per day of hired labour (N).

3 Results

3.1 Descriptive Analysis

3.1.1 Socio-Economic Characteristics of Youth in the Study Area

The descriptive statistics of respondents is shown in Table 1. The result shows that

Table 1 Descriptive Statistics and socio-economic characteristics of youth in the Study area

Variables	Min	Max	Mean	Std. Var
Age	18.00	35.000	32.023	5.321
Gender	0.000	1.000	0.642	0.492
Formal education	0.000	16.000	10.394	0.224
Marital status	0.000	1.000	0.721	0.401
No. of non-farm occupations	0.000	2.000	0.484	0.621
Membership of Social Group	0.000	14.000	3.093	0.884
ICT	0.000	1.000	0.086	0.361
Land ownership	0.000	1.000	0.204	0.456
Access to Agric. Prog	0.000	5.000	0.658	1.162
Perceived Price of fert	0.000	1.000	0.634	0.429
Hours	0.000	12.000	3.067	0.429
Purpose of farming	0.000	1.000	0.843	0.452
Access to Credit	0.000	1.000	0.105	0.326
Avg. wage rate per day	0.000	6.000	1230.5	894.630
Last season Income	0.000	0.002	14,400.0	0.006
No of Extension visit	0.000	45.000	4.8560	8.274

Source Computed by Authors (2016)

the average age of youths in the study area was about 32 years. This means that most youths are in their active age. An average period of formal education stood at 10.4 years. This indicates that most youths in the area were educated, and there is high possibility of adopting agricultural innovations. About 64.20% of the respondents were male youths. The result also showed that 72.10% of youths interviewed were married. Social capital formation among youths was low in the study area, as shown by an average of three years in social organizations. Only 8.60% of the youths sampled had accessed to ICT facilities. The results also showed that about 20.40% of youths owned farm land. The rest acquired farm lands probably through lease and borrowed arrangements among others. About 63.40% of youths in the sampled area perceived that fertilizer price was high. An average of five hours was spent daily in the farm by youths in the area. In addition, about 84.30% of youths engaged in farming activities for commercial purposes. Credit accessibility was very poor among youths in the area. The result further revealed that only 10.50% of the youths had access to credit facilities. An average cost of hired labour stood at N1230.5. Previous farming season income averaged N14400, while extension agent visits 4 times per season on average.

Table 2 Youths' distribution according to rate of participation in farming activities

Participation	Frequency	Percentage
Yes	252	84,00
No	48	16,00

Source Field survey (2016)

3.1.2 Rate of Youths' Participation in Farming Activities

Following Nkonya et al. (1997) and Nnadi and Akwiwu (2008), the rate of youths' participation in farming activities was measured by calculating the percentage of those that participate and those that did not participate.

As shown in Table 2, the results reveal that 84% of the youths participated in farming activities, while 16% did not. The high involvement of youth in farming activities could be attributed to the availability of land in the area and the dependence on land for existence in the study area. The results are in agreement with the findings of Nnadi and Akwiwu (2008) that most youths in the rural area of Imo State, Nigeria are much involved in farming activities.

3.1.3 Farming Activities in Which Youths Participated

Table 3 shows that youths were highly involved in various farming activities with highest participation in planting (76.00%), fertilizer application (65.67%), and weeding (52.33%). Youths were moderately engaged in harvesting (44.00%) and land clearing (42.33%); lowly engaged in transportation (36.67%), processing (32.67%),

Table 3 Farming activities in which youths participated

Activities	Frequency	Percentages
Planting	228	76.00
Fertilizer application	197	65.67
Weeding	157	52.33
Harvesting	132	44.00
Land clearing	127	42.33
Transportation	110	36.67
Processing	98	32.67
Marketing	89	29.67
Staking	86	28.67
Clearing of Pens	59	19.67
Feeding of Birds	44	14.67
Collecting of Fodder	42	14.00
Compounding of poultry Feed	33	11.00

Source Field Survey (2016)

marketing (29.67%), staking (28.67%), clearing of pens (19.67%), feeding of birds (14.67%), collecting of fodder (14.00%), and very lowly involved in compounding of poultry feeds (11.00%). The findings implied that youths participate more in crop production and farm labour supply than livestock production. Such findings are in line with the work of Akpan et al. (2015) and Nnadi and Akwiwu (2008) who reported in their studies that youths were interested in crop production more than livestock, probably due to the short gestation period of the crop varieties produced, which ensures quick turnover. Moreover, livestock production could be more capital-intensive than crop production, hence the preference for crop production by most youths.

3.2 Estimates of Regression Models

3.2.1 Estimates of Logit Regression on Factors Determining the Youth's Decision to Participate in Farming Activities

Empirical results revealed that the log odd coefficients of years of youths in social organization, access to ICT, nature of land ownership, and youth access to State-owned agricultural programmes are positive and statistically significant with respect to the decision or probability of youths' participation in farming activities in the study area. The odd interpretation implies that for every unit increase in years of youth in social organization, the log odd in favour of youths' decision to participate in farming activities increases by 1.204 or about 12.40% compared to youths who are not in a social organization. Similarly, increase in youth access to ICT facilities will result in about 1.760 or about 17.60% increase in the log odd in favour of youths' decision to participate in farming activities compared to rural youths who do not have access to these facilities (Table 4).

Also, increase in youths' access to State-owned agricultural programmes will lead to about 1.277 or 12.77% increase in the log odd in favour of youths' decision to participate in farming activities compared to those who do not have access to such programmes. In a similar manner, about 1.641 or 16.41% increase in the log odd in favour of youth decision to participate in farming activities will occur for a unit increase in farm land owned by rural youth compared to those who do not owned farm lands. The results imply that increase in years of rural youth in a social organization, access to ICT, nature of land ownership, and access to State-owned agricultural programmes will increase their chances to make a positive decision to participate in farming activities. Using marginal effect results, a unit increase in years of rural youth in a social organization, access to ICT facilities, nature of land ownership, and access to State-owned agricultural programmes among youths in the study area will increase the chance or probability of youths' decision to participate in farming activities by 0.0534, 0.2014, 0.1223, and 0.0568 respectively.

The positive determinants of probability of youths' participation in farming activities satisfied a priori expectations. For instance, the increase in years of membership of a social organization promotes the social capital formation or networking among

Table 4 Estimates of Logit Regression on factors determining the youth's decision to participate in farming activities

Variable	Coefficient	Log add coeddicient	Marginal effect	Z-test
Constant	−1.284	–	–	−1.342
Age	0.025	1.034	0.0064	1.301
Gender	0.723	0.296	−0.2062	−3.426***
Education (Years)	0.064	0.820	−0.0123	−2.170**
Marital status	−0.823	0.321	−0.1718	−2.323**
Yrs of Youth in social org	0.167	1.204	−0.0534	3.642***
Access to ICT	1.760	2.507	0.2014	2.040**
Nature of land ownership	0.521	1.641	0.1223	1.863*
Number of non-farm occupations	0.143	1.139	0.0433	0.535
Perceived price of fertilizer	−0.034	0.983	−0.0088	−0.143
Access to state owned agric. Prog	0.235	1.277	0.0563	1.792*
Log likelihood	−164.256	Log ratio test	(10)	65.162***
McFadden R²	0.1352	Correction prediction 79.80%	79.80%	
Akaike Criterion	372.862	Schwarz Criterion	425.304	

Source Computed by authors, data from field survey 2016
*, ** and *** significant levels at 10%, 5% and 1% respectively

youths. Knowledge, ideas, and experiences are often shared among members of a social group. The social interaction among members helped sustain their belief and confidence in their occupations. Groups that shared the same occupation will easily encourage one another to stay put in their occupation. Also, increase in the use of ICTs promotes social interactions among peers and between youths and experienced aged farmers as well as scientists. These results suggest that the increase use of ICTs among youths will likely bring about resource-use efficiency. This can be achieved through exchange of information and exposure of the youths to the latest technology in their areas of activity. In addition, the increase of farm land ownership increased the probability of youths' participation in farming activities in the study area. An area with a high population density will likely have farmland encroachment and resulting in a pressure on agricultural land. This result perhaps suggests that rural youths' decision might be conditioned on the fact that the increase in land ownership among rural youths will likely reduce the cost of production and probably expand the level of farm's profit and constitute a glue that holds the youths perpetually in farming. Similarly, the increase in youths' participation in the State-owned agricultural programmes increases their probabilities of engagement in farming activities.

This could likely be linked to available or anticipated incentives in such programmes. The results on membership of social organization and land ownership corroborate with the research report of Chikezie et al. (2012) and Onemolease and Alakpa (2009).

On the contrary, the marginal effect and the log odd coefficient of youth decision to participate in farming activities with respect to gender, education level, and marital status were negatively signed and statistically significant at 1%, 5% and 10% respectively. This means that as the number of male youths, years of formal education, and marital status increase, the probability of youths engaging in farming activities reduces. With respect to log odd coefficients, the results imply that a unit increase in the number of male youths reduces the odd of increase involvement of male youths in farming activities by 0.296 times compared to a unit increase in number of female youths. Using the marginal effect with respect to gender, the result implies that a unit increase in number of male youths will result in 0.2062 reductions in the probability of male youths' participation in farming activities. This result implies that male youths are more vulnerable to rural–urban migration compared to the female counterpart. This finding could be as the result of economic, social, and environmental factors as asserted by Akpan (2010).

Similarly, a unit increase in the formal education of youths reduces the odd of increase involvement of rural youths in farming activities by 0.820 times compared to a reduction in years of formal education. The results suggest that as youth acquired more years of formal education, they move out of farming and from rural areas in search of 'white collar' jobs in urban areas. The finding indicates that the absent of educational facilities in rural areas, where farming activities are mostly practiced, is a serious factor that prevent youths' involvement in farming activities. As pointed out by Akpan (2010), wage differential between rural and urban areas is one of the motivating factors for youth abandoning farming activities in rural areas. This result is also in agreement with the research findings of Nnadi and Akwiwu (2008) and Chizekie et al. (2012). In a similar manner, the relationship between marital status of youths and the decision or probability to participate in farming activity is inversely related. That is, a number increase in married youth reduces the log odd of youth decision to involve in farming activities in the study area by 0.321 times compared to celibacy. The likely reason for this result could be the difficulty to sustain a family in the area. This could be also linked to low income from farming, lack of health institutions and infrastructures/amenities necessary for family wellbeing. Similar findings had been reported by Chizekie et al. (2012).

3.2.2 Estimates of Poisson Model on Factors Determining the Number(s) of Hours Spent by Rural Youth Farmers in Agricultural Activities

The diagnostic statistics of Poisson regression equation as presented in (Table 5) show that the McFadden R^2 is about 0.0742, which implies that all the explanatory variables included in the model were able to explain about 7.00% variability in the number(s) of hours spent by young farmers in the study area. The value of the

Table 5 Poisson estimates on determinants of hours spent by youth in farming activities

Variable	Coefficient	Standard error	Marginal effect	Z-test
Constant	−1.2570	0.1522	–	6.110***
Age	0.0223	0.0031	0.0335	2.429**
Education (Years)	−0.0124	0.0048	0.0435	−2.342**
Last season farm income	−2.8700	1.2400	−0.0340	−2.321**
No. of times in contact with an ext. agent	−0.0319	0.0039	0.0605	7.234***
Membership of social group	0.0254	0.0061	0.0836	1.763*
Purpose of farming	0.2116	0.0623	0.6132	3.439***
Nature of Land ownership	−0.0739	0.0534	−0.2430	−1.710*
Access to credit facilities	−0.1740	0.0639	−0.0405	−2.521**
Marital status	−0.0390	0.0652	0.2311	−0.431
Avg. wage rate/day	−2.8392	2.7420	−1.1042	−1.0364
Log likelihood	−721.083		Normality test	12.314***
McFadden R^2	0.0742		Schwarz Criterion	1534.603
Akaike Criterion	1316.83		Chi square	16.432***

Source Computed by authors, data from field survey, 2016
*, ** and *** significant levels at 10%, 5% and 1% respectively

normality test attested to the normal distribution of the error term generated in the Poisson regression. The Chi square test (16.432***) is statistically significant at 1% probability level, implying that the estimated Poisson regression has a goodness of fit. Empirical results showed that the age of youths in the study area has a positive relationship with the number of hours spent per day in agricultural activities. The results imply that a unit increase in the youth age will lead to about 3.35% increase in the hours spent per day in farming activities. This means that the number of hours spent by youth per day in farm activities in the study area increases with the increase in age of the youths (Table 5).

Similarly, the number of hours a young farmer has contact with extension agent(s) positively influences hours spent per day in agricultural activities. For instance, a unit increase in extension agent visit will result in 6.05% increase in hours spent in farming activities. This result suggests that a strong extension system can encourage youth to spend more hours in farming activities in the study area. In the same manner, membership in a social organization positively affects hours spent by youths in farming activities. Result revealed that a one-year increase in membership of a social group by youths in will result in 8.36% increase in the number of hours spent per day by them in farming activities. This result showed the importance of social capital formation among youth in the study areas.

The purpose of youth engagement in farming activities also has a strong positive correlation with the number of hours spent by them in farming activities. That is, for every commercial-oriented purpose of youth engagement in farming activities, there is 61.32% increase in the number of hours spent per day in farming activities. The result satisfies the a priori expectation, as a commercial-oriented farmer is expected to spend more hours in farming activities.

On the other hand, the coefficient of youth education, farm income of the previous farming season, land ownership, and access to credit have significant inverse relationship with the number of hours spent per day by youth in farming activities. The result for education suggests that increase in years of formal education will lead to about 4.35% reduction in the average hours spent per day by youths in farming activities. This means that an increase in years of formal education of youth will expose them to better opportunities and high-wage rate jobs in urban areas, and this will motivate them to abandon farming activities. This result is in line with the findings of Chikezie et al. (2012) and Akpan (2010).

Similarly, a unit increase in the previous season farm income of youths reduces the number of hours spent by rural youths in farming activities at 3.40% per day. This means that as the previous farm income of youth increases, current hours spent per day reduces. Onemolease and Alakpa (2009) have reported similar results. The results also revealed that access to credit facilities has a negative association with hour spent per day by rural youths in farming activities. This connotes that a unit increase in access to credit facilities by youth reduces the number of hours spent per day in farming activities by 40.50%. This means that there is an increase tendency of agricultural diversification when access to credit increases among youths. The results suggest that most youths perceived agricultural production as not profitable enough or yield low returns as compared to non-agro-based businesses. Another possible reason for the results could be the conditions attached to such credit facilities; this might motivate youths to diversify investment in order to avoid agricultural activities related risks.

4 Conclusion

This study analysed the expanded factors (pull, push, and economic-based factors) determining youths' decision and actual participation in farming activities in Ogun State, Nigeria. Youth participation in farming activities must be seen as one reliable way of managing food self-sufficiency and poverty as revealed by the results indicating that the rate of youth participation stood at 84% in the study area. The implication of this for Nigeria is that, if the Government continues with feeble efforts and policy summersault on youths' unemployment and involvement in farming, the nation is likely to pay for it with wide food demand–supply gaps that may turn the nation to net food importer in the nearest future, thus resulting in increased food insecurity. The various farming activities in which youths are involved include land clearing, planting, fertilizer application, weeding, collection of fodder, cleaning of pens, etc.

In summary, the study has discovered that year(s) in social organization(s), access to ICTs, nature of land ownership, and access to State-owned agricultural programmes are positive and significant drivers of youths' decision or probability to participate in farming activities in rural areas. On the other hand, rural youths' decision to participate in farming activities is negatively affected by gender composition of rural youths, years of formal education, and marital status. In addition, the number of hours spent by rural youths in farming activities is mostly influenced by age, the number of extension visits per farming season, purpose of youths' involvement in farming activities, and the magnitude of social capital formation among them. The antagonistic factors to number of hours spent by youths in agricultural activities include: increase in acquisition of formal education; increase of access to agricultural credit; increase in the previous farm's income; and increase in self-owned farm land.

5 Recommendations

Based on the above findings, it is recommended that:

- Youths should be given the right incentives by government to intensify their participation in farming activities. Such incentives include among others provision of soft loans and tractor hiring services, and facilitating land acquisition.
- There should be more extension agent visits. This will help increase the rate at which youths adopt new innovations and participate in farming activities.
- Youths should form more cooperative groups since membership of social groups will increase participation and hours spent in farming activities.
- Also, government and extension agents should intensify the provision of amenities in rural areas to make life more comfortable for the youths. These will not only encourage them to stay back, but also stimulate greater participation in full-time farming; hence help fostering the transition toward food self-sufficiency.
- There should be communal support through land donation to young farmers.

References

Ajaegbu OO (2012) Rising youth unemployment and violent crime in Nigeria, *American* J Soc Issues Hum 2(5):315–321. ISSN: 2276 6928. http://www.ajsih.org

Akpabio IA (2012) Youth employment and agricultural Development: the inextricable Siamese twins. 34th inaugural lecture series of the University of Uyo, Nigeria

Akpan SB (2010) Encouraging youth involvement in agricultural production and processing in Nigeria. Policy Note No. 29: International Food Policy Research Institute, Washington, D.C

Akpan SB, Patrick IV, James SU, Agom DI (2015) Determinants of decision and participation of rural youth in agricultural production: a case study of youth in southern region of Nigeria. RJOAS 7(43):35–48

African Union (2019) African youth charter. Addis Ababa. https://au.int./en/treaties

Alhaji BM, Rusmawati BS (2019) Non-income poverty and socio-economics characteristics: evidence from rural nigeria. In: Oortiwjn D (ed) Urban and rural poverty. Nova Science Publishers, New York

Central Bank of Nigeria (CBN) Website (2014) Accessed on the 5th of July, 2015. http://www.cenbank.org/.

Chikezie NP, Omokore DF, Akpoko JG, Chikaire J (2012) Factors influencing rural youth adoption of cassava recommended production practices in onu-imo local government area of imo state, Nigeria

Economic Watch (2013) Nigeria Economic Statistics and Indicators, Accessed 29 Oct 2013 from http://www.economywatch.com/economicStatistics/country/Nigeria/

Gunnar R (2018) Food Self-Sufficiency- Does it make sense? Garden Earth. https://www.Resilience.org

Jibowo AA, Sotomi AO (1996).The youth in sustainable rural development. a study of youth programme in odeda local government area of ogun state, Nigeria, In: Adedoyin SF, Ilonsu JOY (eds) Sustainable development in rural Nigeria, Proceedings of the 8th annual conference of nigeria rural sociological association

Nnadi F, Akwiwu C (2008) Determinants of youths" participation in rural agriculture in imo state, Nigeria. J Appl Sci 8(2):328–333

Nkonya E, Schroeder T, Norman D (1997) Factors affecting adoption of improved seed and fertilizer in Northern Tanzania. J Agric Econ 18:1–12

Ogun state Government (2016) Ogun State Population. Ogunstate.gov.ng/og

Okafor EE (2011) Youth unemployment and implications for stability of democracy in Nigeria. J Sustain Dev Africa 13(1):358–373

Olaoye OJ, Adekoya BB, Ezeri GNO, Omoyinmi GAK, Ayansanwo TO (2007) Fish hatchery production trends in ogun state: 2001–2006. J Field Aquat Stud Aquafield 3:29–40

Oloruntoba A, Adegbite DA (2006). Improving agricultural extension services through university outreach initiatives: a case of farmers in model villages in ogun state, Nigeria. J Agric Educ Ext 12

OGADEP (2005) Annual in-house review fisheries report. Ogun State Agricultural Development Programme (OGADEP) Fisheries Unit 10(4):273–283

Onemolease E, Alakpa S (2009) Determinants of adoption decisions of rural youths in the Niger delta region of Nigeria. J Soc Sci 20(1):61–66

Sarah S (2014) Encouraging youth involvement in agriculture and agribusiness. Foodtank Newslett. https://foodtank.com

Thakur A, Jasral L (2018) A logit model to predict innovativeness among mobile telecom service users. Global Bus Rev. https://doi.org/10.1177/09721509177133

UN General Assembly Resolutions (2008) Definition of Youth. http://www.un.org/documents/resga.htm

World Bank (2019) Population Data for Nigeria. https://data.worldbank.org.

World Bank (2016) Poverty reduction in Nigeria in the last decade. Country Report No- NGA World Bank, Washington, DC

Postface

At the end of this volume, some key points can be identified. First of all, the issue of food security in a context of climate change is really challenging with complex ramifications. It has certainly to be addressed from a trans-disciplinary viewpoint, mobilizing knowledge from several fields, to allow following changes occurring, generally, at a very rapid pace. Moreover, it implies that bridges between researchers and their thematic fields have to be built, to take into consideration the complexities of food security, and the nexus involving water, energy, land, and food, without omitting labour. The examples addressed in the book are also very interesting as they allow adding context elements to the complexity of this nexus in developing countries, encompassing situations in North and sub-Saharan Africa, Middle East, and South Asia.

One of the main solutions presented in the volume to try to alleviate the challenges induced by climate change for the agricultural sector is to promote climate-smart practices. This has to be one of the pillars of the future of extension services and programs, particularly in areas where smallholder farmers are the most important actors in food supply chains. Climate-smart agriculture has certainly to be a top item in the agenda of decision makers, planners, and development agencies, meaning that consequent policies and programs aiming at promoting farmers' skills have to be designed. These policies and programs should entail a paradigm shift towards efficiency rather than production. Thy should have the potential to mobilize all the stakeholders to address the questions of managing scarce resources and reducing negative externalities, by implementing good governance throughout supply chains. For the specific question of water productivity, proactive approaches, anticipating future risks, have to promoted, certainly needing innovative tools and participative options. Such approaches seem inevitable to avoid the entire collapse of previous production models, which may not be viable and sustainable in the future decades, given the ongoing and expected impacts of climate change on both renewable water availability and rising temperatures.

Another striking point from the examples presented in the volume is the possibility to rely on smart practices such as digitalization to improve the agricultural sector's performance and enhance the whole governance of food chains. It is now widely acknowledged that new information and communication technologies, databases' management and Internet of Things (IoT) can provide valuable tools to enhance the monitoring of production processes as well as the management of complex chains where different actors, most of all smallholders in remote rural areas, are involved. Short message systems (SMS), at a very low cost can therefore be used to spread valuable information from a decision source (farmers' co-operatives, public extension services, meteorological stations, veterinarians, etc.) to farmers, and that can be very helpful to improve their levels of performance. At the scale of the agricultural plot, as well as in hydraulic basins, such tools may allow improved water-use efficiency, early warning alerts to irrigate crops in order to face the effects of a heat wave, and the availability of signals to start harvesting products. Such tools might also be very useful to monitor herds' diets conception, heat detection, and this can only enhance livestock profitability.

The volume is also presenting examples pertaining to the effects of Covid-19 pandemic on the agricultural sector, which acknowledge that this pandemic has provided an occasion to recognize the vital roles of agriculture and food chains in social resilience, especially in times of crisis. Related activities not only create wealth and job opportunities but also are crucial in ensuring the management of territories as well as supplying food. At a time where many employment opportunities collapsed during the global lockdown, agriculture and food systems related activities kept on almost uninterruptedly. Moreover, they even provided jobs to people who became without occupation, because of the lockdown. The pandemic has also clearly revealed the limits of food systems totally dependent on an unregulated globalization. It has soon become obvious that food sources produced locally have much more value than products imported from very far locations, inducing an important carbon footprint and unnecessary costs. Moreover, the pandemic has also reinvigorated old protectionist measures, implying that major staple food (particularly cereal and legumes' grains) export countries have issued limitations to avoid stockpiles' depletion. Such measures—which are almost similar to the ones adopted during the last global food crisis 2007–2008—have, therefore, alerted decision-makers and even the civil society organizations on the crucial issue of food security. In water-stressed countries, where imports are often mandatory to ensure the supply of staple food, some experts have even exploited the situation created by the Covid-19 pandemic to advocate the return of the concept of food sovereignty in political agendas.

Altogether, these ideas strongly converge towards the rehabilitation of scientific knowledge—together with the valuation of traditional knowledge and practices—as a milestone in future public policies devoted to agriculture and sustainable food systems. Given the ongoing challenges induced by global warming as well as the decreased water availability in many regions of the Global South, and the decarbonation of human activities it entails, future policies have to be sound, taking into consideration the complexities of the challenges at stake. Since agriculture and food systems' activities remain the first employment sectors at a global scale, with more than 1.5 billion persons concerned, a renewed interest in the sustainability of these jobs has to be promoted. This not only focuses on the resources needed to ensure the resilience of such activities, like renewable water sources, but it has also to deal with a decent remuneration for the people involved in securing a day-to-day supply of food to the whole humanity. The scope of this volume has been significantly addressing these issues. Let's hope that it may contribute by its innovative thoughts to implement fairer and more resilient food systems in developing countries, by allowing better incomes to people involved in food production and distribution, as well as ensuring a balanced use of natural resources, in a time of growing uncertainty due to climate change implications and higher food price volatility.

Agadir, Morocco	Behnassi, Mohamed
Riyadh, Kingdom of Saudi Arabia	Baig, Mirza Barjees
Rabat, Morocco	Sraïri, Mohamed Taher
Riyadh, Kingdom of Saudi Arabia	Alsheikh, Abdulmalik
Riyadh, Kingdom of Saudi Arabia	Abu Risheh, Ali Wafa
	Editors

Biographical Notes of Authors

Abbas, Syed Ghazanfar
Dr. Abbas graduated as Agricultural Engineer from the University of Agriculture, Faisalabad, Pakistan in 1980 and entered into private sector employment where he worked during five years in the tractor manufacturing industry of Pakistan, precisely for the after-sale-service and establishment of Dealer's network in the country. In 1985, he joined the Pakistan Agricultural Research Council (PARC) under the Ministry of National Food Security and Research, until his retirement. During this period, he completed his M.Phil. in "Agricultural Water Management" and PhD in "Use of Computers in Agriculture" from Massey University, Palmerston North, New Zealand. Dr. Abbas has been engaged during his professional career in various United Nations assignments, including Team Leader (Farm Mechanization) under the Oil for Food Program in Iraq. Besides that, he has also taken some assignments for the United Nations Industrial Development Organization (UNIDO) in Afghanistan. His specialization has always been in the areas of using smart agricultural techniques easily and practically adoptable by farmers. He is one of those who advocates smaller farming equipment for easier operational techniques instead of larger and complicated farm machineries. Dr. Abbas is a campaigner of using alternative techniques to undo the climatic changes causing Global warming.

Abu Risheh, Ali Wafa A.
Eng. Abu Risheh is the Executive Director of the Prince Sultan International Prize for Water (PSIPW), at the Prince Sultan Institute for Environmental, Water and Desert Research, King Saud University, KSA. His professional interests include issues related to Environment, Water and Desert. He uses the most updated tools like GIS to purse his research work. He is instrumental to prepare an atlas to identify and assess natural resources and their location. It would be a ready reference resource for academia, researchers, planners, policy makers to conduct their research and launch development plans. He also oversees a research journal on water to promote water science in KSA and beyond. He shares his expertise at the King Saud University

(KSU) and other scientific fora in the country and internationally. He loves to share his wisdom with the fellow scientists. Due to his vision, the Prince Sultan Institute has been able to execute numerous projects of national importance. He constantly collaborates with many international research institutes to shape national issues according to international perspectives.

Al-Zahrani, Khodran H.
Dr. Al-Zahrani is a Professor at the Department of Agricultural Extension and Rural Society, College of Food and Agriculture Sciences, King Saud University, Riyadh, Saudi Arabia. He obtained his B.Sc and M.Sc from the College of Food and Agriculture Sciences, King Saud University and his Ph.D. from the Ohio State University, USA. He has conducted research on major issues like water and climate change faced by KSA. He has done several studies to explore the shortcomings and challenges faced by the national extension service in the country and did produce recommendations for improving the extension system and enhancing the efficiency of its Staff. His professional interests include: sustainable use of water, conservation of natural resources, combating food waste, and innovations in extension education. He is the author of two books and has published extensively in his areas of interests. He has represented KSA at many international fora and has delivered many scholarly talks.

Alaagib, Sharafeldin B.
Dr. Alaagib works as researcher in the College of Food and Agricultural Sciences in the University of King Saud, Riyadh, KSA. His areas of interest include food security, agricultural investment, environmental degradation, and pollution. He has published more than thirty papers in prestigious regional and international journals especially in the following areas: technical competence of broiler poultry farms, pricing of agricultural economic resources in KSA, the economic analysis of the use of pesticides in Saudi agriculture and its impact on food security and the environment, use of desalination water in agriculture. He also participated in several studies with the Ministry of Environment, Water and Agriculture in KSA, the King Abdullah Institute for Economic Studies and Research, and the Riyadh Economic Forum.

Alary, Veronique
Dr. Alary is an Agricultural Economist in the International Center of Research in Agronomy for Development (CIRAD), based at the International Center for Agricultural Research in the Dry Areas (ICARDA). She holds a Ph.D and a Doctoral Direction Habilitation in Economics. She has carried out research on household viability, risk management, and vulnerability in rural areas over the last 25 years in several developing countries (Cameroon, India, Mali, Tunisia, Egypt, and Morocco). During the last 10 years, she has coordinated different projects on the contribution of livestock to the household viability in North-African countries.

Alrwis, Khaled N.
Dr. Alrwis received his PhD from Oklahoma State University, Oklahoma, United States. He is currently a Professor in the College of Food and Agricultural Sciences

at King Saud University, Riyadh, KSA where he teaches economics of agricultural production and management of agricultural enterprises, supervised many masters and doctoral students, and sentenced many theses. Dr. Alrwis serves as the Head of the Department of Agricultural Economics for several years and supervisor of the King Abdullah Chair for Food Security. He served as the supervisor of the Riyadh Economic Forum at the Chamber of Commerce and Industry in Riyadh, and President of the Saudi Economic Association for several sessions, and worked as a consultant at the Royal Court and a member of the Food and Drug Authority. Dr. Alrwis has published more than thirty papers in prestigious regional and international journals especially in the following areas: technical competence of broiler poultry farms in the central region of KSA, the current and proposed crop composition in light of water security considerations in KSA, the economic importance of the agricultural sector in KSA, study and analysis of the demand structure for KSA's imports of red meat, pricing of agricultural economic resources in KSA, the economic analysis of the use of pesticides in Saudi agriculture and its impact on food security and the environment. He also undertook several studies with the Ministry of Environment, Water and Agriculture in KSA, the King Abdullah Institute for Economic Studies and Research, and the Riyadh Economic Forum.

Azhar, Bismah
Ms Bismah Azhar is a graduate of Social Sciences with specialization in Development Studies. Moreover, Ms. Azhar has a Master degree in Project Management with specialization in NGO Management from Bahria University Islamabad Campus, Pakistan. Through her academic and professional career, Ms. Azhar has a keen interest in writing research essays. As a junior researcher in development studies, she wrote various research essays revolving around the areas of women empowerment, global development experiments, development projects, child labour, globalization and development, development policy, law and development, governance and development, climate change and global warming, etc. During her service at the Institute of Rural Management (IRM), she prepared a *Training and Recruitment Proposal* for Khushali Bank Limited (KBL) and a Report on *How Effective is Income Generation for Women Empowerment?* Currently she is pursuing her passion of academic writing as a freelancer.

Bedrani, Slimane
Dr. Bedrani is a Professor of Economics at the National Higher School of Agronomy, Algiers and Associate researcher at CREAD. His areas of interest include agricultural economics and public policies evaluation.

Benabdelouahab, Tarik
Mr. Benabdelouahab is a researcher at the National Institute of Agricultural Research, Research Center of Tadla, Morocco. He has published nationally and internationally in reputable journals on natural resource management, remote sensing, and GIS issues.

Bin Kamal, Abdullah

Mr. Bin Kamal is a Social Scientist cum Educationist. He has a Bachelor degree in Social Sciences with specialization in Development Studies and a Postgraduate degree in Project Management with specialization in NGO Management from Bahria Univesity Islamabad Campus, Pakistan. Currently, he is managing two portfolios: Student Advisor and Program Coordinator, Public Health Program in Bahria University Islamabad Campus. His has an expertise in education administration, curriculum development, coordination, liaison, event management, and quality assurance related work. Apart from the technical skills, Mr. Bin Kamal has a passion for academic writing and research. He has a dissertation on *"Bodybuilding and Steroids: Are They Worth the Risk?"*, from which he has presented two papers in national conferences. Moreover, two of his research papers are in pipeline to be published in near future. Recently, he has published a research article titled *"Portrayal of Women in Media: A Symbol of Empowerment or Oppression?* in BU Prodigy 1st Issue.

Boudhar, Abdelghani

Dr. Boudhar is a Professor of Hydrology and Remote Sensing in the Faculty of Science and Technology, Sultan Moulay Slimane University, Beni Mellal, Morocco. He has several scientific publications related to land and water management. He is a member of the Research Team on Water and Remote Sensing at the same faculty.

Bouzid, Amel

Dr. Bouzid holds a PhD in Agricultural Economics from the National Higher School of Agronomy, Algiers, Algeria. She is a researcher at CREAD where she leads the Agriculture, Territories and Environment Division. His areas of interest include the analysis of agricultural and agrifood chains, innovation in agriculture as well as the social life-cycle analysis of agricultural products.

Cheikh M'hamed, Hatem

Cheikh M'hamed is an Agronomist at the National Institute for Agronomic Research of Tunisia (INRAT). He has a Ph.D in Crop Production Sciences from the National Agronomic Institute of Tunis (INAT). As agronomist, he is extensively and permanently involved in the conception and implementation of on-farm and on-station agronomy experiments. His research activities focus on sustainable intensification systems, mainly on innovative agricultural practices under conservation agriculture. His research is mostly related to cereals and food legumes cropping systems. Dr. Cheikh M'hamed has also coordinated and participated in several national and international research and research-development projects. Recently, Dr. Cheikh M'hamed is elected for a board member of the African Association for Precision Agriculture (AAPA) with the position of Regional Representative of North Africa. Dr. Cheikh M'hamed has authored more than 40 peer-reviewed journal articles and book chapters.

Dhehibi, Boubaker

Dr. Dhehibi is an agricultural resource economist in the International Center for Agricultural Research in the Dry Areas (ICARDA). He is distinguished for his research and teaching on production economics, economics of climate change, economics of natural resources management, value chain analysis, economics of development, and competitiveness and productive analysis of the agricultural sector in the CWANA region. Prior to joining ICARDA, he has worked at the National Institute for Agricultural Research of Tunisia (INRAT). Dr. Dhehibi has authored more than 60 peer-reviewed journal articles and book chapters. He received his Ph.D. in Economics from the University of Zaragoza, Spain.

Dube, Ahmed Kasim
Mr. Dube is currently pursuing his PhD at the Department of Agricultural Economics, University of Akdeniz, Antalya, Turkey. He is a member staff at Mada Walabu University, Ethiopia. His research interest areas include food security, efficiency and profitability assessment, market participation, and impact of market participation on food security and welfare of smallholder farmers. He has authored and published different publications in these areas.

Devkota, Mina
Dr. Devkota is Agronomist in the International Center for Agricultural Research in the Dry Areas (ICARDA). She had a Ph.D in Agronomy (Conservation Agriculture) from the University of Bonn, Germany. She has more than 15 years of research and development experience in soil, water, and agronomy with focus on resource conservation agriculture-based crop production system from different parts of the world (South and Central Asia and the MENA region). She had published over 40 book chapters and scientific papers in high impact journals. She also had good experience on scale appropriate farm mechanization and scaling out technologies at scale.

Elame, Fouad
Mr. Elame is a researcher in Agricultural Economics in the National Institute of Agricultural Research (INRA), Research Center of Agadir, Morocco. He is working on several research projects on natural resource modelling and performance analysis of agricultural systems.

Falsafi, Peyman
Dr. Falsafi is an Associated Professor on Agricultural Extension Education and has graduated from the University of Tehran in 2003 and now academic member of Agricultural Education and Extension Institute, Ministry of Agriculture, Tehran, Iran. He also has acted as: the Director General of the Agricultural Extension Bureau, (2004–2005); Deputy Minister for Extension and Farming System, (2006–2007); Deputy Minister for the Parliament, International Affairs and African Development Cooperation, (2007–2009); Deputy Minister for Parliamentary and Provincial Affairs, (2009–2010); Advisor to the Minister of Agriculture (2010–2013); Deputy of Agricultural Research, Education and Extension Organization, (2011–2012); Head of Higher Education Institute of Applied Science in the Ministry of Agriculture (2011–2014); Director General of Halal Affairs Bureau: (2014-Continues); and academic

member of Agricultural Education and Extension Institute, Ministry of Agriculture. Dr. Falsafi published more than 50 books, journal articles, conference proceedings, and technical reports.

Frija, Aymen
Dr. Frija is an agricultural economist with ICARDA's Social, Economic, and Policy Research Team. He is also the coordinator of ICARDA's activities in Tunisia and Algeria. His current research interests focus on farm modeling, farm efficiency and productivity analysis, agricultural water management instruments, institutional performance analysis, and the economics of conservation agriculture. Earlier in his career, Dr. Frija was a post-doctoral researcher at Ghent University in Belgium, specializing in agricultural water policy analysis in developing countries. He was also Assistant Professor and Researcher at the College of Agriculture of Mograne, Carthage University in Tunisia. His publications records can be found at Google Scholar, SCOPUS, ORCID, Research Gate.

Gopichandran, Ramachandran
Dr. Gopichandran is a Professor teaching various aspects of mitigation, adaptation, preventive environmental management, and business communication at the NTPC School of Business in India. These are important elements of public policy, pertaining to which Gopichandran has an overall work profile that spans thirty-two years. He has recently contributed to the review of the first and second order drafts of the upcoming sixth assessment report of the IPCC; with a special emphasis on chemical ecology for improved adaptation strategies. Gopichandran's contributions through the Compliance Assistance Programme, OzonAction of the UNEP have been quite significant and consistently so for more than two decades at the regional and global levels. This created the opportunity for him to serve as a Member of the Inter-Ministerial Empowered Steering Committee constituted by the Ministry of Environment, Forest and Climate Change, Government of India, on aspects of the Montreal Protocol. He is a well-known specialist in the areas of science and technology management communication, with a large number of theme-specific editorials and other publications to his credit. He holds two doctoral degrees in the fields of microbial and chemical ecology respectively, with a degree in law and is an Alumnus of the International Visitors Leadership Program of the Department of State, United States of America.

Hadria, Rachid
Mr. Hadria is a researcher at the National Institute of Agricultural Research (INRA), Research Center of Oujda, Morocco. His research interests include crop modeling, remote sensing and GIS applied to natural resources and environment management.

Htitiou, Abdelaziz
Mr. Htitiou is a PhD student at the Faculty of Science and Technology, Sultan Moulay Slimane University, Beni Mellal, Morocco. He is a member of the Research Team of Environment and Remote Sensing at the same faculty.

Idoudi, Zied

Idoudi is an Agricultural Economist with professional expertise in conservation agriculture, agroecology, rural livelihoods assessment, agricultural innovation systems, monitoring and evaluation, knowledge management, and innovations scaling. For the past eight years, he has been involved in different agricultural research for development projects dealing with diversification and sustainable intensification of production systems. He is currently a Technology Scaling Specialist (TSS) at the International Center for Agricultural Research in the Dry Areas (ICARDA) and is based in Tunisia.

Kamal Sheikh, Muhammad

Dr. Kamal Sheikh is a Professor of Development and Project Planning in Pakistan Agricultural Research Council (PARC) Institute of Advanced Studies in Agriculture, Islamabad, Pakistan. He completed his Master's degree in Agricultural Economics from the University of Agriculture Faisalabad, Pakistan and later received PhD degree (1998) from the Development and Project Planning Center (DPPC), University of Bradford, United Kingdom in the field of Development and Planning. He has served in PARC in various technical positions and retired in 1999 as Chief Scientific Officer. Dr. Kamal has 35 years of experience in management of research and development programs, project planning and management, monitoring and evaluation, and impact assessment. He completed several mega R&D projects and contributed significantly in research system re-organization and national priority setting and strategic planning. Working for research grants program, he processed and evaluated 100 s of research grants on emerging issues. His writings include: research articles in peer-reviewed and online journals, book chapters, technical reports, popular articles in newspapers, concept papers and feasibilities, M&E and completion reports. He obtained several merits, awards, and distinctions and is member of nine professional forums. He attended national and international trainings and conferences, workshops, and seminars. In recognition, during 2015–16 he was awarded best scientist of the year.

Kandegama, W. M. Wishwajith W.

Dr. Kandegama is working as a Senior Lecturer (grade I) at the Department of Horticulture and Landscape Gardening, Faculty of Agriculture and Plantation Management, Wayamba University of Sri Lanka, since April 2019. His teaching areas include: HC 22,063-Principals and application of Food Science; HC 31,083-Commercial Vegetable Production; HC 31,102-Medicinal and Herbal Plan Production; HC 32,182-Commercial Nursery Management; HC32232-Food Safety and Quality Assurance; CG 41,011-Professional and Employment Skill Development; and HC 41,292-Application of Biorationals in Agriculture. His Research Interests include: climate change, organic farming, plant protection, food safety, food product development, herbal medicine and commercial horticulture production, nursery management and nanotechnology application on agriculture. In terms of experiences, Dr. Kandegama graduated from the Faculty of Agriculture, University of Peradeniya Sri Lanka with BSc Agriculture (Special) Degree, in 1999. He initiated the career joining with UN project for rural agriculture development in Sri Lanka. After completion of the project, he joined a private company as an Area Manager. In 2010, he possessed the master qualification in Food Safety & Control from London South Bank University. During his stay in UK, he worked as a quality assurance specialist for food processing industries. In 2012, he served as a visiting lecturer at the Faculty of Animal Science and Export Agriculture, Uva wellassa university of Sri Lanka. Later, he earned his PhD with an excellent pass in nanotechnology application on agrochemicals to reduce its environmental hazards in Central China Normal University (CCNU), China. After completion of PhD, he worked as a postgraduate research associate in CCNU until 2018. He has developed a novel eco-friendly nanoencapsulated pesticide formulation and claimed patent rights in China. Once he returned to Sri Lanka, he joined as Director of Post Graduate Studies for DBA Degree programme of Lincoln University College Malaysia. In addition to his academic work in Wayamba University, he is contributing in the implementation of FAO and IFAD projects for food safety and agriculture and community development.

Larbi, Aziz
Dr. Larbi is an Agricultural Engineer from the National School of Agriculture (ENA), Meknes, Morocco. He holds a Doctorate in Environmental Sciences and Management (environmental socio-economics) from the University of Liège, Belgium. He is a Researcher-Professor at the Department of Rural Development Engineering at ENA and Coordinator of the Research Group on 'Rural Societies and Territorial Development'. His work focuses on agricultural and rural development, particularly the social management of natural resources in the context of global and climatic changes.

Lazereg, Messaoud
Mr. Lazereg is a researcher at CREAD since 2011 and is working on agricultural systems in Algeria. He is leading a team working on agricultural systems dynamics and is involved in many research projects such as: food security outlook, value chain, and rural development. His areas of interest include the contract farming, innovation in agriculture, and the dairy sector development.

Lebrini, Youssef
Mr. Lebrini is a PhD student at the Faculty of Science and Technology, Sultan Moulay Slimane University, Beni Mellal, Morocco. He is a member of the Research Team on Environment and Remote Sensing at the same faculty.

Lionboui, Hayat
Dr. Lionboui is a researcher in Agricultural Economics at the National Institute of Agronomic Research (INRA), Morocco. She works at the Department of Economics and Rural Sociology team. She has completed several research projects on agro-economic modelling of water resources, land and water resources management, and efficiency analysis studies.

Louahdi, Nasreddine

Nasreddine is an Agronomist and Director of the ITGC experimental station of Setif (Algeria), a specialist in agricultural machinery. He made a career within the ITGC as an experimenter of culturales techniques and cereal breeding. He is also a specialist in agricultural policies and working as an extension engineer for the development of conservation agriculture and sustainable agricultural practices in Algeria. He is member of several national and international agricultural and rural development projects that encourage the adoption of new agricultural practices.

Muneeb, M. Musthafa

Dr. Muneeb is currently working as Senior Lecturer and Head of the Biosystems Department of South Eastern University of Sri Lanka. He is heavily involved in the Department's development and improvement to another level on routine basis. He received his PhD from the University of Malaya, Malaysia in 2018 with excellence in Ecology and Biodiversity. He received international Graduate Research Assistant Scheme (iGRAS) from the same university. Before that, from 2013–2014, he worked as a Researcher at Ege University, Turkey. Dr. Muneeb got Chinese government scholarship during 2011–2013 to work at Chinese Academy of Agricultural Sciences (CAAS). He completed his MSc in Animal Science from King Saud University, Saudi Arabia, while serving as a researcher at Animal Biotechnology Lab. He earned his Bachelor's degree from the University of Peradeniya, Sri Lanka in agriculture in 2007. Dr. Muneeb has published more than 35 peer-reviewed and indexed journals on various areas including ecology, biodiversity, genetic diversity, environmental conservation, etc. He also has been involved in different journals as a reviewer. He has tried to develop a strong public awareness on natural resource conservation via public forums and newspaper articles too. He has initiated a project on organic home garden among public.

Munir, Muhammad

Dr. Munir is a Gold Medalist, having a rich experience of 39 years working on multifarious professional aspects including Coordination, National and International Liaison, as a Scientist, and in management & senior administrative positions. He has served in different capacities in Government, including the Pakistan Agricultural Research Council (PARC), Islamabad as a Member of the Plant Sciences Division, Member of the Coordination & Monitoring Division, Secretary Council (PARC), National Project Director Olive Promotion in Pakistan, and National Coordinator for International Consortium for Rice–Wheat system in Indo-Gangetic Region. He served during 4 years as a Deputy Director (Policy Planning) in SAARC Agricultural Information Centre (SAIC), Dhaka, Bangladesh representing the Government of Pakistan. He also served as a Consultant for FAO and the Higher Education Commission of Pakistan for an Islamic Development Project. He has a Ph.D from Oregon State University, USA and possesses 33 years of post-Ph.D experience. He is author of 63 research publications in different journals of international repute. He got international exposure by representing the Government of Pakistan in 22 countries and attained specialized professional training from 11 international institutions worldwide. He succeeded to win and successfully executing five project grants worth Rs. 900 million from international and national sources. He is working as member of Senate & Syndicate of the University of Agriculture Faisalabad and on board of studies of different universities. In addition, he is supervising students working on their M.Sc. and Ph.D theses and also working as a member of HEC curriculum committee for Agronomy. He is well-versed in the use of different computer software needed in professional work. He is member of many national and international professional societies.

Najim, M. M. M.
Dr. Najim is currently a Professor at the Faculty of Science, University of Kelaniya, Sri Lanka and the Vice-Chancellor of the South-Eastern University of Sri Lanka since 2015. He earned his B.Sc. degree in Agriculture from the University of Peradeniya, Sri Lanka in 1994 with a first class honours pass. He completed his M. Eng. in Irrigation Engineering and Management from the Asian Institute of Technology, Thailand in 2000 and was honored with the 'Hisamatsu Price' for his outstanding performance in the field of study 'Irrigation Engineering and Management'. He completed his Ph.D. in Water Resources Engineering from the University Putra Malaysia, Malaysia in 2004. Dr. Najim has published extensively (more than 50 full papers out of which some are in Q1 – Q4 journals and more than 70 short communications) on the issues associated with water resources in the national and international journals. He has also presented issues in water resources management and agriculture at various international conferences. His areas of interest include hydrological modeling and stream flow assessment, impacts of climate change on agriculture and aquaculture, agricultural water management, and wastewater agriculture. He has attempted to develop strategies to conserve water resources and promote environmental conservation addressing issues of climate change. He was honoured by many prestigious awards such as Presidential Awards for Scientific Research, NRC Merit Award for Scientific Publication, Award for the popularization of Science, etc. Dr. Najim started his scientific career in 1995 as an academic at the University of Peradeniya, Sri Lanka. He served at the Sabaragamuwa University of Sri Lanka as an academic from 1997–1998, as an academic again at the University of Peradeniya, Sri Lanka from 1998 to 2007, as a Senior Academic and a Professor at University of Kelaniya, Sri Lanka from 2007 to to-date and as the Vice Chancellor of South-Eastern University of Sri Lanka from 2015 to to-date. He serves as a member of many international journals' editorial boards and professional organizations.

Ouhnini, Ahmed
Mr. Ouhnini is an Agroeconomist Engineer specialized in development studies. He graduated from the National School of Agriculture in Meknes in 2014 and holds a Master degree in law, economics and management from the Sorbonne Institute of Development Studies in Paris. Ouhnini is an alumnus of the U.S State Department Middle East Partnership Initiative and began his professional career in the field of consulting, then in socioeconomic research at the Paris School of Economics (PSE) before joining the Policy Center For the New South (PCNS) in 2019 as a Research

Assistant in economics, where he currently leads reflections on rural and agricultural policies in Africa and public policy subjects.

Olowa, Olatomide Waheed

Dr. Olowa is a Chief Lecturer at the Department of Agricultural Education of the Federal College of Education (Technical) Akoka, Lagos. He obtained a second-class honours upper division in Agricultural Science from the Obafemi Awolowo University, Ile-Ife and a Doctorate and Master degrees in Agricultural Economics from the University of Ibadan. He won the best African Economics Research Consortium (Nairobi, Kenya) Ph.D Thesis Research Grant. He was awarded Letters of commendation by the Department as well as the College Management in 2006 and 2017, respectively, for Diligence and productivity. Dr. Olowa is a member of the Nigerian Association of Agricultural Economists (NAAE), Farm Management Association of Nigeria (FAMAN), Agricultural Society of Nigeria (ASN), and Agricultural Education Teachers Association of Nigeria (AETAN). In pursuit of academic advancement, Dr. Olowa has participated in National and international conferences of these professional bodies. He has authored five books and four chapters in springer Nature plus over sixty journal articles in reputable international and university-based journals, most of which can be accessed on self-archive platforms such as Research-gate and Academia. Dr. Olowa was at various times Head, Department of Agricultural Education, Director, Degree Programme and Chairman of several college-based committees. He has served close to two decades teaching and supervising degree and NCE students. In terms of community service, Dr. Olowa is a member of Federal Road Safety Special Marshal, RS 2.17 unit 53 in Lagos and he is a pastor with Deeper Life Bible Church.

Olowa, Omowumi Ayodele

Dr. Olowa obtained the BSc., MSc. and Ph.D. degrees in Agricultural Economics from the University of Ibadan, Nigeria. She has been teaching Agricultural science and related courses for a number of years. Dr. Olowa has published over twenty peer-reviewed papers in reputable national and international journals and conferences and authored and co-authored more than five books. Dr. Olowa is currently a Lecturer at the Department of Agricultural Education, Federal College of Education (Technical) Akoka, Lagos. Dr. Olowa has a second-class upper degree in Agricultural Science from the Obafemi Awolowo University, Ile-Ife and a Doctorate in Agricultural Economics from the University of Ibadan having also completed a Master degree in Agricultural Economics from the same University. With more than a decade expertise experience in teaching, research and leadership, Dr. Olowa has over 60 publications among books, book chapters and scientific papers published both at national and international levels. Presently, Dr. Olowa is working with the Federal College of Education (Technical) Akoka assuming responsibilities of teaching and supervising both Degree and NCE Programmes in Agricultural Education.

Ozkan, Burhan

Dr. Ozkan is a Professor of Agricultural Economics at the Akdeniz University, Antalya, Turkey. He holds a MSc from Reading University, United Kingdom and

PhD from Cukurova University, Adana, Turkey. His principal areas of research are agricultural business management, production economics, energy economics, and food security. He is the founder of the Department of Agricultural Economics at the Akdeniz University and chaired the Department for more than 15 years. He is currently chair of the Agribusiness Management Division within the Department.

Prasanna, Ariyarathna
Mr. Prasanna is a doctoral student pursuing DBA research degree in the University of Kelaniya, Sri Lanka. He holds two post-graduate degrees: Executive Master of Business Administration degree (2011), and Master of Engineering degree in Electronic and Telecommunication (2003), followed by his Bachler of Science degree on Electrical and Electronics (1997). Currently, he has been employed by the United Nations International Children's Emergency Fund (UNICEF) as an international professional working in the area of operations management with special focus on quality assurance of office governance, enterprise risk management, business continuity planning and operations, and business processes analysis and simplifications. The present research interest of Prasanna is the use of modern ICTs in industry value chain governance and coordination with a special interest for the agriculture industry.

Qureshi, Ajmal Mahmood
Prof. Qureshi serves as a Senior Associate at the Asia Center, Faculty of Arts and Sciences, Harvard University, Cambridge, MA, USA. Previously, he has served at the United Nations as the FAO Resident Representative (Ambassador) for over seven years in China, multiple accreditations to North Korea and Mongolia, regarded as one of the most challenging assignments in the UN System. He has represented the Organization to the host government, diplomatic missions, bilateral and multilateral donors, international organizations, NGOs and other stakeholders. His services have been recognized at the highest administrative and political levels. He was awarded the title of 'Senior Advisor and Professor at Chinese Academy of Agricultural Sciences'. He successfully implemented a large array of technical assistance projects throughout the country including Tibet. In addition, through a large number of regional projects, Chinese expertise was used in training and capacity building of countries in the Asia Pacific Region. He was the pioneer to launch of Food Security Project in Sichuan Province that was named as an ideal model and was being replicated in over hundred countries. He also assisted in the preparation of China's Agenda 21, holding of the Roundtable on Agenda 21 and was the keynote speaker on the "Environmental Issues". In the restructuring of the FAO, Prof. Qureshi contributed to the preparation of the Organization's strategic plan in the context of China's priorities. As Chairman of the Donors Group in China, Prof. Qureshi helped mobilize millions of dollars for the agriculture, forestry and fisheries sectors of China. For over eight years, at the Ash Centre of Democratic Governance and Innovation at Harvard Kennedy School, he has worked on governance and democracy issues. Prof. Qureshi has served in North Korea, Mongolia, Uganda and Represented the United Nations to the host governments, diplomatic missions, bilateral and international organizations and all other stakeholders. He was the Member of the High-Level Steering Committee advising the

President of Uganda on the implementation of the Plan for Modernization of Agriculture. Through his offices, he had close interactions with the multilateral donors; inter alia, USAID, World Bank, UN System, Asian Development Bank, SIDA, AUSAID and NORAD. As Head of International Cooperation, he led the country's delegation to the Governing Boards of Rome-based UN Organizations: FAO, WFP and IFAD. Prof. Qureshi also organized and assisted in holding numerous international events like: FAO Ministerial level Regional Conference for Asia and Pacific in Islamabad, attended by 28 Ministers. He has the honour to serve as the Consul General (Head of Mission) of Pakistan in Istanbul, Turkey. He received his education in Pakistan, France and the USA with major focus on political science, international relations, and development economics. At the Harvard Kennedy School, he dealt with issues of governance, democracy and innovation; and under the leadership training programs, inter-acted and collaborated with senior officials from China, Viet Nam and Indonesia. He has published extensively in Harvard journals, bearing on China and DPRK's Food Security. He has also been named as the 'Most Distinguished Alumni' at the Boston University. He is also the Member of the Board of Directors of the United Nations Association of Greater Boston (UNAGB).

Rathnayake, Rathnayake Mudiyanselage Praba Jenin
Ms. Rathnayake worked as a Demonstrator at Wayamba University of Sri Lanka. She is an agriculture graduate from the same University. She demonstrated on crop genetics, post-harvest practices, horticultural crop-based farming, modern farming practices, integrated crop management, and industrial food quality. She undertook studies on post-harvest losses of horticultural crops and research on horticultural crop-based product development. Having a great interest for industries, she received exposure on exportations in Sri Lanka. The perceptions from Sri Lankan economy and background information of farmers have inspired her contribution to the chapter. Currently, she studies waste management and man-made influences on the world. Eco-friendly agricultural practices are mandatory to shift the farming practices towards sustainability and reduce the negative impacts of climate change. She intends to investigate further rural farmer communities in Sri Lanka and the impact of modern day agriculture along with future scenarios.

Reed, Michael R.
Dr. Reed is Emeritus Professor of Agricultural Economics at the University of Kentucky, USA. He holds a Ph.D. in economics from Iowa State University (1979); a Doctor Honoris Causa (Honorary Ph.D) from Bucharest University of Agricultural Sciences and Veterinary Medicine (Romania); and an Honorary Ph.D. from the Faculty of Business Administration, Maejo University (Thailand). Dr. Reed's principal area of research is international trade in agricultural products, including the effects of macroeconomic policies and exchange rates on U.S. food exports, international commodity price dynamics, consumer demand in various countries, and the effects of competition patterns on world agricultural trade patterns.

Rekik, Mourad
Dr. Rekik holds a Ph.D. in animal production from the University of Reading, United Kingdom. He is a Principal Livestock Scientist working at the International Center for Agricultural Research in the Dry Areas (ICARDA) with more than 25 years of experience in animal reproduction and small ruminants' production and management in drylands. His expertise includes sheep and goat reproduction and its interaction with nutrition, health, and genetics. His current research interests focus on boosting the resilience and productivity of livestock production systems at the household level and attenuating the impact of environmental and economic stressors. Dr. Rekik is author of more than 120 peer-reviewed journal publications, book chapters, and conference papers.

Rudiger, Udo
Rudiger is an Agricultural Innovation Specialist, with more than 25 years of experience in agricultural development and R4D projects. Before joining the International Center for Agricultural Research in the Dry Areas (ICARDA) in 2016, Rudiger worked for the International Fertilizer Development Center (IFDC) and GIZ in Togo, Benin, Rwanda, Burundi, Uganda, RDC and Burkina Faso where he managed and advised agricultural value chain and natural resource management projects. He studied international agriculture at the University of Witzenhausen in Germany, receiving his Master's degree in 1989.

Straquadine, Gary S.
Dr. Straquadine serves as the Interim Chancellor, Vice Chancellor for Academic Programs, and Vice Provost at the Utah State University—Eastern, USA. He did his Ph.D. from Ohio State University, USA. Presently he also leads the applied sciences division of the USU Eastern campus. He is responsible for faculty development and evaluation, program enhancement, and accreditation. In addition to his heavy administrative assignments, he manages to find time to teach some undergraduate and graduate courses and supervise the research projects of his graduate students. Being an extension educator, he has a passion for the economic development of the communities through education and has also successfully developed significant relations with agricultural leadership in the private and public sectors. He has also served as the Chair, Agricultural Comm, Educ, and Leadership, at Ohio State University, USA. Before accepting the present position as the Vice Provost, he served on many positions as the Department Head; Associate Dean; Dean and Executive Director, USU-Tooele Regional Campus, and the Vice-Provost (Academic). His professional interests include extension education, sustainable agriculture, food security, statistics in education, community and international development, the motivation of youth, and outreach educational programs. He has also helped several under developing countries improving their agriculture and educational programs. In the USA and at Utah State, he is seen as an administrator, educator, extension expert, community developer, and an International Development Professional.

Biographical Notes of Authors

Sujirtha, N. Vishnukumar

Ms. Sujirtha is presently a Lecturer at the Department of Biosystems Technology of South Eastern University of Sri Lanka. She is presently pursuing her Ph.D in Nutritional Economics at Massey University, New Zealand. She is a recipient of the AHEAD scholarship for Ph.D funded by the World Bank. She graduated with a Bachelor degree in Agriculture with a first-class pass from Eastern University of Sri Lanka, Sri Lanka in 2014. Following that, she earned a Master degree in Food Science and Technology from University of Peradeniya in 2019. Ms. Sujirtha has published nearly 12 full papers and 18 abstracts in international and national conferences and journals in the areas including food and nutrition, food security, and plant science.

Tripathi, Gireesh Chandra

Dr Tripathi serves as the Deputy Director General (Academics) of the NTPC School of Business, India. He secured his graduate and master degrees in civil engineering, followed by a doctoral degree in management from Indian Institute of Technology, Delhi. He also chose to specialize in areas of financial management. He has taught extensively about principles and practice of financial management in several centres of higher learning. Presently, he leads investigations on forms and functions of energy markets in close association with the Ministry of Power, Government of India. This logically extends from a core energy perspective with implications for energy security and sustainability. He is highly networked with communities of energy professionals in many geographies and is a well-known capacity building specialist in the stated areas.

Yousif, Imad Eldin A.

Dr. Yousif received his PhD from Humboldt University of Berlin, Germany. He is an Associate Professor at the College of Food and Agricultural Sciences, King Saud University, Riyadh, KSA. His research interests include: socioeconomic analysis, agricultural trade, food security and poverty alleviation policies, international commodity markets, liberalization policies, and the World Trade Organization (WTO). He is teaching the principles of agricultural economics, environmental economics, environmental tourism economics, and food consumption economics for undergraduate, and farm management and environmental economics for postgraduates. He has supervised seven M.Sc. research and acting as external and internal examiner for M.Sc. students. In addition, Dr. Yousif is a consultant in agricultural trade policies and development with AOAD, AAAD, and the World Bank.

Zarkik, Afaf

Ms. Zarkik is an Engineer specialized in energy transition and environmental studies. She holds a Master's degree in energy strategies from the School of Mines in Paris and a Bachelor of Science degree in engineering with a major in environmental science from Al Akhawayn University in Ifrane. Zarkik was an analyst in an oil and gas mergers and acquisitions consulting firm in Paris, a venture capital analyst in a cleantech venture fund in Paris, and an asset management analyst with top consulting firms. She joined the Policy Center for the New South (PCNS) in 2020 as a Research Assistant in economics where she currently leads reflections on climate change and sustainable development policies.